T0297401

Clinical Challenges in Therapeutic Drug Monitoring

Clinical Challenges in Therapeutic Drug Monitoring

Special Populations, Physiological Conditions, and Pharmacogenomics

Edited By

William Clarke, PhD, MBA

Associate Professor of Pathology

Johns Hopkins University, Baltimore, MD

Amitava Dasgupta, PhD

Professor of Pathology and Laboratory Medicine

University of Texas Medical School at Houston, Houston, TX

AMSTERDAM • BOSTON • HEIDELBERG • LONDON • NEW YORK • OXFORD
PARIS • SAN DIEGO • SAN FRANCISCO • SINGAPORE • SYDNEY • TOKYO

Elsevier
Radarweg 29, PO Box 211, 1000 AE Amsterdam, Netherlands
The Boulevard, Langford Lane, Kidlington, Oxford OX5 1GB, UK
50 Hampshire Street, 5th Floor, Cambridge, MA 02139, USA

Notices

Knowledge and best practice in this field are constantly changing. As new research and experience broaden our understanding, changes in research methods, professional practices, or medical treatment may become necessary.

Practitioners and researchers must always rely on their own experience and knowledge in evaluating and using any information, methods, compounds, or experiments described herein. In using such information or methods they should be mindful of their own safety and the safety of others, including parties for whom they have a professional responsibility.

To the fullest extent of the law, neither the Publisher nor the authors, contributors, or editors, assume any liability for any injury and/or damage to persons or property as a matter of products liability, negligence or otherwise, or from any use or operation of any methods, products, instructions, or ideas contained in the material herein.

Medicine is an ever-changing field. Standard safety precautions must be followed, but as new research and clinical experience broaden our knowledge, changes in treatment and drug therapy may become necessary or appropriate. Readers are advised to check the most current product information provided by the manufacturer of each drug to be administered to verify the recommended dose, the method and duration of administrations, and contraindications. It is the responsibility of the treating physician, relying on experience and knowledge of the patient, to determine dosages and the best treatment for each individual patient. Neither the publisher nor the authors assume any liability for any injury and/or damage to persons or property arising from this publication.

British Library Cataloguing-in-Publication Data
A catalogue record for this book is available from the British Library

Library of Congress Cataloging-in-Publication Data
A catalog record for this book is available from the Library of Congress

ISBN: 978-0-12-802025-8

For Information on all Elsevier books,
visit our website at http://www.elsevier.com/

Publisher: Mara Conner
Acquisition Editor: Tari Broderick
Editorial Project Manager: Jeffrey Rossetti
Production Project Manager: Melissa Read
Designer: Victoria Pearson

Typeset by MPS Limited, Chennai, India

Contents

LIST OF CONTRIBUTORS xiii

PREFACE xv

CHAPTER 1 Overview of Therapeutic Drug Monitoring 1
William Clarke

1.1 Introduction 1
1.2 Principles of TDM 1
1.3 Clinical Areas in Which TDM is Routine Practice 2
1.3.1 Epilepsy 3
1.3.2 Organ Transplantation 5
1.3.3 Cardiology (Antiarrhythmic Drugs) 7
1.3.4 Psychiatry 8
1.3.5 Infectious Disease 9
1.3.6 Oncology 11
1.4 Conclusions 12
References 13

CHAPTER 2 Immunoassays and Issues With Interference in Therapeutic Drug Monitoring 17
Amitava Dasgupta

2.1 Introduction 17
2.2 Immunoassay Platforms Used in TDM 18
2.3 Interference of Bilirubin, Hemolysis, and Lipemia 20
2.4 Interferences in Digoxin Immunoassays 21
2.4.1 DLIS Interferences in Digoxin Immunoassays 22
2.4.2 Effect of Digibind and DigiFab on Digoxin Immunoassays 23
2.4.3 Potassium-Sparing Diuretics and Digoxin Immunoassays 24
2.4.4 Herbal Supplements and Digoxin Immunoassays 25
2.5 Issues of Interferences With Immunoassays for Immunosuppressants 26
2.5.1 Metabolite Interferences in Cyclosporine Immunoassays 26

2.5.2 Metabolite Interferences in Tacrolimus Immunoassays 27
2.5.3 Metabolite Interferences in Sirolimus and Everolimus Immunoassays 27
2.5.4 Metabolite Interferences in Mycophenolic Acid Immunoassays 28
2.6 Interferences in Immunoassays for Anticonvulsants 29
2.6.1 Interferences in Phenytoin Immunoassays 29
2.6.2 Interferences in Carbamazepine Immunoassays 33
2.6.3 Interferences in Phenobarbital and Valproic Acid Immunoassays 35
2.6.4 Interferences in Immunoassays for Newer Anticonvulsants 35
2.7 Interferences in Immunoassays for TCAs 35
2.7.1 Interference of Phenothiazines and Their Metabolites in Immunoassays for TCAs 36
2.7.2 Interference of Carbamazepine in Immunoassays for TCAs 36
2.7.3 Interference of Cyproheptadine and Quetiapine With Immunoassays for TCAs 37
2.7.4 Interference of Miscellaneous Drugs With Immunoassays for TCAs 38
2.8 Conclusions 38
References 39

CHAPTER 3 Application of Chromatography Combined With Mass Spectrometry in Therapeutic Drug Monitoring 45
Madhuri Manohar and Mark A. Marzinke

3.1 Introduction 45
3.2 Liquid Chromatography 48
3.3 Mass Spectrometry 51
3.3.1 Ion Source 51
3.3.2 Mass Analyzers 53
3.3.3 Detectors 55
3.4 Preanalytical Stage 55
3.5 Application of LC-MS/MS Methods in TDM 56
3.5.1 Immunosuppressants 56
3.5.2 Anticonvulsants 57
3.5.3 Antidepressants 58
3.5.4 Antifungal Drugs 59
3.5.5 Others Drug Classes 61
3.6 LC-MS/MS Limitations 61
3.7 Conclusions 63
References 63

CHAPTER 4 Monitoring Free Drug Concentration: Clinical Usefulness and Analytical Challenges 71
Amitava Dasgupta

4.1 Introduction 71

4.2 Drug–Protein Binding 72
4.3 Drugs Requiring Free Drug Monitoring 73
4.4 Conditions in Which Monitoring Free Anticonvulsants is Necessary 75
4.4.1 Clinical Utility of Monitoring Free Phenytoin Concentrations 76
4.4.2 Clinical Utility of Monitoring Free Valproic Acid Concentration 78
4.4.3 Clinical Utility of Monitoring Free Carbamazepine Concentrations 80
4.5 Mechanisms of Elevated Free Anticonvulsant Levels in Various Pathophysiological Conditions 80
4.5.1 Mechanism of Elevated Free Anticonvulsant Concentrations in Uremia 80
4.5.2 Mechanism of Elevated Free Anticonvulsant Concentrations in Liver Diseases 81
4.5.3 Mechanism of Elevated Free Anticonvulsant Concentrations in AIDS 82
4.5.4 Mechanism of Elevated Free Anticonvulsant Concentrations in Pregnancy 83
4.6 Elevated Free Anticonvulsant Due to Drug–Drug Interactions 83
4.6.1 Displacement of One Anticonvulsant by Another Anticonvulsant from Protein Binding Sites 83
4.6.2 Displacement of Anticonvulsant from Protein Binding Site by Various Drugs 85
4.7 Free Drug Monitoring of Drugs Bound to AGP 86
4.8 Monitoring Free Concentrations of Immunosuppressants 87
4.8.1 Monitoring Free Mycophenolic Acid 87
4.9 Special Situation: Monitoring Free Digoxin 90
4.10 Monitoring Free Concentrations of Protease Inhibitors 91
4.11 Saliva and Tears for Determination of Free Drug Concentration 92
4.12 Methods for Monitoring Free Drug Concentration 92
4.13 Conclusions 94
References 94

CHAPTER 5 Therapeutic Drug Monitoring of Newer Antiepileptic Drugs 101
Gwendolyn A. McMillin and Matthew D. Krasowski

5.1 Introduction 101
5.2 Therapeutic Drug Monitoring of Antiepileptic Drugs 104
5.2.1 Therapeutic Ranges and Specimen Types for TDM of Newer Antiepileptic Drugs 104
5.2.2 Analytical Methods Used in TDM of Antiepileptic Drugs 108
5.2.3 Monitoring of Free Drug Fractions 110
5.2.4 Monitoring for Special Populations 110
5.2.5 Pharmacogenetics of Antiepileptic Drugs 111

5.3 Newer Generation of Antiepileptic Drugs 112
5.3.1 Clobazam 113
5.3.2 Eslicarbazepine Acetate 114
5.3.3 Ezogabine (Retigabine) 114
5.3.4 Felbamate 115
5.3.5 Gabapentin 116
5.3.6 Lacosamide 116
5.3.7 Lamotrigine 116
5.3.8 Levetiracetam 118
5.3.9 Oxcarbazepine 118
5.3.10 Perampanel 119
5.3.11 Pregabalin 119
5.3.12 Rufinamide 120
5.3.13 Stiripentol 120
5.3.14 Tiagabine 121
5.3.15 Topiramate 122
5.3.16 Vigabatrin 122
5.3.17 Zonisamide 122
5.4 Conclusions 123
References 123

CHAPTER 6 Therapeutic Drug Monitoring of Antiretrovirals 135
Mark A. Marzinke
6.1 Introduction 135
6.2 Rationale for Therapeutic Drug Monitoring of Antiretroviral Agents 137
6.3 Antiretroviral Drug Classes 137
6.3.1 Nucleoside/Nucleotide Reverse Transcriptase Inhibitors 137
6.3.2 Non-Nucleoside Reverse Transcriptase Inhibitors 140
6.3.3 Integrase Strand Transfer Inhibitors 141
6.3.4 Protease Inhibitors 143
6.3.5 Other Antiretroviral Classes 147
6.4 Analytical Methodologies 148
6.5 Clinical Trials 149
6.6 Challenges and Limitations of Therapeutic Drug Monitoring for Antiretroviral Agents 152
6.6.1 Recommendations 152
6.7 Conclusions 153
References 154

CHAPTER 7 Therapeutic Drug Monitoring in Infants and Children 165
Uttam Garg and Clinton Frazee
7.1 Introduction 165
7.2 Basic Concepts of TDM as It Relates to Infants and Children 166
7.2.1 Absorption 166
7.2.2 Distribution 166

7.2.3 Metabolism ... 167
7.2.4 Excretion ... 168
7.3 Pharmacokinetic Calculations ... 168
7.4 Sample Collection, Analysis, and Interpretation ... 171
7.5 Specific Drug Classes ... 173
7.5.1 Antibiotics ... 173
7.5.2 Anticonvulsants ... 176
7.5.3 Immunosuppressive Drugs ... 177
7.5.4 Antidepressants ... 178
7.5.5 Cardiovascular Drugs ... 178
7.5.6 Bronchodilators ... 179
7.5.7 Antineoplastic Drugs ... 180
7.6 Conclusions ... 181
References ... 181

CHAPTER 8 Therapeutic Drug Monitoring in Pregnancy ... 185
Sarah C. Campbell, Laura M. Salisbury, Jessica K. Roberts, Manijeh Kamyar, Jeunesse Fredrickson, Maged M. Costantine and Catherine M.T. Sherwin

8.1 Introduction ... 185
8.2 TDM in Pregnancy ... 186
8.3 What Drugs Warrant TDM? ... 186
8.4 Steps Involved in the TDM Process ... 187
8.4.1 Decision to Request a Drug Concentration ... 188
8.4.2 Collection of the Sample ... 188
8.4.3 Laboratory Measurement of Drug Concentrations in Specimens ... 190
8.4.4 Communication of Lab Result ... 191
8.4.5 Clinical Interpretation of Drug Levels ... 191
8.4.6 Implementation and Therapeutic Management ... 192
8.5 Pregnancy-Induced Physiologic, Pharmacokinetic, and Pharmacodynamic Changes in Drug Disposition ... 192
8.5.1 Pregnancy-Induced Physiologic Changes ... 193
8.5.2 Pregnancy-Induced Changes in Drug Metabolism ... 195
8.6 Drugs Used in Pregnancy With Recommended TDM ... 196
8.6.1 Antiarrhythmics ... 197
8.6.2 Antibiotics ... 198
8.6.3 Anticoagulants ... 198
8.6.4 Antiepileptics ... 198
8.6.5 Antidepressants/Antimanics/Antipsychotics ... 201
8.6.6 Antiretrovirals ... 201
8.6.7 Bronchodilators ... 202
8.6.8 Immune Modulators ... 203
8.7 Limitations of TDM in Pregnancy ... 203
8.8 Conclusions ... 206
References ... 206

CHAPTER 9 Therapeutic Drug Monitoring in Older People 213
Andrew J. McLachlan

9.1 Introduction 213
9.2 Clinical Pharmacology of Medicines in Older People: Implications for TDM 214
9.2.1 Age-Related Changes in Pharmacokinetics 215
9.2.2 Age-Related Changes in Pharmacodynamics 217
9.3 Which Drugs and When Should TDM Be Used in Older People? 218
9.4 Studies Investigating TDM in Older People 220
9.5 Pharmacodynamic Monitoring to Guide Dosing in Older People 223
9.6 Cost-Effectiveness of TDM in Older People 223
9.7 Big Data and Best Evidence to Support TDM in Older People 224
9.8 Conclusions 225
Acknowledgments 225
References 225

CHAPTER 10 Therapeutic Drug Monitoring in Obese Patients 231
Ventzislava Hristova and William Clarke

10.1 Introduction 231
10.2 PK in Obesity 234
10.2.1 Absorption 234
10.2.2 Distribution 234
10.2.3 Clearance 235
10.2.4 Metabolism 235
10.2.5 Renal Elimination 236
10.3 Chemotherapy in Obese Patients 237
10.4 Antimicrobial Treatment in Obese Patients 239
10.5 PK After Bariatric Surgery 240
10.6 Conclusions 242
References 242

CHAPTER 11 Special Issues in Therapeutic Drug Monitoring in Patients With Uremia, Liver Disease, and in Critically Ill Patients 245
Kamisha L. Johnson-Davis and Amitava Dasgupta

11.1 Introduction 245
11.2 Monitoring Free Drug Concentrations in Patients With Uremia, Liver Disease, and in Critically Ill Patients 246
11.3 Altered Drug Disposition in Patients With Kidney Disease 247
11.3.1 Altered Clearance of Renally Cleared Drugs in Patients With Kidney Disease 248
11.3.2 Altered Clearance of Nonrenally Cleared Drugs in Patients With Kidney Disease 249

11.4 Altered Drug Disposition in Hepatic Disease 250
11.5 Altered Drug Disposition in Critically Ill Patients 252
11.6 Application of TDM in Critically Ill Patients 255
11.6.1 TDM of Antimicrobial Drugs in Critically Ill Patients 255
11.6.2 TDM of Digoxin in Critically Ill Patients 256
11.6.3 TDM of Sedatives and Analgesics in Critically Ill Patients 257
11.7 Conclusions 257
References 257

CHAPTER 12 Issues of Pharmacogenomics in Monitoring Warfarin Therapy 261
Jennifer Martin and Andrew Somogyi

12.1 Introduction 261
12.2 Potential for Pharmacogenomics of Warfarin 262
12.3 Pharmacology 263
12.3.1 CYP2C9 Status 263
12.3.2 VKORC Status 264
12.3.3 Other Genetic Mutations 265
12.3.4 Nongenetic Factors Affecting Warfarin Dosing 266
12.3.5 Vitamin K 268
12.3.6 Diet 268
12.3.7 Age 268
12.3.8 Gender 269
12.3.9 Anthropometric Variables 269
12.4 Clinical Relevance 270
12.5 Cost-Effectiveness of Pharmacogenomics Testing in Warfarin Therapy 272
12.6 New Medications to Replace Warfarin 273
12.7 Conclusions 274
References 275

CHAPTER 13 Alternative Sampling Strategies for Therapeutic Drug Monitoring 279
Sara Capiau, Jan-Willem Alffenaar and Christophe P. Stove

13.1 Introduction 279
13.2 The Ideal Alternative Matrix for Therapeutic Drug Monitoring 280
13.3 The Correlation Between Alternative Matrix and Systemic Levels 282
13.4 Alternative Specimen: Dried Blood Spots 284
13.4.1 Applications of Dried Blood Spots in Therapeutic Drug Monitoring 290
13.5 Alternatives to Classical Dried Blood Spots 296
13.5.1 Volumetric Dried Blood Spots 296
13.5.2 Dried Plasma Spots 297
13.5.3 Other Dried Matrix Spots 297

13.6 Paper Spray Ionization for the Analysis of Dried Blood Spots 298
13.7 Alternative Specimen: Oral Fluid 299
13.7.1 Applications of Oral Fluid for Therapeutic Drug Monitoring 302
13.8 Alternative Specimen: Tears 307
13.8.1 Application of Tears for Therapeutic Drug Monitoring 308
13.9 Alternative Specimen: Interstitial Fluid 310
13.9.1 Application of Interstitial Fluid for Therapeutic Drug Monitoring 311
13.10 Alternative Specimen: Hair 312
13.10.1 Application of Hair Specimens for Therapeutic Drug Monitoring 314
13.11 Alternative Specimen: Exhaled Breath 316
13.11.1 Application of Exhaled Breath for Therapeutic Drug Monitoring 317
13.12 Alternative Specimen: Sweat 318
13.12.1 Application of Sweat for Therapeutic Drug Monitoring 319
13.13 Alternative Specimen: Nasal Mucus 319
13.14 Toward Routine Implementation 319
13.15 Conclusions 322
Acknowledgment 322
References 323

CHAPTER 14 Integrating Therapeutic Drug Monitoring in the Health Care Environment: Therapeutic Drug Monitoring and Pharmacists 337
William Clarke

14.1 Introduction 337
14.2 Team-Based Care in the Modern Health Care Environment 337
14.3 Hypertension 338
14.4 Diabetes and Hypertension 338
14.5 Cardiovascular Diseases 339
14.6 Pharmacists and Cost-Effective Care 339
14.7 Impact of Health Care Information Technology on TDM 340
14.8 Practical Examples of Integrating Pharmacy for TDM and Patient Care 341
14.8.1 Anticoagulation 342
14.9 Epilepsy 344
14.9.1 Antiinfective Management 346
14.10 Conclusions 349
References 349

INDEX **351**

List of Contributors

Jan-Willem Alffenaar Department of Clinical Pharmacy and Pharmacology, University Medical Center Groningen, University of Groningen, Groningen, The Netherlands

Sarah C. Campbell Division of Clinical Pharmacology, Department of Pediatrics, University of Utah, Salt Lake City, UT, United States

Sara Capiau Laboratory of Toxicology, Department of Bioanalysis, Faculty of Pharmaceutical Sciences, Ghent University, Ghent, Belgium

William Clarke Department of Pathology, Johns Hopkins University School of Medicine, Baltimore, MD, United States

Maged M. Costantine Department of Obstetrics and Gynecology, University of Texas Medical Branch, Galveston, TX, United States

Amitava Dasgupta Department of Pathology and Laboratory Medicine, University of Texas Health Science Center at Houston, Houston, TX, United States

Clinton Frazee Department of Pathology and Laboratory Medicine, Children's Mercy Hospitals and Clinics, University of Missouri School of Medicine, Kansas City, MO, United States

Jeunesse Fredrickson Department of Obstetrics and Gynecology, Maternal–Fetal Medicine, University of Utah School of Medicine, Salt Lake City, UT, United States

Uttam Garg Department of Pathology and Laboratory Medicine, Children's Mercy Hospitals and Clinics, University of Missouri School of Medicine, Kansas City, MO, United States

Ventzislava Hristova Department of Pathology, Johns Hopkins University School of Medicine, Baltimore, MD, United States

Kamisha L. Johnson-Davis Department of Pathology, University of Utah Health Sciences Center, Salt Lake City, UT, United States; ARUP Institute for Clinical and Experimental Pathology, Salt Lake City, UT, United States

Manijeh Kamyar Department of Obstetrics and Gynecology, Maternal–Fetal Medicine, University of Utah School of Medicine, Salt Lake City, UT, United States

Matthew D. Krasowski Department of Pathology, University of Iowa Hospitals and Clinics, Iowa City, IA, United States

Madhuri Manohar Division of Clinical Pharmacology, Department of Medicine, Johns Hopkins University School of Medicine, Baltimore, MD, United States

Jennifer Martin School of Medicine and Public Health, University of Newcastle, Newcastle, NSW, Australia

Mark A. Marzinke Division of Clinical Pharmacology, Department of Medicine, Johns Hopkins University School of Medicine, Baltimore, MD, United States; Division of Clinical Chemistry, Department of Pathology, Johns Hopkins University School of Medicine, Baltimore, MD, United States

Andrew J. McLachlan Faculty of Pharmacy and Centre for Education and Research on Ageing, University of Sydney and Concord Hospital, Sydney, Australia; National Health and Medical Research Council, Centre of Research Excellence in Medicines and Ageing, University of Sydney, Sydney, Australia

Gwendolyn A. McMillin Department of Pathology, School of Medicine, University of Utah, Salt Lake City, UT, United States; ARUP Institute for Clinical and Experimental Pathology, ARUP Laboratories, Inc., Salt Lake City, UT, United States

Jessica K. Roberts Division of Clinical Pharmacology, Department of Pediatrics, University of Utah, Salt Lake City, UT, United States

Laura M. Salisbury Department of Nursing, Clinical Pharmacology and Pediatric Nursing, Eagle Gate College, Salt Lake City, UT, United States

Catherine M.T. Sherwin Division of Clinical Pharmacology, Department of Pediatrics, University of Utah, Salt Lake City, UT, United States

Andrew Somogyi Discipline of Pharmacology, School of Medicine, University of Adelaide, Adelaide, Australia

Christophe P. Stove Laboratory of Toxicology, Department of Bioanalysis, Faculty of Pharmaceutical Sciences, Ghent University, Ghent, Belgium

Preface

Therapeutic drug monitoring in patient care is useful for individual dosage adjustment and for avoiding drug toxicity for certain drugs with a narrow therapeutic window. However, only relatively few drugs require therapeutic drug monitoring, and most of these drugs are used for treating chronic conditions, with the exceptions of vancomycin and aminoglycosides, which are used over a relatively short time for treating a life-threatening infection. In general, approximately 25 drugs are routinely monitored in clinical laboratories, and for most of these drugs there are readily available immunoassays that can be adopted on automated chemistry analyzers. These drugs include cardioactive drugs, anticonvulsants, antiasthmatics, tricyclic antidepressants, immunosuppressants, certain antibiotics, and certain anticancer drugs, most commonly methotrexate. There are approximately another 25–30 drugs that are monitored less frequently, and for most of these drugs immunoassays are not commercially available.

There are many critical issues in the field of therapeutic drug monitoring that currently are not thoroughly covered in a single book. The goal of this book is to provide up-to-date information on critical issues in therapeutic drug monitoring, including special requirements of therapeutic drug monitoring in selected populations (infants and children, pregnant women, elderly patients, and obese patients), as well as issues with free drug monitoring and interferences in using immunoassays for therapeutic drug monitoring. Although there are excellent books on pharmacogenomics, this book covers warfarin pharmacogenomics, which is probably the most important topic in pharmacogenomics of therapeutics. Therefore, this book focuses on special topics in therapeutic drug monitoring that will be useful not only to clinicians and toxicologists but also to medical students and nurses.

Chapter 1 "Overview of Therapeutic Drug Monitoring" provides an overview of therapeutic drug monitoring, and Chapter 2 "Immunoassays and Issues With Interference in Therapeutic Drug Monitoring" discusses the limitations of immunoassays used in therapeutic drug monitoring. The gold standard of therapeutic drug monitoring is liquid chromatography combined with tandem mass spectrometry. This important methodology in therapeutic drug monitoring is discussed in Chapter 3"Application of Chromatography Combined With Mass Spectrometry in Therapeutic Drug Monitoring." Monitoring free drug is essential for strongly protein-bound drugs, and this topic is covered in-depth in Chapter 4 "Monitoring Free Drug Concentration: Clinical Usefulness and Analytical Challenges." Monitoring of newer antiepileptics is discussed in Chapter 5 "Therapeutic Drug Monitoring of Newer Antiepileptic Drugs," and the emerging field of therapeutic drug monitoring of antiretrovirals is covered in Chapter 6 "Therapeutic Drug Monitoring of Antiretrovirals." Therapeutic drug monitoring in special patient populations is thoroughly discussed in Chapter 7 "Therapeutic Drug Monitoring in Infants and Children," 8 "Therapeutic Drug Monitoring in Pregnancy," 9 "Therapeutic Drug Monitoring in Older People," 10 "Therapeutic Drug Monitoring in Obese Patients," and 11 "Special Issues in Therapeutic Drug Monitoring in Patients With Uremia, Liver Disease, and in Critically Ill Patients." The role of pharmacogenomics in monitoring of warfarin is addressed in Chapter 12 "Issues of Pharmacogenomics in Monitoring Warfarin Therapy." Therapeutic drug monitoring in alternative specimens is gaining popularity, and this important topic is addressed in Chapter 13 "Alternative Sampling Strategies for Therapeutic Drug Monitoring." The important role of pharmacists in integrating therapeutic drug monitoring in clinical service is discussed in Chapter 14 "Integrating Therapeutic Drug Monitoring in the Health Care Environment: Therapeutic Drug Monitoring and Pharmacists."

Many special issues addressed in this book regarding therapeutic drug monitoring are currently discussed only in peer-reviewed publications. Experts in respective subspecialities of therapeutic drug monitoring generously offered their valuable time to write individual chapters on these special topics so that readers have easy access to these topics in a single book for their convenience. We thank all contributors for their chapters. Without them taking time from their busy work schedules to write these chapters, this book would not have been possible. Lastly, Bill and I thank our wives for putting up with us during the long hours we spent writing chapters and editing this book. We hope this book will serve as a single resource covering all special issues related to therapeutic drug monitoring. If readers find this book useful, the hard work of all contributors and our hard work as editors will be duly rewarded.

Respectfully,

William Clarke

Baltimore, MD

Amitava Dasgupta

Houston, TX

CHAPTER 1

Overview of Therapeutic Drug Monitoring

William Clarke
Department of Pathology, Johns Hopkins University School of Medicine, Baltimore, MD, United States

CONTENTS

1.1 Introduction 1
1.2 Principles of TDM 1
1.3 Clinical Areas in Which TDM is Routine Practice ... 2
1.3.1 Epilepsy 3
1.3.2 Organ Transplantation 5
1.3.3 Cardiology (Antiarrhythmic Drugs) ... 7
1.3.4 Psychiatry 8
1.3.5 Infectious Disease 9
1.3.6 Oncology 11
1.4 Conclusions .. 12
References 13

1.1 INTRODUCTION

The vast majority of drugs are managed in a very standard way: The drugs are dosed on a unit per body mass (eg, mg/kg) basis, and then the dosage is adjusted based on the clinical response as empirically assessed by a physician. This is often described as "titration to clinical effect." For a small subset of drugs, the clinical effects (pharmacodynamics) can be assessed objectively through laboratory measurements—for instance, Coumadin and assessment of coagulation via international normalized ratio measurement, or statin drugs and blood lipid levels. For an even smaller subset of drugs, the pharmacokinetic–pharmacodynamic relationship is not predictable from the dose, and the pharmacokinetics are highly variable between individuals; for these drugs, management can be particularly challenging. It is for these drugs that therapeutic drug monitoring (TDM) is most effective.

1.2 PRINCIPLES OF TDM

TDM in practice is performed by collection of a blood sample at a known time relative to administration of the last (or next) dose. The concentration of drug and/or metabolite is measured in the sample and compared to a target range or predicted pharmacokinetics for the drug. In order for TDM to be effective and necessary, several criteria must be met: (1) The drugs must have a narrow therapeutic index (the difference between the minimum effective concentration and the toxic concentration, relative to the pharmacokinetic variability), (2) the relationship between the drug dose and concentration in blood must be highly variable and/or not predictable, (3) the relationship between blood concentration and clinical or toxic effect must be well-defined, (4) there should be serious consequences for under- or overdosing,

W. Clarke & A. Dasgupta (Eds): Clinical Challenges in Therapeutic Drug Monitoring. DOI: http://dx.doi.org/10.1016/B978-0-12-802025-8.00001-5

and (5) the result of TDM testing must be interpretable and actionable—there should be an effect on clinical outcomes. It should answer a clinical question.

There are some drugs that meet most of the criteria for TDM but for which measurement of drug concentrations is not particularly useful. For instance, when the drug is administered as a prodrug, the biologically active form of the drug is a metabolite of the administered substance. Therefore, measurement of the administered drug does not help with guidance of therapy, and the active metabolite must be measured (eg, irinotecan and SN-38). In other cases, the drug is converted to its active form intracellularly, so blood measurements are not reflective of the therapeutic activity (eg, antiretroviral drugs and peripheral blood mononuclear cells). In addition, drugs for which tolerance can be developed (eg, narcotics and pain management) prevent the utility of TDM because the effective blood concentration range is a moving target and not stable within an individual patient.

Although the primary function of TDM is to allow for adjustment of the drug dose, there are other applications of TDM results (eg, questions to be answered). For example, TDM in antiretroviral management in patients with HIV is not necessary for optimization of dose; however, it is vitally important to confirm adherence to the prescribed drug regimen in the context of increasing viral loads and apparent therapeutic failure. The TDM results can be used to assist in the determination of whether the patient has developed viral resistance to the prescribed drugs or whether the patient has simply stopped taking the drugs. Another question that can be answered is whether drug is being absorbed at all. For instance, gastrointestinal (GI) inflammation could prevent drugs administered orally from entering the circulation. It is important to remember that TDM is a tool for assessing the clinical presentation of the patient. Most therapeutic target intervals were not derived from large clinical studies but, rather, are based on the observations and data from a single site. The absence of a therapeutic interval does not mean that TDM lacks utility, provided that there are specific criteria for interpretation of the result and it will aid in clinical decision-making.

1.3 CLINICAL AREAS IN WHICH TDM IS ROUTINE PRACTICE

Certain pharmacotherapies in patients with specific clinical conditions require routine TDM. Most of these drugs are used for treating chronic conditions such as epilepsy, but certain toxic antibiotics, such as aminoglycosides and vancomycin, require TDM if used for more than 2 days. These clinical situations and drugs that are subjected to TDM are discussed in this section.

1.3.1 Epilepsy

For treatment of epilepsy, TDM has significant value because the drugs act on the central nervous system (CNS) to treat the seizures, but they also can cause toxic CNS effects. In some cases, the drugs can cause seizures, making it difficult to differentiate subtherapeutic effects from toxic effects. TDM is useful to quickly determine whether drug concentrations are too low or too high upon initiation of therapy. In addition, once a baseline concentration has been established for patients on chronic therapy, blood levels can be used to investigate loss of seizure control or unexpected toxicity in a stable patient.

Phenytoin and fosphenytoin (prodrug for phenytoin) are widely used for treating epilepsy in a broad range of populations. Phenytoin has been in use for epilepsy since 1936; fosphenytoin was developed in 1996. The drug has a high bioavailability, but the rate of absorption is highly dependent on formulation—in some cases, absorption can occur up to 48 h after administration. There is significant pharmacokinetic variability because phenytoin is susceptible to drug–drug interactions and is highly protein bound. The typical target interval for this drug is 10–20 μg/mL, with CNS toxicity typically observed with concentrations greater than 30 μg/mL [1].

Carbamazepine is a drug that was introduced for treatment of epilepsy in 1962. This drug is CYP450 metabolized, and it is affected by both inhibitor and inducers of this enzyme family. Interestingly, its inductive capabilities lead to a drug–drug interaction with itself for patients on chronic therapy, giving increased blood concentrations. The therapeutic interval is defined as 4–12 μg/mL [2], and blood concentrations greater than 12 μg/mL are associated with adverse CNS effects.

Primidone is a barbiturate drug used primarily for treatment of epilepsy in children; it is also used in the treatment of essential tremor. Primidone is primarily metabolized hepatically to phenobarbital, and given that both parent drug and primary metabolite are biologically active, both must be monitored when primidone is used for antiepileptic therapy. The therapeutic target for primidone is 5–12 μg/mL, and adverse effects are seen at concentrations greater than 15 μg/mL [3]. Clinical toxicity is primarily due to accumulation of phenobarbital, and it is manifested as reduced respiratory function and CNS depression. Primidone is CYP450 metabolized and thus sensitive to drug–drug interactions.

Phenobarbital is another barbiturate drug used for treatment of epilepsy. As noted previously, it is also the primary biologically active metabolite of primidone; it can be used alone as a therapeutic compound or in conjunction with other antiseizure medications. The therapeutic interval is 15–40 μg/mL, and blood concentrations of greater than 40 μg/mL may cause significant

CNS depression [4]. Phenobarbital has a significantly longer half-life than many drugs that are monitored, and it is thus not as sensitive to timing of collection as other drugs.

Valproic acid is a drug that has long been used for treatment of epilepsy (since 1962), but it is also used for the treatment of bipolar disorder, migraines, and neuropathic pain. As with phenytoin, valproic acid is highly protein bound, and measurement of the free (or unbound) drug can be used for patient management. The therapeutic interval is defined as 50–100 μg/mL, and concentrations greater than 100 μg/mL are associated with significant toxic effects, including hepatic, GI, and CNS toxicity [5].

Lamotrigine is one of a newer generation of antiepileptic drugs; it was introduced for treatment of epilepsy in the early 1990s. Currently, it is not only used for its antiseizure effects but also used as a mood stabilizer in the treatment of bipolar disorder and clinical depression. The therapeutic interval is generally reported as 2.5–15.0 μg/mL (C_0 concentration) [6]. Adverse effects from elevated lamotrigine concentrations include vision abnormalities and GI toxicity. TDM is particularly useful for adherence monitoring.

Levetiracetam is another newer-generation antiepileptic drug used for treatment of partial myoclonic, tonic–clonic, and partial seizures. It is also used for the treatment of manic states in bipolar disorder and for migraine headaches. Levetiracetam has a relatively short half-life (~6 or 7 h) and is cleared renally, so any renal dysfunction or acute kidney injury may warrant TDM and/or dose adjustment. The generally reported therapeutic interval for C_0 samples is 12–46 μg/mL; no toxic threshold has been established [7]. Clinical toxicities associated with increased levetiracetam concentrations include general anemia, neutropenia, and significant drowsiness (somnolence). These toxicities can sometime occur even when the drug concentration is within the therapeutic interval.

Gabapentin is another of the newer antiepileptic drugs that was introduced in the early 1990s for treatment of seizures. Since its introduction, it has also found use for treatment of neuropathic pain, restless leg syndrome, anxiety, bipolar disorder, and insomnia. Gabapentin does not undergo any hepatic metabolism and is cleared almost completely by renal excretion—the elimination half-life is approximately 5–7 h in patients with normal renal function. The generally reported therapeutic interval for C_0 concentrations is 2–20 μg/mL [8]. Toxic effects of high gabapentin concentration include extreme fatigue, drowsiness, dizziness, and ataxia (loss of full control of body movements).

Oxcarbazepine is a prodrug that is metabolized to a biologically active metabolite, 10-hydroxy-10,11-dihydrocarbamazepine, known more commonly as monohydroxycarbamazepine (MHC). MHC is responsible for the antiseizure activity of the drug, and it is the component that is measured in the blood for TDM. The half-life for MHC is longer than that of oxcarbazepine (8–10 h vs 1–2.5 h). The therapeutic interval for MHC is reported generally as 3–35 μg/mL [9]. Toxicity associated with MHC includes dizziness, drowsiness, GI toxicity, tremor, ataxia, and abnormal gait. These toxicities can sometimes occur even when the drug concentration is within the therapeutic interval.

1.3.2 Organ Transplantation

Organ transplantation is an extremely complex medical procedure, and immunosuppression is central to controlling the body's response to the transplanted organ to ensure a successful xenograft. Management of these powerful drugs requires a balance of ensuring enough drug exposure to prevent rejection while keeping the concentration of the drug low enough to avoid toxic effects. TDM is an essential tool in the process, allowing rapid titration of blood concentrations of the drug to maximize immunosuppression and avoid acute rejection while minimizing adverse events from exposure to the drug.

Tacrolimus is a calcineurin inhibitor, and it is the most widely used immunosuppressive drug used in transplantation. It can be administered either intravenously or orally, and it exhibits significant interindividual variability. Tacrolimus is monitored in whole blood rather than serum or plasma, and a general therapeutic interval is reported as 5–15 ng/mL for C_0 concentrations [10], although in practice target concentration ranges are more narrow and dependent on the type of organ transplanted, as well as the time from transplantation. Adverse effects of elevated concentrations of tacrolimus in blood include nephrotoxicity, neurotoxicity, hypertension, and nausea.

Cyclosporine A (CsA) is a calcineurin inhibitor that has been available for longer than tacrolimus but is not as widely used. CsA is available in both intravenous and oral forms, with variable absorption and distribution; it is also highly protein bound. CsA is found in both plasma and red blood cells, but measurement of the drug primarily occurs in whole blood samples because the drug distributes into red cells in vitro after collection as the temperature decreases. The general target interval for C_0 concentrations is 100–400 ng/mL [11], although the target in clinical practice is dependent on multiple factors, including the type of transplant, time from transplantation, method of analysis, and coadministered drugs. There is evidence that C_2 (peak concentration 2 h after administration) is more closely correlated with clinical outcomes [12]. CsA is metabolized in the liver by CYP3A4, so it demonstrates significant interindividual pharmacokinetic variability and is susceptible to drug–drug interactions. Adverse effects of high CsA concentrations include renal toxicity, as well as liver or CNS effects.

Sirolimus is an interleukin-2 (IL-2)-inhibiting drug that targets the mTOR (mammalian target of rapamycin) receptor, which interrupts the cell cycle. Sirolimus is primarily metabolized by CYP3A4/5, so it exhibits significant pharmacokinetic variability and is susceptible to drug–drug interactions. The half-life of sirolimus is relatively long (48–72 h) compared to those of the other immunosuppressants, so it takes longer to achieve steady-state concentrations after initiation of therapy or with a dosage change. Generally, the therapeutic interval for sirolimus is 4–12 ng/mL, but as with the other drugs, it is dependent on transplant type and comedication, in addition to other variables [13]. Sirolimus is measured in whole blood samples rather than serum or plasma due to its significant distribution into red blood cells. The benefit of this drug is that it does not have renal toxicity, making it appealing for use in kidney transplantation. However, significant toxic effects are still possible, including leukopenia, thrombocytopenia, and hypercholesterolemia.

Everolimus is a structural analog of sirolimus, and as such they share the same target and mechanism of action (mTOR inhibition). The primary difference between everolimus and sirolimus is that everolimus has a shorter half-life (18–36 h) and its toxic effects are less significant. However, it is still important to manage this drug using TDM for optimal therapeutic effect while minimizing toxicities. Everolimus is also measured in whole blood samples for TDM. The target therapeutic interval is 3–15 ng/mL, with modifications based on coadministered medications, transplant type, and time from transplantation [14].

Mycophenolic acid (MPA) is an inhibitor of inosine monophosphate dehydrogenase, which inhibits cell group by inhibition of purine synthesis. MPA is the active compound; however, the drug is actually administered as a prodrug such as mycophenolate mofetil (CellCept) or mycophenolate sodium (Myfortic), which is an extended-release formulation. The drug is used in conjunction with either calcineurin or mTOR inhibitors as adjuvant therapy for immunosuppression. The drug has a relatively short half-life (9–18 h) and is not distributed into red blood cells like the other commonly used immunosuppressant drugs. Important differences for MPA include the following: (1) The measurement of the drug is in plasma rather than whole blood; and (2) a peak or C_0 measurement is not sufficiently correlated with the area under the curve (AUC), so a single measurement is not sufficient. It is recommended that AUC with limited sampling be used for TDM of this compound [15]. The primary toxicity for MPA is GI toxicity; there is some debate about whether TDM is really needed for MPA or whether the dose can just be reduced if GI toxicity is observed.

1.3.3 Cardiology (Antiarrhythmic Drugs)

Historically in cardiology, certain therapeutic agents have been used to control arrhythmia in patients with abnormal heart physiology. However, these historic therapeutic agents displayed two important tendencies. First, each of the agents demonstrated significant pharmacokinetic variability. Second, although these drugs can control arrhythmia when the dose is optimized, they can also cause arrhythmia when concentrations in blood are too high. Based on these characteristics, it was recognized that TDM is a pivotal tool in optimal management of these agents.

Procainamide is an antiarrhythmic drug used for control of both atrial and ventricular arrhythmias, and *N*-acetylprocainamide (NAPA) is the primary metabolite that also has antiarrhythmic activity (via a different mechanism). The half-life of procainamide is approximately 3 or 4 h, and the half-life for NAPA is approximately 6 h; both compounds exhibit significant interindividual variability. The therapeutic target for procainamide is 4–8 μg/mL, and that for NAPA is 10–20 μg/mL; when considered together, the therapeutic target interval for procainamide and NAPA combined is 5–30 μg/mL [16]. Adverse effects associated with high concentrations of procainamide and NAPA include hypotension, induced arrhythmia, and widening of QRS intervals (associated with chronically high exposure).

Lidocaine is a drug commonly used for emergent treatment of acute or life-threatening arrhythmias. It is highly protein bound (to $\alpha_1$acid glycoprotein) and is metabolized in the liver. Because lidocaine is used in acutely ill patients, and $\alpha_1$acid glycoprotein is an acute phase reactant, it is common for the free/unbound fraction of the drug to change rapidly along with the patient condition, thus changing the effective concentration of the drug. However, despite this, the free fraction of lidocaine is not commonly monitored. In addition, the hepatic metabolism of the drug means that the pharmacokinetics can be significantly impacted by changes in liver perfusion or liver injury, which is particularly pertinent in an acutely ill population. The therapeutic interval for lidocaine is reported as 1.5–5 μg/mL [17]. Toxic effects from high lidocaine concentrations include bradycardia, hypotension, and CNS dysfunction.

Quinidine is a drug used for both atrial and ventricular arrhythmias, but it is less commonly used due to GI side effects present even when blood concentrations are within the therapeutic range, in addition to significant toxic effects associated with high concentrations in the blood. Blood concentrations are best measured as C_0 levels, and the therapeutic interval is reported as 2–5 μg/mL [18]. Concentrations greater than 5 μg/mL are associated with hypotension, ventricular tachycardia or fibrillation, cinchonism,and QT interval elongation on an electrocardiogram. Interestingly, the QT interval elongation is a desired effect for patients with Brugada syndrome, making it a preferred treatment in that population [19]. Several metabolites of quinidine have been shown to have biologic activity, but they are not routinely monitored.

Digoxin is a cardiac glycoside used in treatment of cardiac arrhythmia and also congestive heart failure. One of the primary toxic effects of the drug at higher than therapeutic concentrations is cardiac arrhythmia that is difficult to differentiate from the clinical indication for the drug in the first place. The therapeutic interval for digoxin is reported as 0.8–2.0 ng/mL [20]. Toxic symptoms in addition to arrhythmia include GI toxicity and neurologic symptoms. Digoxin assays can encounter significant interference from digoxin-like immunoreactive factors or Digibind (Fab fragments specific for digoxin used for treatment of overdose); these interferences are discussed in detail in Chapter 2, "Immunoassays and Issues with Interference in Therapeutic Drug Monitoring".

1.3.4 Psychiatry

In the treatment of psychiatric disorders such as depression or schizophrenia, pharmacotherapy is commonly managed by titration to clinical effect. This approach is taken because of the lack of robust relationship between dose administered (or blood concentration) and the clinical response to therapy. In addition, it is difficult to investigate these relationships because the clinical endpoint of response is somewhat subjective and can vary based on the variability of how the patients communicate their experience and also the way the clinicians perceive their interaction with the patients. However, for some drugs, the adverse effects of the drug can be objectively measured and correlated with blood concentration of the drug. In these cases, TDM can be a useful tool for patient management.

Lithium is a drug (administered as lithium salts) that is widely used for treatment of bipolar disorder; specifically, it is used for control of the manic phase in this disorder. The blood concentration of lithium can be affected by changes in other physiologically relevant electrolytes. For instance, increased intake of sodium can enhance lithium excretion, whereas decreased physiologic sodium concentration can lead to reduced lithium excretion. Lithium is cleared from circulation only by the kidneys, so decreased renal function can lead to toxic accumulation of the drug. The therapeutic interval for lithium is generally reported as 0.4–1.2 mmol/L, with toxic effects seen at concentrations greater than 1.5 mmol/L [21]. Clinically significant side effects of high lithium concentrations include renal failure and excessive water and electrolyte loss. Chronic lithium administration can also impact thyroid function (up to 35% of patients treated with lithium develop hypothyroidism), so thyroid function should be monitored regularly for patients on lithium therapy [22].

Tricyclic antidepressants include amitriptyline, nortriptyline, imipramine, desipramine, and clomipramine and are used for treatment of major depression. Although a robust relationship between blood concentration and therapeutic clinical effect is difficult to define, there is a well-described relationship between blood concentration of tricyclic antidepressant drugs and life-threatening cardiac arrhythmia. It is because of this significant toxicity and potential for suicide using the prescribed therapeutic drug that tricyclic antidepressants are not widely used at present. However, in certain patient populations, these drugs are still preferred, and when they are used, TDM is necessary to ensure patient safety. Blood concentrations greater than 500 ng/mL are generally reported as toxic and are associated with increased risk of cardiac arrhythmia [23]. For other antidepressant drugs, such as selective serotonin reuptake inhibitors and selective norepinephrine reuptake inhibitors, the toxic side effects of the drugs are not so severe, and without a well-defined therapeutic interval, TDM of these drugs is not commonly performed.

More recently, TDM of antipsychotic drugs has emerged as an area of interest for TDM in psychiatry, particularly for monitoring adherence to therapy. Although the same challenges exist for these drugs as for antidepressants, in terms of drug exposure predicting clinical success in treatment, there is growing interest in using drug monitoring to assess adherence to therapeutic regimens. Common requests for TDM in this area include clozapine, olanzapine, risperidone, and aripiprazole. Of course, to assess whether a patient is taking the drug does not require a blood sample or TDM; one recent study demonstrated the potential of using urine metabolites for assessment of adherence in patients treated with aripiprazole [24]. However, TDM is still not standard of practice for most psychiatric patients.

1.3.5 Infectious Disease

In the treatment of infectious disease, the biologic activity of the drug is directed against the microorganism responsible for the infection and not toward the host (the person taking the drug). As such, the therapeutic index is quite wide, and although many of the drugs have significant pharmacokinetic variability and it is important to have the drug concentration (exposure) higher than the minimal effective concentration (MIC) for eradication of the microorganism, the lack of toxic effects against the host/patient allows for a high dose relative to the MIC in order to ensure sufficient blood concentrations for microbicidal effect. However, some antimicrobial drugs do have both significant pharmacokinetic variability and significant clinical toxicity for the patient. In these cases, TDM is a valuable tool for management of the drugs.

Aminoglycoside antibiotics include gentamicin and tobramycin (effective against Gram-positive and Gram-negative bacteria) as well as amikacin (used primarily against Gram-negative bacteria). These drugs are characterized by their short half-lives (1 or 2 h) and lack of oral bioavailability—they are all administered intravenously. The aminoglycoside antibiotics are cleared by the kidney, and they exhibit significant interindividual pharmacokinetic variability. Traditional dosing of these drugs includes dosing every 6–8 h, with peak monitoring target concentrations of 5–10 μg/mL for gentamicin and tobramycin and 20–25 μg/mL for amikacin [25]. However, due to the antibiotic residual effect [26], where the antibiotic effects of the drug are related to the peak exposure concentration but toxic effects are related to the duration of exposure, daily dosing regimens (or pulse dosing) are more common. In this therapeutic paradigm, all of the doses for a day are now combined into a single administered dose for the entire day. When this approach is taken, the peak therapeutic target intervals are no longer applicable, and TDM is performed using C_0 monitoring to ensure that the drug is being cleared. In this approach, C_0 levels should be less than 2 μg/mL. Significant toxic effects are commonly observed when these drugs are used, including irreversible ototoxicity and renal toxicity in the form of renal tubular damage.

Vancomycin is a glycopeptide antibiotic used for treatment of many Gram-positive bacterial infections, and it also demonstrates treatment efficacy for some Gram-negative bacteria as well. It works particularly well for methicillin-resistant *Staphylococcus aureus* infection. As with the aminoglycosides, vancomycin does not have oral bioavailability and is administered as an intravenous infusion. The half-life of vancomycin is 4–6 h, and it is typically given twice per day. It is renally cleared with significant interindividual variability; TDM is performed using trough levels with a target concentration interval of 10–20 μg/mL [27]. Toxic effects of vancomycin include ototoxicity and nephrotoxicity, although these toxic effects are more commonly seen when the drug is coadministered with drugs that have the same toxicity profile such as the aminoglycoside antibiotics.

Voriconazole is a triazole antifungal drug that is used to treat invasive fungal infections (eg, candidiasis or aspergillosis) and also for prophylaxis in severely immunocompromised patients. Voriconazole is administered both orally and intravenously, with an oral bioavailability of 96%. It is metabolized in the liver by the cytochrome P450 system and thus demonstrates significant interindividual pharmacokinetic variability and is susceptible to a number of drug–drug interactions. The reported therapeutic target interval is 1.0–5.5 μg/mL, and adverse side effects include GI toxicity, headache, peripheral edema, and visual disturbances. The primary toxicity associated with blood concentrations greater than 6.0 μg/mL is liver toxicity [28].

Although the antimicrobial drugs listed previously are those most commonly managed using TDM, there are some emerging applications in the area of infectious disease. Because the drug posaconazole is more frequently used to treat antifungal infections, there are those that suggest it should be monitored similarly to voriconazole and for the same reasons. In addition, interest has recently increased regarding the application of TDM for antituberculosis drugs. β-Lactam drugs exhibit significant pharmacokinetic variability, and a number of studies suggest that TDM may be of benefit in patients administered those drugs as well.

1.3.6 Oncology

In the treatment of cancer, most drugs are dosed based on body surface area normalization, and then the patients are monitored for significant toxicity or clinical effect—this is similar to the classic titrate to clinical effect paradigm. However, in cancer treatment, the additional parameter of maximum tolerated dose (MTD) must be considered. Because the drugs are known to be toxic (they are cytotoxic drugs after all), part of the clinical trials during drug development involves assessment of dose-limiting toxicity relative to the administered dose in a small number of patients in order to determine the maximum amount of drug that should be given. When patients are receiving the MTD and not exhibiting symptoms of toxicity, treatment is continued without dose adjustment until it is determined whether the treatment is affecting the cancer. However, it is important to note that many chemotherapy drugs have significant pharmacokinetic variability so that when MTD is given to the patient and no toxic effects are observed, it is not certain that the patient is getting the correct dose; it may be that the patient is actually getting less drug than needed (or that he or she can tolerate). Based on this, some have advocated for the concept of maximum tolerated exposure [29], which would require blood concentration measurements. TDM is not routine in the management of chemotherapy.

Methotrexate is a folic acid antagonist drug that initially was used as an immunosuppressant drug in transplantation but has found more widespread use in the treatment of malignancy or autoimmune disease. As with many other drugs that require TDM, methotrexate exhibits significant interindividual variability and is susceptible to drug–drug interactions. In the treatment of autoimmune disorders, low-dose methotrexate is used and typically TDM is not needed for management of these patients. However, in treatment of malignancy, the risk of adverse effects on normal cells is much greater, and the drug is commonly used with leucovorin (a folic acid derivative) to prevent damage to noncancer cells. The amount of leucovorin needed is dependent on the rate of methotrexate clearance, so methotrexate concentrations are measured over multiple days. The target concentration intervals are 5–10 μM at 24 h postdose, 0.5–1.0 μM at 48 h postdose, and 0.05–0.1 μM at 72 h postdose [30]. When methotrexate concentrations are greater than those time-dependent target intervals, additional leucovorin is needed. The adverse effects of high methotrexate levels include hepatotoxicity, leukopenia, and ulcerative stomatitis.

Busulfan is an alkyl sulfonate drug that functions as a nonspecific alkylating neoplastic agent. It was initially approved as a treatment for chronic myeloid leukemia (CML); however, it is currently primarily used as a conditioning agent along with cyclophosphamide, fludarabine, or total body irradiation prior to bone marrow transplantation. Busulfan can be given as either an oral or an intravenous formulation; the kinetics is more predictable when given intravenously. TDM-based dosing of busulfan is not based on a single measurement but instead based on 6-h AUC. The general target AUC is between 900 and 1500 μmol/min. When the AUC is less than 900 μmol/min, the risk of incomplete bone marrow ablation is increased, along with the chances of graft rejection or graft-versus-host disease; in these cases, the dose should be increased. AUCs greater than 1500 μmol/min are associated with liver toxicity known as sinusoidal obstructive syndrome, also known as venoocclusive disease, which can result in significant organ damage and poor clinical outcomes [31].

Although TDM has been sparsely utilized in oncology with the exception of methotrexate and busulfan, there is an emerging body of evidence suggesting that it can be a useful tool for managing other chemotherapy drugs. The largest number of studies supporting TDM in the scientific literature support optimization of 5-fluorouracil (5-FU) management, including one of the few randomized control trials published in 2008 [32]. However, despite the large number of studies suggesting the utility of 5-FU monitoring, it has not become standard of practice. Recently, clinical studies have demonstrated the potential for TDM to reduce toxicity in patients receiving taxanes (docetaxel and paclitaxel) for treatment of cancer [33]. Finally, several studies have suggested that TDM for management of imatinib in the settings of CML [34] and GI stromal tumors [35] has potential to positively impact patient outcomes. However, despite the emerging evidence, TDM remains far from routine implementation in oncology settings.

1.4 CONCLUSIONS

As the concept of personalized or precision medicine matures, there are many opportunities for TDM moving forward. Increasing focus on special populations, whether they are age-related (eg, pediatric or geriatric medicine), disease-related (eg, transplantation or cystic fibrosis), or health and wellness-defined (eg, obesity), opens interesting opportunities for the use of blood concentrations of drugs to manage patients. Advances in technology allow for increased pharmacogenetic testing, highlighting the complementary relationship of TDM and pharmacogenetics for personalized therapeutic drug management. Whereas development of commercial assays for TDM has stagnated, whether due to the relative small size of potential markets or the perception that the role of pharmacokinetics in personalized medicine will decrease based on advances in genomics, research in the area of assay development using noncommercial platforms is rapidly growing, especially in the area of liquid chromatography–tandem mass spectrometry. In addition, changes in healthcare infrastructure are impacting how the laboratory can be integrated into the medical team responsible for patient care. TDM is now a mature field in laboratory medicine, with a number of well-established applications and a history of almost 50 years, but with recent advances in healthcare information systems and technologies for rapid measurement of drug concentrations in biological matrices, the real advances to optimize TDM practices for improved patient outcomes are just beginning.

References

[1] Patsalos PN, Berry DJ, Bourgeois BF, Cloyd JC, Glauser TA, Johannessen SI, et al. Antiepileptic drugs-best practice guidelines for therapeutic drug monitoring: a position paper by the sub-commission on therapeutic drug monitoring, ILAE Commission on Therapeutic Strategies. Epilepsia 2008;49:1239–76.

[2] Brodie MJ, Richens A, Yeun AW. Double-blind comparison of lamotrigine and carbamazepine in newly diagnosed epilepsy. UK Lamotrigine/Carbamaepin Monotherapy Trial Group. Lancet 1995;345:476–9.

[3] Booker HE, Hosokowa K, Burdette RR, Darcey B. A clinical study of serum primidone levels. Epilepsia 1970;11:395–402.

[4] Schmidt D, Einicke I, Haenel FT. The influence of seizure type on the efficacy of plasma concentrations of phenytoin, phenobarbital, and carbamazepine. Arch Neurol 1986;43:263–5.

[5] Henriksen O, Johannessen SI. Clinical and pharmacokinetic observations on sodium valproate—a 5-year follow-up study in 100 children with epilepsy. Acta Neurol Scand 1982;65:504–23.

[6] Morris RG, Black AB, Harris AL, Batty AB, Sallustio BC. Lamotrigine and therapeutic drug monitoring: retrospective survey following the introduction of a routine service. Br J Clin Pharmacol 1998;46:547–51.

[7] Leppik IE, Rarick JO, Walczak TS, Tran TA, White JR, Gumnit RJ. Effective levetiracetam doses and serum concentrations: age effects. Epilepsia 2002;43(Suppl. 7):240.

[8] Sivenius J, Kalviainen R, Ylinen A, Riekkinen P. A double-blind study of gabapentin in the treatment of partial seizures. Epilepsia 1991;32:539–42.

[9] Striano S, Striano P, Di Nocera P, Italiano D, Fasiella C, Ruosi P, et al. Relationship between serum mono-hydroxy-carbazepine concentrations and adverse effects in patients with epilepsy on high-dose oxcarbazepione therapy. Epilepsy Res 2006;69:170–6.

[10] Agarwal YP, Cid M, Westgard S, Parker TS, Jaikaran R, Levine DM. Transplant patient classification and tacrolimus assays: more evidence of the need for assay standardization. Ther Drug Monit 2014;36:706–9.

[11] Oellerich M, Armstrong VW, Kahan B, Shaw L, Holt DW, Yatscoff R, et al. Lake Louise Consensus Conference on cyclosporine monitoring in organ transplantation: report of the consensus panel. Ther Drug Monit 1995;17:642–54.

[12] Pape L, Lehnhardt A, Latta K, Ehrich JH, Offner G. Cyclosporin A monitoring by 2-h levels: preliminary target levels in stable pediatric kidney transplant recipients. Clin Transplant 2003;17:546–8.

[13] Morris RG, Salm P, Taylor PJ, Wicks FA, Theodossi A. Comparison of the reintroduced MEIA assay with HPLC-MS/MS for the determination of whole-blood sirolimus from transplant recipients. Ther Drug Monit 2006;28:164–8.

[14] Mabasa VH, Ensom MH. The role of therapeutic monitoring of everolimus in solid organ transplantation. Ther Drug Monit 2005;27:666–76.

[15] Shaw LM, Nowak I. Mycophenolic acid: measurement and relationship to pharmacologic effects. Ther Drug Monit 1995;17:685–9.

[16] Wierzchowiecki M, Jasinski K, Trojanowicz R, Ochotny R, Tomaszkiewicz T. Pharmacokinetic and pharmacodynamics studies of procainamide given intermittently intravenously in patients with severe ventricular arrhythmias. Cor Vasa 1978;20:176–83.

[17] Alderman EL, Kerber RE, Harrison DC. Evaluation of lidocaine resistance in man using intermittent large-dose infusion techniques. Am J Cardiol 1974;34:342–9.

[18] Halkin H, Vered Z, Milman P, Rabinowitz B, Neufeld HN. Steady-state serum quinidine concentration: role in prophylactic therapy following acute myocardial infarction. Isr J Med Sci 1979;15:583–7.

[19] Pellegrino PL, Di Biase M, Brunetti ND. Quinidine for the management of electrical storm in an old patient with Brugada syndrome and syncope. Acta Cardiol 2013;68:201–3.

[20] Horn JR, Christensen DB, deBlaquiere PA. Evaluation of a digoxin pharmacokinetic monitoring service in a community hospital. Drug Intell Clin Pharm 1985;19:45–52.

[21] Frings CS. Lithium monitoring. Clin Lab Med 1987;7:545–50.

[22] Shine B, McKnight RF, Leaver L, Geddes JR. Long-term effects of lithium on renal, thyroid, and parathyroid function: a retrospective analysis of laboratory data. Lancet 2015;386:461–8.

[23] Preskorn SH, Dorey RC, Jerkovich GS. Therapeutic drug monitoring of tricyclic antidepressants. Clin Chem 1988;34:822–8.

[24] McEvoy J, Millet RA, Dretchen K, Morris AA, Corwin MJ, Buckely P. Quantitative levels of aripiprazole parent drug and metabolites in urine. Psychopharmacology (Berl) 2014;231:4421–8.

[25] Begg EJ, Barclay ML, Kirkpatrick CM. The therapeutic drug monitoring of antimicrobial agents. Br J Clin Pharmacol 2001;52(Suppl. 1):35S–43S.

[26] Stankowicz MS, Ibrahim J, Brown DL. Once-daily aminoglycoside dosing: an update on current literature. Am J Health Syst Pharm 2015;72:1357–64.

[27] Kullar R, Leonard SN, Davis SL, Delgado Jr G, Pogue JM, Wahby KA, et al. Validation of the effectiveness of a vancomycin nomogra in achieving target trough concentrations of 15–20 mg/L suggested by the vancomycin consensus guidelines. Pharmacotherapy 2001;31:441–8.

[28] Chu HY, Jain R, Xie H, Pottinger P, Fredricks DN. Voriconazole therapeutic drug monitoring: retrospective cohort study of the relationship to clinical outcomes and adverse events. BMC Infect Dis 2013;13:105.

[29] Beumer JH, Chu E, Salamone SJ. Body-surface area-based chemotherapy dosing: appropriate in the 21st century? J Clin Oncol 2012;30:3896–7.

[30] Dombrowsky E, Jayaraman B, Narayan M, Barrett JS. Evaluating performance of a decision support system to improve methotrexate pharmacotherapy in children and young adults with cancer. Ther Drug Monit 2011;33:99–107.

[31] Malar R, Sjoor F, Rentsch K, Hassan M, Gungor T. Therapeutic drug monitoring is essential for intravenous busulfan therapy in pediatric hematopoietic stem cell recipients. Pediatr Transplant 2011;15:580–8.

[32] Gamelin E, Delva R, Jacob J, Merrouche Y, Raoul JL, Pezet D, et al. Individual fluorouracil dose adjustment based on pharmacokinetic follow-up compared with conventional dosage: results of a multicenter randomized trial of patients with metastatic colorectal cancer. J Clin Oncol 2008;26:2099–105.

[33] Kraff S, Nieuweboer AJ, Mathijssen RH, Baty F, de Graan AJ, van Schaik Rh, et al. Pharmacokinetically based dosing of weekly paclitaxel to reduce drug-related neurotoxicity based on a single sample strategy. Cancer Chemother Pharmacol 2015;75:975–83.

[34] Yu H, Steeghs N, Mijenhuis CM, Schellens JH, Beijnen JH, Huitema AD. Practical guidelines for therapeutic drug monitoring of anticancer tyrosine kinase inhibitors: focus on the pharmacokinetic targets. Clin Pharmacokinet 2014;53:305–25.

[35] Sawaki A, Inaba K, Nomura S, Kanie H, Yamada T, Hayashi K, et al. Imatinib plasma levels during successful long-term treatment of metastatic-gastrointestinal stromal tumors. Hepatogastroenterology 2014;61:1984–9.

CHAPTER 2

Immunoassays and Issues With Interference in Therapeutic Drug Monitoring

Amitava Dasgupta
Department of Pathology and Laboratory Medicine, University of Texas Health Science Center at Houston, Houston, TX, United States

CONTENTS

2.1 Introduction 17

2.2 Immunoassay Platforms Used in TDM 18

2.3 Interference of Bilirubin, Hemolysis, and Lipemia 20

2.4 Interferences in Digoxin Immunoassays 21

2.4.1 DLIS Interferences in Digoxin Immunoassays 22

2.4.2 Effect of Digibind and DigiFab on Digoxin Immunoassays 23

2.4.3 Potassium-Sparing Diuretics and Digoxin Immunoassays 24

2.4.4 Herbal Supplements and Digoxin Immunoassays 25

2.5 Issues of Interferences With Immunoassays for Immunosuppressants..26

2.5.1 Metabolite Interferences in Cyclosporine Immunoassays 26

2.1 INTRODUCTION

Usually serum or plasma is used for therapeutic drug monitoring (TDM) except for immunosuppressant drugs such as cyclosporine, tacrolimus, sirolimus, and everolimus, for which TDM is conducted using whole blood. However, the concentration of the immunosuppressant mycophenolic acid is determined using serum or plasma. Different types of assays are used in clinical laboratories for TDM. Historically, concentrations of various anticonvulsants, such as phenytoin, carbamazepine, phenobarbital, and primidone, in serum or plasma were measured using gas chromatography. Several bioassays were also available for monitoring certain antibiotics. In the 1960s, extensive research took place to develop various assays for therapeutic drugs using high-performance liquid chromatography. In the 1970s, immunoassays were available for the determination of concentrations of various drugs in serum and plasma, thus revolutionizing the field of TDM. However, immunoassays are not commercially available for all drugs currently monitored in clinical practice—for example, antiretroviral agents. For monitoring such drugs, sophisticated techniques such as high-performance liquid chromatography combined with tandem mass spectrometry (LC-MS/MS) are usually used. In general, chromatography-based methods (including LC-MS/MS) for TDM are superior approaches compared to immunoassays because such methods are relatively free from interferences. For immunosuppressants such as cyclosporine, tacrolimus, sirolimus, everolimus, and mycophenolic acid where immunoassays are commercially available, chromatographic methods, particularly LC-MS/MS methods, are considered as the "gold standard" due to significant metabolite cross-reactivity in immunoassays for the parent drug.

W. Clarke & A. Dasgupta (Eds): Clinical Challenges in Therapeutic Drug Monitoring. DOI: http://dx.doi.org/10.1016/B978-0-12-802025-8.00002-7

2.5.2 Metabolite Interferences in Tacrolimus Immunoassays 27
2.5.3 Metabolite Interferences in Sirolimus and Everolimus Immunoassays 27
2.5.4 Metabolite Interferences in Mycophenolic Acid Immunoassays 28

2.6 Interferences in Immunoassays for Anticonvulsants............29
2.6.1 Interferences in Phenytoin Immunoassays 29
2.6.2 Interferences in Carbamazepine Immunoassays 33
2.6.3 Interferences in Phenobarbital and Valproic Acid Immunoassays..........35
2.6.4 Interferences in Immunoassays for Newer Anticonvulsants.................35

2.7 Interferences in Immunoassays for TCAs..............................35
2.7.1 Interference of Phenothiazines and Their Metabolites in Immunoassays for TCAs ..36
2.7.2 Interference of Carbamazepine in Immunoassays for TCAs ..36
2.7.3 Interference of Cyproheptadine and Quetiapine With Immunoassays for TCAs ..37
2.7.4 Interference of Miscellaneous Drugs With Immunoassays for TCAs ..38

2.8 Conclusions 38

References 39

2.2 IMMUNOASSAY PLATFORMS USED IN TDM

Most TDM testing is now performed by immunoassays in which specimens (serum or plasma) can be used without any pretreatment, and they are run on fully automated, continuous, random access systems. For analysis of immunosuppressants (except mycophenolic acid) using whole blood specimen, a pretreatment phase may be required. The immunoassays require very small amounts of sample (mostly $<100\ \mu L$), reagents are stored in the analyzer, and most analyzers have stored calibration curves on the system. In immunoassays, the analyte is detected by its binding with a specific binding molecule, which in most cases is an analyte-specific antibody (or a pair of specific antibodies).

With respect to assay design, there are two formats of immunoassays: competition-based and immunometric (commonly referred as "sandwich"). Competition-based immunoassay is the method of choice for analytes with small molecular weight, requiring a single analyte-specific antibody. In contrast, sandwich immunoassays are mostly used for analytes with larger molecular weight, such as proteins or peptides, and use two different specific antibodies. Because most TDM immunoassays involve analytes of small molecular size, these assays employ the competition-based format. In this format, the analyte molecules in the specimen compete with analyte (or its analogues), labeled with a suitable tag provided in the reagent, for a limited number of binding sites provided by, for example, an analyte-specific antibody (also provided in the reagent). Thus, in these types of assays, the higher the analyte concentration in the sample, the lower amount of label that can bind to the antibody to form the conjugate. If the bound label provides the signal, which in turn is used to calculate the analyte concentration in the sample, the analyte concentration in the specimen is inversely proportional to the signal produced. If the free label provides the signal, then signal produced is proportional to the analyte concentration. The signal is mostly optical—absorbance, fluorescence, or chemiluminescence. There are several variations in this basic format. The assay can be homogeneous or heterogeneous. In the former, the bound label has different properties than the free label, and no separation between the bound and free label is needed before measuring the signal. In heterogeneous assay format, bound label must be separated from the free label. Examples of commonly used formats of immunoassays used in TDM include fluorescence polarization immunoassay (FPIA), enzyme-multiplied immunoassay technique (EMIT), cloned enzyme donor immunoassay (CEDIA), microparticle enzyme immunoassay (MEIA), enzyme-linked immunosorbent assay (ELISA), particle-enhanced turbidimetric inhibition immunoassay (PETNIA), chemiluminescent microparticle immunoassay (CMIA), and antibody-conjugated magnetic immunoassay (ACMIA).

Older immunoassays used for TDM utilized polyclonal antibodies, and such assays were subject to many interferences. With development and use of more specific monoclonal antibodies against a specific analyte, the magnitude of interferences in immunoassays has been significantly reduced. There are several benefits of monoclonal antibodies over polyclonal ones, including the following: (1) The characteristics of polyclonal antibodies are dependent on the animal producing the antibodies, and when the source individual animal is changed, the resultant antibody may be quite different; and (2) because polyclonal antibodies constitute many antibody clones, these antibodies are less specific compared to monoclonal antibodies for the analyte. Sometimes, instead of using the whole antibody, fragments of the antibody, generated by digestion of the antibody with peptidases (eg, Fab, Fab′, or their dimeric complexes), are used as a reagent.

Although the immunoassay methods are now widely used, there are significant limitations of this analytical technique. Antibody specificity is the major concern with regard to an immunoassay. Many endogenous metabolites of the analyte (drug) may have a very similar structural recognition motif as the analyte. There may also be other molecules unrelated to the analyte but producing comparable recognition motif as the analyte. These molecules are generally called cross-reactants. When present in the sample, these molecules may produce both positive and negative interference in the relevant immunoassay [1–3]. Other components in a specimen, such as bilirubin, hemoglobin, or lipid, may interfere in the immunoassay by interfering with the assay signal, thus producing incorrect results. A third type of immunoassay interference involves endogenous human antibodies in the specimen, which may interfere with components of the assay reagent such as the assay antibodies or the antigen labels. Such interference includes the interference from heterophilic antibodies or various human antianimal antibodies. However, interference of heterophilic antibody in immunoassays used for TDM is reported rarely.

Theoretically, all immunoassays used for TDM could be subject to interferences. However, in reality, digoxin immunoassays are most commonly reported to have issues with interferences, followed by issues of metabolite interferences in immunoassays used for monitoring immunosuppressant drugs including cyclosporine, tacrolimus, sirolimus, everolimus (all monitored in whole blood), and mycophenolic acid (monitored in serum or plasma). Interference of carbamazepine-10,11-epoxide in carbamazepine immunoassays may be a concern, depending on the immunoassay. Fosphenytoin interference in phenytoin immunoassays in uremic patients may also be a concern. Immunoassays for tricyclic antidepressants (TCAs) are also subjected to interferences from metabolites and structurally similar drugs. However, immunoassays for other common drugs, including vancomycin and aminoglycosides, are subjected to little interference.

2.3 INTERFERENCE OF BILIRUBIN, HEMOLYSIS, AND LIPEMIA

Most of the interferences from bilirubin, hemoglobin, or lipids occur during the optical detection process of the assay. Bilirubin absorbs approximately 450–460 nm, whereas hemoglobin absorbs 340–560 nm with an absorbance peak at 541 nm (oxyhemoglobin). Bilirubin and hemoglobin may also interfere in assays through unintended side reactions. The package inserts from most commercial assay kits normally list the effects of these interfering substances. Most manufacturers follow the EP7-P protocol from the National Committee for Clinical Laboratory Standardization (NCCLS; currently called the Clinical Laboratory Standards Institute or CLSI). The protocol recommends testing interference of bilirubin at 20 mg/dL, hemoglobin at 500 mg/dL, and lipid at 1000 mg/dL.

Elevated bilirubin causes interference proportional to its concentration. The interference of bilirubin in TDM assays is mainly caused by bilirubin absorbance at 454 or 461 nm. In addition, bilirubin may interfere in colorimetric ELISA that uses alkaline phosphatase label and *p*-nitrophenol phosphate substrate (405 nm). However, if the assay is enzymatic or colorimetric, bilirubin may also interfere by reacting chemically to the reagents [4]. Interference of high bilirubin in colorimetric assay for acetaminophen has been reported, but immunoassays are free from such interference [5]. Polson et al. observed false-positive acetaminophen results in serum specimens using colorimetric assays when bilirubin concentrations exceeded 10 mg/dL. However, immunoassays (FPIA and EMIT), as well as the more sophisticated gas chromatography combined with mass spectrometric method, were free from such interferences [6].

Bilirubin also interferes with the colorimetric salicylate assay (Trinder salicylate assay) based on the ability of salicylate to form a colored complex with ferric ion, and the colored complex can be measured at 560 nm. However, bilirubin also forms a colored complex with the reagent that absorbs at 600 nm. The salicylate assay on the Beckman Synchron LX analyzer utilizes a primary wavelength at 560 nm and secondary wavelength at 700 nm. Therefore, a high concentration of bilirubin causes positive interference with serum salicylate determination using this assay. The Trinder salicylate reagent is used in salicylate assays for both Hitachi 917 and Synchron analyzers, but high bilirubin causes negative interference in salicylate determination using the Hitachi 917 analyzer (primary wavelength 546 nm; no secondary wavelength). Interestingly, the FPIA for salicylate for application on the AxSYM analyzer (Abbott Laboratories, Abbott Park, IL) is free from bilirubin interference [7].

Hemolysis can occur in vivo, during venipuncture and blood collection, or during processing of the sample. Hemoglobin interference depends on its concentration in the sample. Serum appears hemolyzed when the hemoglobin concentration exceeds 20 mg/dL [8]. However, icteric serum may contain a higher concentration of hemoglobin before hemolysis can be noticed. Hemoglobin interference is caused not only by the spectrophotometric properties of hemoglobin but also by its participation in chemical reaction with sample or reactant components [9]. The absorbance maxima of the heme moiety in hemoglobin are at wavelengths of 540–580 nm. However, hemoglobin begins to absorb at approximately 340 nm, with absorbance increasing at 400–430 nm. The iron atom in the center of the heme group is the source of such absorbances. Of the many variants of hemoglobin, methemoglobin and cyanmethemoglobin also absorb at 500 and 480 nm, respectively. Methods that use the absorbance properties of NADH or NADH (340 nm) may thus be affected by hemolysis. When hemoglobin is oxidized to methemoglobin, the absorbance at 340 nm decreases [4].

Lipids in serum or plasma exist as complexed with proteins called lipoproteins. The lipoprotein particles with high lipid contents are micellar and are the main source of assay interference. Unlike bilirubin and hemoglobin, lipids normally do not participate in chemical reactions and mostly cause interference in assays due to their turbidity by scattering light. Because scattered light does not follow the Lambert–Beer law of absorbance, scattering normally reduces absorbance producing false results (positive or negative, depending on the reaction principle). Among the plasma lipoproteins, chylomicrons and very low-density lipoprotein (VLDL) particles only scatter light. VLDL exists in three size classes: small (27–35 nm), intermediate (35–60 nm), and large (60–200 nm). Only the latter two sizes of VLDL scatter light. Lipemic interference is most pronounced with spectrophotometric assays, less important with fluorometric methods, and rarely interferes with chemiluminescent methods. Thus, assays that use turbidimetry for signal are the ones most affected by lipid interference [10].

2.4 INTERFERENCES IN DIGOXIN IMMUNOASSAYS

Digoxin, a cardiac glycoside, is an old drug, but despite introduction of many relatively new cardioactive drugs, digoxin is still regularly used for treating patients with cardiac illnesses. Today, digoxin is isolated from a species of foxglove plant *Digitalis lantana*, and TDM of digoxin is essential in the management of patients receiving digoxin. Immunoassays are commonly used for TDM of digoxin in clinical laboratories, but immunoassays are also subjected to interferences from endogenous digoxin-like immunoreactive substances; Digibind/DigiFab; digoxin metabolites; spironolactone, potassium canrenoate, and their common metabolite canrenone; as well as a variety of herbal supplements, including Chinese medicine Chan Su and Lu-Shen-Wan,

Table 2.1 Interferences in Digoxin Immunoassays

Interfering Substance	Comments
DLIS	DLIS, endogenous substances elevated in volume-expanded patients, only interfere with polyclonal antibody-based older digoxin assays. MEIA digoxin assay showed negative interference. However, newer monoclonal antibody-based assays are free from DLIS interference
Spironolactone	Minimal interference even with polyclonal antibody-based digoxin immunoassays, but newer monoclonal antibody-based assays are free from such interferences
Potassium canrenoate (not used in the United States)	Higher magnitude of interference with polyclonal antibody-based assays compared to spironolactone, but newer monoclonal antibody-based assays are virtually free from interferences of potassium canrenoate
Digibind and DigiFab	Digibind/DigiFab used in treating severe digoxin overdose, interferes with most digoxin immunoassays. Free digoxin monitoring is recommended in patients overdosed with digoxin and being treated with Digibind or DigiFab
Chinese medicines (bufalin)	Chan Su and Lu-Shen-Wan contain bufalin, which interferes with serum digoxin measurements using both polyclonal and more specific monoclonal antibody-based assays. However, the magnitude of interference is much higher in polyclonal antibody-based digoxin assays compared to monoclonal antibody-based assays
Herbal supplements Active ingredients: oleandrin for oleander extract and convallatoxin for lily of the valley extract	Herbal supplements containing oleander and lily of the valley extracts interfere with serum digoxin measurements using both polyclonal and more specific monoclonal antibody-based assays. However, the magnitude of interference is much higher in polyclonal antibody-based digoxin assays compared to monoclonal antibody-based assays

oleander-containing herbal products, and lily of the valley extract-containing herbal products. In general, the magnitude of interference is much higher in digoxin assays containing a polyclonal antibody against digoxin compared to digoxin assays that utilize a specific monoclonal antibody against digoxin. Interference of digoxin metabolites on newer monoclonal antibody-based immunoassays is not clinically significant. Major interferences in digoxin immunoassays are listed in Table 2.1.

2.4.1 DLIS Interferences in Digoxin Immunoassays

The presence of endogenous digoxin-like immunoreactive substances (DLIS) was first described in a volume-expanded dog in 1980 [11]. After publication of that initial report, many investigators confirmed the presence of endogenous DLIS in serum and other biological fluids in volume-expanded patients not limited to patients with uremia, liver disease, essential hypertension, transplant recipients, eclampsia, pregnant women, and preterm infants [12,13]. DLIS is not a single compound but may be a complex mixture of several compounds that have common steroid-like structure [14].

DLIS showed significant positive interference with FPIA for application on the TDx analyzer, but the interference was negative with the MEIA digoxin assay, also marketed by Abbott Laboratories for application on the AxSYM analyzer. However, taking advantage of strong protein binding of DLIS and only 25% protein binding of digoxin, interference of DLIS in the MEIA can be eliminated by monitoring free digoxin concentration in protein-free ultrafiltrate. Due to the high concentration of digoxin in protein-free ultrafiltrate (approximately 75% of total digoxin concentration), immunoassays designed for total digoxin can also be used for measuring free drug digoxin, and there is an insignificant matrix effect from lack of proteins in the protein-free ultrafiltrate used for measuring free digoxin concentrations. Protein-free ultrafiltrate can be easily prepared by centrifuging serum for approximately 20–25 min in a Centrifree Micropartition filter with a molecular cutoff of 30 kDa [15].

Newer, more specific monoclonal antibody-based digoxin assays are virtually free from DLIS interferences. Nevertheless, due to the narrow therapeutic range, discordance in digoxin values measured using two different assays may occur. Jones and Morris analyzed digoxin values in 36 plasma samples by sending aliquots to two different laboratories using different digoxin immunoassays (CEDIA DRI digoxin, Microgenics Corporation, Fremont, CA; and DGNA digoxin assay, Dade Behring, now Siemens Diagnostics, Deerfield, IL). The authors observed clinically significant discordance in 39% of these samples and commented that DLIS interference may explain only some of the discordance [16].

2.4.2 Effect of Digibind and DigiFab on Digoxin Immunoassays

Digibind and DigiFab are Fab fragments of antidigoxin antibody used in treating life-threatening acute digoxin overdose. Digibind was the first antidote available for treating digoxin overdose in the United States and was marketed in 1986 by Glaxo Wellcome Company. In 2001, the US Food and Drug Administration (FDA) approved DigiFab, the second product for treating digoxin overdose. Both Digibind and DigiFab interfere with serum digoxin measurement using immunoassays. McMillin et al. studied the effects of Digibind and DigiFab on 13 different digoxin immunoassays. Significant interference was observed with both Digibind and DigiFab, although the magnitude of interference was somewhat less with DigiFab. The magnitude of interference varied significantly with each method, whereas IMMULITE, Vitros, Dimension, and Access digoxin methods showed the highest interference. Minimal interferences were observed with FPIA, MEIA, Synchron, and CEDIA methods. The authors also commented that monitoring free digoxin (in the protein-free ultrafiltrate) eliminates this interference because both

Digibind and DigiFab, have an approximate molecular weight of 46 kDa, and are absent in the protein-free ultrafiltrate [17].

Case Report: A 35-year-old woman intentionally swallowed 100 Lanitop tablets (methyl digoxin) in a suicide attempt. Methyl digoxin is rapidly converted into digoxin after oral administration. On admission, approximately 19 h after ingestion, her serum digoxin level was 7.4 ng/mL. She was treated immediately with 80 mg of Digibind followed by continuous infusion at a rate of 30 mg/h. A total of 395 mg of Digibind was administered. The total serum digoxin level increased significantly after initiation of Digibind therapy and peaked at 125 ng/mL, but the free serum digoxin level immediately decreased to nontoxic levels immediately after initiation of Fab therapy. Her symptoms of digoxin toxicity as well as nausea and vomiting resolved within 3 h of initiation of therapy, and the patient was discharged from the hospital after 3 days [18].

2.4.3 Potassium-Sparing Diuretics and Digoxin Immunoassays

Potassium-sparing diuretics such as spironolactone, potassium canrenoate, and eplerenone, not only pharmacokinetically interact with digoxin but also may interfere with serum digoxin measurements using various digoxin immunoassays. Spironolactone is used clinically for treating primary aldosteronism, essential hypertension, congestive heart failure, as well as edema, and it may be used along with digoxin therapy. After oral administration, spironolactone is rapidly and extensively metabolized to several metabolites including canrenone, which is an active metabolite. Potassium canrenoate is also metabolized to canrenone, but this drug is not approved for clinical use in the United States, although it is used clinically in Europe and other countries throughout the world. Because of structural similarity between spironolactone and related compounds with digoxin, these substances interfere with serum digoxin assays, especially assays utilizing polyclonal antibody against digoxin. Although interference of spironolactone and related compounds with FPIA digoxin immunoassay is clinically significant, this assay is no longer available.

Steimer et al. reported for the first time the negative interference of canrenone in serum digoxin measurement using MEIA (Abbott Laboratories). Misleading subtherapeutic concentrations of digoxin as measured on several occasions led to falsely guided digoxin dosing, resulting in serious digoxin toxicity in patients [19]. In a follow-up study, Steimer et al. reported that spironolactone, potassium canrenoate, and their common metabolite, canrenone, can cause both positive interference and negative interference in serum digoxin measurement using immunoassays. Positive interference was observed using FPIA, ACA (Dade Behring, Deerfield, IL) or Elecsys (Roche Diagnostics, Indianapolis, IN). Digoxin values were falsely lower (negative interference) if measured by MEIA, IMx (both from Abbott Laboratories), and Dimension digoxin assays (Dade Behring). The magnitude of interference was more significant with potassium canrenoate, where the concentration of its metabolite canrenone could be significantly higher compared to spironolactone therapy [20].

The relatively new digoxin assays that utilize more specific monoclonal antibodies are free from interference of spironolactone, potassium canrenoate, and their common metabolite, canrenone. For example, digoxin immunoassays manufactured by Abbott Laboratories for application on ARCHITECT clinical chemistry platforms (*c*Dig; PETINIA) and ARCHITECT immunoassay platforms (*i*Dig; CMIA) are completely free from such interferences [21]. The relatively new luminescent oxygen channeling technology (LOCI) digoxin assay introduced by the Siemens Diagnostics is also free from interferences from spironolactone and related compounds [22]. Yamada et al. reported that another potassium-sparing diuretic, eplerenone, interferes with FPIA and ACMIA (on Dimension analyzer; Siemens Diagnostics) but has no cross-reactivity with the MEIA digoxin assay [23].

2.4.4 Herbal Supplements and Digoxin Immunoassays

In the United States, complementary and alternative medicines are classified as dietary supplements and sold pursuant to the Dietary Supplement Health and Education Act of 1994. Interestingly, herbal supplements interfere only with digoxin immunoassays, and the magnitude of interference depends on the antibody specificity of the assay. As expected, polyclonal antibody-based digoxin immunoassays are more affected by these supplements than are specific monoclonal antibody-based digoxin immunoassays. Significant interferences of Chinese medicines Chan Su and Lu-Shen-Wan and oleander-containing herbal products with various digoxin assays (both monoclonal and polyclonal antibody based) have been reported. The interference of Chan Su and Lu-Shen-Wan to serum digoxin measurement can be positive (falsely elevated digoxin concentrations) or negative (falsely lower digoxin concentration), depending on the assay design. Although Beckman assay (Synchron LX system) and Roche assay (Tina-quant) showed positive interference in the presence of Chan Su, the MEIA digoxin assay on the AxSYM platform showed negative interference of Chan Su in serum digoxin measurement [24].

The oleanders are evergreen ornamental shrubs that grow in the southern United States from Florida to California, Australia, India, Sri Lanka, China, and other areas of the world. Although oleander causes positive interference with most digoxin assays, the MEIA digoxin assays demonstrated negative interference [25]. Lily of the valley that grows in the northern United States is a poisonous plant and contains convallatoxin. Due to structural similarity with digoxin, this toxin interferes with serum digoxin measurement using immunoassays [26].

2.5 ISSUES OF INTERFERENCES WITH IMMUNOASSAYS FOR IMMUNOSUPPRESSANTS

Immunoassays are available for TDM of cyclosporine, tacrolimus, sirolimus (monitored in whole blood), and mycophenolic acid (monitored in serum), but such immunoassays are subjected to interferences mostly from metabolites of the parent drug. Therefore, chromatographic methods, especially LC-MS/MS, are considered the gold standard for TDM of immunosuppressants.

For TDM of immunosuppressants, several immunoassay platforms are available commercially, including ACMIA, CEDIA, CMIA, EMIT, quantitative microsphere system (QMS), and PETINIA. For whole blood-based assays, extraction of the parent drug by an appropriate organic solvent prior to assay is necessary except for ACMIA assay, which does not require any specimen pretreatment and relies on online mixing and ultrasonic lysing of whole blood followed by exposure to β-galactosidase–antibody conjugate, removal of free conjugate using analyte-coated magnetic particles, and detection via a spectrometric β-galactosidase reaction.

2.5.1 Metabolite Interferences in Cyclosporine Immunoassays

Bias between immunoassays and chromatography-based methods in the analysis of various immunosuppressants has been well documented. Compared to the chromatographic method, mean cyclosporine concentrations have been found to be approximately 12%, 13%, 17%, 22%, and 40% higher for Dimension ACMIA, Syva EMIT, CEDIA PLUS, FPIA on the TDx, and FPIA on the AxSYM, respectively [27–31]. The FPIA assay for application on the TDx analyzer (Abbott Laboratories) is no longer commercially available. More recently, Abbott Laboratories marketed a CMIA for application on the ARCHITECT analyzer. Wallemacq et al. reported findings from multicenter evaluation of Abbott ARCHITECT cyclosporine assay using seven clinical laboratories. Values obtained by the immunoassay compared well with corresponding values obtained by LC-MS/MS due to minimal cross-reactivity of AM1 and AM9, the two major cyclosporine metabolites (up to 2.2% cross-reactivity, respectively) [32]. Soldin et al. evaluated the performance of a new ADVIA Centaur cyclosporine immunoassay that requires a single-step extraction and observed excellent correlation between cyclosporine values obtained by the LC-MS/MS assay and the Centaur cyclosporine assay [33]. However, falsely elevated blood cyclosporine levels due to the presence of endogenous antibody have been reported with the ACMIA cyclosporine assay run on the Dimension RXL analyzer. De Jonge et al. reported a falsely elevated cyclosporine level of 492 ng/mL in a 77-year-old patient. However, using LC-MS, the cyclosporine level was undetectable. Treating specimen with polyethylene glycol and remeasuring cyclosporine in the supernatant by the same ACMIA assay demonstrated no detectable level of cyclosporine, confirming that the interfering substance was a protein, most likely heterophilic antibody [34].

2.5.2 Metabolite Interferences in Tacrolimus Immunoassays

Although metabolite cross-reactivity is the major reason for falsely elevated tacrolimus levels using immunoassays, several studies have reported false-positive tacrolimus concentrations in patients with low hematocrit values and high imprecision at tacrolimus value less than 9 ng/mL with the MEIA assay for application on the AxSYM platform. However, other immunoassays studied, such as EMIT, were not affected [35].

Immunoassays for tacrolimus are affected by the cross-reactivity from tacrolimus metabolites. Westley et al. observed a 33.1% bias with the CEDIA assay and 20.1% bias with the MEIA assay when tacrolimus values measured by these two assays were compared with tacrolimus values determined by LC-MS/MS in renal transplant recipients [36]. Bazin et al. evaluated CMIA ARCHITECT tacrolimus assay and observed an average bias of 20% between values determined by the CMIA assay and LC-MS/MS [37]. Like cyclosporine, the ACMIA tacrolimus assay is also affected by rheumatoid factors and endogenous heterophilic antibodies. Altinier et al. described the interference of heterophilic antibody in the ACMIA tacrolimus assay. A sample of a patient showed tacrolimus values in the range of 4.9–12.5 ng/mL even after interruption of the treatment. The authors confirmed that the elevated tacrolimus levels were due to the presence of heterophilic antibody by treating samples with heterophilic blocking tubes and protein G resin that removed such interference [38].

2.5.3 Metabolite Interferences in Sirolimus and Everolimus Immunoassays

Sirolimus immunoassays are also subjected to significant metabolite interference. Morris et al. compared result from MEIA sirolimus assay with the values obtained using a specific LC-MS/MS assay and observed a mean bias of 49.2% [39]. Schmidt et al. evaluated the CMIA sirolimus assay for application on the ARCHITECT analyzer and concluded that the assay cross-reacts only with sirolimus metabolites F4 and F5 but hematocrit has no effect on the assay. In a multisite clinical trial, the authors observed an average of 14%, 25%, and 39% mean bias with the LC-MS/MS method in three different sites that evaluated the CMIA assay. The authors concluded that the CMIA assay showed positive bias in sirolimus values compared to values determined by more specific LC-MS/MS assays [40]. Holt et al. reported a mean positive bias of 21.9% with the CMIA assay when sirolimus values were compared with LC-MS/MS values [41].

The QMS everolimus assay received FDA approval for clinical use in 2011. The average bias in everolimus values determined by the QMS everolimus assay and corresponding values obtained by a specific LC-MS/MS method was 11% based on comparison of 90 specimens obtained from patients receiving everolimus [42]. Based on a study of 250 specimens from 169 patients, Hoffer et al. reported that the mean everolimus concentration determined by the QMS everolimus assay (using the ARCHITECT ci4100 analyzer) was 6.3 ng/mL, which was significantly higher than the average everolimus concentration of 4.8 ng/mL as determined by the LC-MS/MS method (31.2% average positive bias with the QMS assay). In addition, 69% of specimens, when analyzed by the QMS everolimus assay, showed supratherapeutic concentrations (>8 ng/mL), whereas everolimus values determined by the LC-MS/MS method were within therapeutic range [43]. Sallustio et al. observed an average bias of greater than 30% in everolimus concentration as determined by the Seradyn FPIA (not FDA approved) and a specific LC-MS/MS for everolimus and concluded that further investigation is needed before this assay can be used for routine therapeutic monitoring of everolimus [44].

2.5.4 Metabolite Interferences in Mycophenolic Acid Immunoassays

Mycophenolic acid is the only immunosuppressant monitored in serum or plasma and as such no sample pretreatment is needed prior to immunoassay. Hosotsubo et al. studied the analytical performance of EMIT mycophenolic acid immunoassay and observed no interference from the major metabolite mycophenolic acid glucuronide. In addition, the EMIT assay also correlated well with a high-performance liquid chromatography combined with ultraviolet detection (HPLC-UV) method for determination of mycophenolic acid [45]. Another study investigated the analytical performance of Roche total mycophenolic acid assay for application on the Cobas Integra and the Cobas 6000 analyzer by comparing these methods with a specific LC-MS/MS method for determination of mycophenolic acid in specimens obtained from liver transplant recipients. The authors did not observe any significant bias between values obtained by Roche total mycophenolic acid immunoassay and values determined by LC-MS/MS [46]. However, Westley et al. observed significant bias in mycophenolic acid determined by another immunoassay (CEDIA assay on a Hitachi 911 analyzer) and values obtained by a chromatographic method (HPLC-UV); the average positive bias was 18% with the CEDIA immunoassay [47]. Shipkova et al. observed a mean positive bias of 15% with samples obtained from patients after heart transplantation, but mean bias was 41.7% and 52.3% in specimens obtained from kidney transplant recipients and liver transplant recipients, respectively, using the CEDIA mycophenolic acid immunoassay compared to values obtained by the more specific HPLC-UV method [48]. Recently, the PETINIA assay for mycophenolic acid for application on the Dimension EXL analyzer became commercially available, but this method showed an average 12% positive bias compared to a reference chromatographic method [49]. Major interferences in immunoassays available for monitoring various immunosuppressants are listed in Table 2.2.

2.6 INTERFERENCES IN IMMUNOASSAYS FOR ANTICONVULSANTS

In general, classical anticonvulsants such as phenytoin, phenobarbital, valproic acid, and carbamazepine are routinely monitored in clinical laboratories using immunoassays, and issues of interferences with such immunoassays are discussed in this section. Although there is a reduced need of TDM for newer anticonvulsants, some of these anticonvulsants, such as lamotrigine, zonisamide, and topiramate, may require some monitoring. In general, gabapentin, pregabalin, tiagabine, and vigabatrin are not good candidates for TDM. TDM of levetiracetam and pregabalin is justified in patients with renal impairment. Monitoring the active metabolite of oxcarbazepine (10-hydroxycarbazepine) has some justification. Usually, chromatographic techniques are employed for TDM of these newer anticonvulsants. These methods are usually free from interferences. However, there are commercially available immunoassays for lamotrigine, zonisamide, and topiramate. Interferences in immunoassays used for TDM of various anticonvulsants are listed in Table 2.3.

2.6.1 Interferences in Phenytoin Immunoassays

Phenytoin (diphenylhydantoin) was first introduced as an anticonvulsant agent in 1938, and it is one of the most widely used anticonvulsant drugs. Immunoassays are used in clinical laboratories for TDM of phenytoin, and in

Table 2.2 Issues of Interferences with Immunoassays for Various Immunosuppressants

Immunoassay for Immunosuppressant	Comments
Cyclosporine	In general, metabolite cross-reactivity in various immunoassays may cause 10–20% positive bias in cyclosporine values measured by immunoassays (CEDIA, EMIT, etc.) compared to LC-MS/MS values. CMIA assay showed relatively small bias
	ACMIA assay is the only assay affected by interference from endogenous heterophilic antibodies
Tacrolimus	CEDIA tacrolimus assay shows an average 33.1% bias and CMIA assay shows an average 20% positive bias compared to chromatographic methods
	MEIA tacrolimus assay may show false-positive value due to low hematocrit (<25%)
	ACMIA assay is the only assay affected by interference from endogenous heterophilic antibodies
Sirolimus	CMIA sirolimus assay shows bias ranging from 14% to 39% compared to chromatographic methods
	MEIA sirolimus assay shows average bias of 49.2% compared to chromatographic method
Everolimus	QMS everolimus assay shows average 11% positive bias compared to chromatographic method in one report, but another report indicates 31.2% average positive bias
	FPIA assay (not approved by the FDA) shows an average 30% positive bias compared to chromatographic methods
Mycophenolic acid	CEDIA mycophenolic acid assay shows significant cross-reactivity with mycophenolic acid acyl glucuronide metabolite, and positive bias may range from 15% to 52.3% compared to chromatographic methods
	EMIT 2000 mycophenolic acid assay shows negligible cross-reactivity with metabolites, and values compared well with LC-MS/MS values
	Relatively new PETINIA assay shows average 12% positive bias compared to a reference chromatographic method

ACMIA, antibody-conjugated magnetic immunoassay from Dade Behring (now Siemens Diagnostics); CEDIA, cloned enzyme donor immunoassay developed by Microgenics Corporation (now Thermo Fisher); CMIA, chemiluminescent microparticle immunoassay developed by Abbott Laboratories; EMIT, enzyme-multiplied immunoassay technique developed by Syva (now in Siemens Diagnostics product line; Roche Diagnostics also marketed the EMIT assay); PETINIA, particle-enhanced turbidimetric inhibition immunoassay.

general, these assays are robust, with few reported interferences. One of the potential interferences in immunochemical measurements of phenytoin is cross-reactivity of the major phenytoin metabolite 5-(*p*-hydroxyphenyl)-5-phenylhydantoin (HPPH) and its glucuronide conjugate with phenytoin immunoassays. HPPH is the primary metabolite of phenytoin, and it is readily conjugated to glucuronide (HPPG), which is excreted in urine. It is estimated that 60–90% of the administered dose of phenytoin can by recovered in the urine as HPPG [50]. This cross-reactivity is particularly important in patients with renal insufficiency because of the increased concentration of metabolites. A 1981 study showed significant interference of metabolites with phenytoin immunoassays [51]. However, newer

Table 2.3 Issues of Interferences with Immunoassays for Various Anticonvulsants

Immunoassay for Anticonvulsant	Comments
Phenytoin	Metabolite cross-reactivity is clinically nonsignificant in newer phenytoin assays based on more specific monoclonal antibodies
	Fosphenytoin cross-reacts with various phenytoin immunoassays. Therefore, phenytoin should be measured after complete in vivo conversion of fosphenytoin into phenytoin (2 h after IV or 4 h after IM injection)
	True phenytoin levels may be falsely elevated in uremic patients receiving fosphenytoin. This is due to oxymethylglucuronide metabolite derived from fosphenytoin in sera of uremic patients
Carbamazepine	Carbamazepine-10,11-epoxide cross-reactivity varies from 0% (Vitros) to 96% (PETINIA) in various immunoassays
	Hydroxyzine and cetirizine interfere only with PETINIA assay
	10-hydroxy-10,11-dihydro-carbamazepine, a metabolite of oxcarbazepine only affects EMIT carbamazepine assay
Phenobarbital	Minimally affected by interference
Valproic acid	Minimally affected by interference
Lamotrigine	Seradyn QMS lamotrigine assay showed an average 21% positive bias when values were compared with a chromatographic method
	Another QMS lamotrigine assay (Thermo Scientific) showed average 15.6% positive bias compared to a HPLC-UV method. However, the positive bias was concentration dependent, with higher bias (up to 49.5%) observed when lamotrigine concentrations were relatively low (<2 mg/L)

monoclonal antibody-based immunoassays are less affected by phenytoin metabolites. Tutor-Crespo et al. compared phenytoin concentrations determined by two immunoassays (FPIA and EMIT) with values determined by HPLC for determination of phenytoin using specimens from 36 patients receiving phenytoin. These patients had renal function ranging from normal to severe renal insufficiency (glomerular filtration rate 10–102 mL/min of creatinine clearance). The authors used a deviation of 15% in value as clinically significant, and they concluded that immunoassays provided accurate results in therapeutic monitoring of phenytoin in patients with renal insufficiency [52]. Datta et al. studied the analytical performance of a turbidimetric assay on the ADVIA 1650 analyzer and reported that the phenytoin assay had very low cross-reactivity (5–8%) with the HPPH metabolite and virtually no cross-reactivity with oxaprozin. Oxaprozin interference in some older phenytoin immunoassays had been reported [53] (Table 2.4).

Another drug that must be considered for cross-reactivity with phenytoin assays is fosphenytoin (5,5-diphenyl-3-[(phosphonooxy)methyl]-2,4-imidazolidine-dione disodium salt). Fosphenytoin is a prodrug of phenytoin that is rapidly

Table 2.4 Common Interferences in Immunoassays for TCAs

Interfering Substance	Comments
Metabolites of tricyclic antidepressants (TCAs)	TCAs with tertiary amine structures are metabolized to secondary amines that have almost 100% cross-reactivity with immunoassays
Phenothiazines	Parent drugs and metabolites may cross-react with TCA immunoassay even at therapeutic concentration
Carbamazepine	May cause false-positive TCAs using immunoassays even at therapeutic concentration
Cyproheptadine	No interference with therapeutic concentration but may interfere if concentration is greater than 400 ng/mL
Quetiapine	May interfere with dosage of 600 mg quetiapine per day
Hydroxyzine and cetirizine	Cetirizine is a metabolite of hydroxyzine, and cetirizine is also used as antihistamine. After therapeutic use, these drugs showed no interference with TCA immunoassay but with overdose may cause interference
Diphenhydramine	May cause interference only in overdosed patients
Buflomedil	No interference after therapeutic use but in overdosed patients (>13 μg/mL; therapeutic 1–4 μg/mL) interference in TCA immunoassays may be observed

converted into phenytoin after administration. Fosphenytoin, unlike phenytoin, is readily water-soluble and can be administered via intravenous (IV) or intramuscular (IM) routes. Significant cross-reactivity of fosphenytoin in various immunoassays has been reported, and it is recommended that specimens for determination of phenytoin concentrations should not be obtained for patients on fosphenytoin until at least 2 h after IV infusion or 4 h after IM injection. Also, incubating 1 mL of specimen with 10 μL of alkaline phosphatase enzyme (Sigma Chemical Company) converts any fosphenytoin present in the specimen to phenytoin within 5 min at room temperature. This procedure eliminates interference of fosphenytoin in phenytoin immunoassays. The authors observed complete conversion of fosphenytoin to phenytoin by alkaline phosphatase in heparin, EDTA, and citrated plasma [54].

Roberts et al. studied in detail falsely elevated phenytoin values when measured by immunoassays compared to HPLC in patients with renal failure. The authors observed falsely increased phenytoin results up to 20 times higher than the HPLC results using AxSYM, TDx Phenytoin II (Abbott Laboratories), ACS:180 (Bayer Diagnostics, Tarrytown, NY), and Vitros assays (currently, TDx Phenytoin II and ACS:180 phenytoin assays are not commercially available). Interestingly, no fosphenytoin was detected in any of these specimens by HPLC [55]. Annesley et al. identified a unique immunoreactive oxymethylglucuronide metabolite derived from fosphenytoin in sera of uremic patients and demonstrated that this unusual metabolite was responsible for the cross-reactivity [56].

2.6.2 Interferences in Carbamazepine Immunoassays

Carbamazepine is an anticonvulsant drug that is structurally similar to TCAs and is used in the treatment of generalized tonic–clonic, partial, and partial–complex seizures. It was approved for treatment of epileptic patients in the United States in 1974 and subsequently approved for use in children older than age 6 years in 1979. Similar to phenytoin, carbamazepine is one of the most widely used anticonvulsant drugs. The active as well as major metabolite of carbamazepine is carbamazepine-10,11-epoxide, which represents 10–20% of the parent drug concentration at the steady state. Many factors, including renal insufficiency, polytherapy, genetic predisposition, and drug–drug interactions, may disproportionately increase the concentration of active epoxide metabolite. Tutor-Crespo et al. commented that in patients with moderate to severe renal insufficiency, the relative proportion of epoxide with respect to carbamazepine is significantly increased, and in such patients carbamazepine concentration obtained by EMIT or other assays with low cross-reactivity with the epoxide may have inadequate diagnostic efficiency because the pharmacological activities of carbamazepine and the epoxide are the same [57].

Lamotrigine is often used along with carbamazepine to treat epilepsy. In one study of nine patients (five male and four female, ages 19–31 years), the authors observed that after introduction of lamotrigine (median daily dosage 200 mg), the mean serum epoxide concentration was increased by 45% compared to prelamotrigine level. In four patients, such increases in the epoxide concentration caused clinically significant toxicity [58]. Quetiapine also interacts with carbamazepine and increases the concentration of epoxide metabolite possibly by inhibiting the epoxide hydroxylase enzyme [59]. It has been well documented that valproic acid increases carbamazepine toxicity due to accumulation of epoxide metabolite in the serum, and serum carbamazepine levels may be within the reference range in these patients [57]. Valpromide, valnoctamide, and progabide also inhibit epoxide hydrolase, thus increasing concentrations of carbamazepine-10,11-epoxide. Inhibition of carbamazepine metabolism and elevation of plasma carbamazepine to potential toxic concentrations can also be due to cotherapy with stiripentol, remacemide, acetazolamide, macrolide antibiotics, isoniazid, metronidazole, verapamil, diltiazem, cimetidine, danazol, or propoxyphene [60].

The cross-reactivity of carbamazepine epoxide in carbamazepine immunoassays has been studied extensively by many investigators. The cross-reactivity of carbamazepine-10,11-epoxide with different immunoassays for carbamazepine may vary between 0% (Vitros) and 94% (Dade Dimension) [61]. Parant et al. also reported high cross-reactivity of PETINIA carbamazepine assay with carbamazepine-10,11-epoxide and negligible cross-reactivity with the EMIT 2000 assay [62]. In general,

PETINIA assay actually measures the total concentration of the parent drug and the epoxide metabolite, whereas carbamazepine assays with low cross-reactivities with epoxide measure actual carbamazepine value. McMillin et al. demonstrated significant discordance in carbamazepine values obtained by two immunoassays, one with low cross-reactivity with epoxide (ADVIA Centaur carbamazepine assay) and the other with high cross-reactivity with epoxide (PETINIA). The authors observed that cross-reactivity of epoxide in the PETINIA assay varied from 93% to 101%, with an average cross-reactivity of 95.7%, whereas cross-reactivity of epoxide varied from 5% to 7.6% in the ADVIA Centaur assay, with an average cross-reactivity of only 6.3%. In one specimen collected from a patient taking carbamazepine, the discordance was 115%, which was clinically very significant. The carbamazepine concentration using the ADVIA Centaur assay was 6.0 μg/mL, whereas the carbamazepine concentration observed by the PETINIA assay was 12.9 μg/mL. The true carbamazepine concentration determined by LC-MS/MS was 6.4 μg/mL. The discordant between carbamazepine values determined by two assays was due to the presence of a high amount of epoxide in the specimen (epoxide concentration 8.1 μg/mL as determined by LC-MS/MS). The authors also demonstrated that carbamazepine concentration as determined by the ADVIA Centaur assay correlated better with the values obtained by LC-MS/MS than with values obtained by the PETINIA assay. In general, PETINIA overestimates carbamazepine concentration by 12% [63].

Hydroxyzine is a commonly prescribed first-generation antihistamine with sedative properties. Cetirizine is a metabolite of hydroxyzine but also available as a drug that is used in treating allergies. Although hydroxyzine and cetirizine are structurally unrelated to carbamazepine, Parant et al. documented two cases in which hydroxyzine in serum caused false-positive carbamazepine levels using the PETINIA assay. A 22-year-old female with a hydroxyzine concentration of 1.77 μg/mL and cetirizine concentration of 2.1 μg/mL showed an apparent carbamazepine level of 5.3 μg/mL. Parant et al. also demonstrated cross-reactivity of the PETINIA assay with oxatomide and other benzhydryl piperazine drugs. However, EMIT 2000 assay for carbamazepine showed no cross-reactivity [64].

Oxcarbazepine is structurally similar to carbamazepine and is also used in the treatment of epilepsy. In some cases, both drugs and their metabolites may be present in patients who are transitioning from one therapeutic regimen to the other. In a study regarding whether oxcarbazepine or its metabolites cross-react with an EMIT carbamazepine assay, it was shown that from a clinical perspective, only the 10-hydroxy-10,11-dihydro-carbamazepine metabolite of oxcarbazepine had any significant cross-reactivity with the assay, whereas there was no significant interference from the parent drug oxcarbazepine [61].

2.6.3 Interferences in Phenobarbital and Valproic Acid Immunoassays

Phenobarbital is the oldest member of the anticonvulsants and, in general, immunoassays for phenobarbital are robust and relatively free from interferences, including metabolite interference. Valproic acid is an eight-carbon branched-chain fatty acid indicated for patients with absence, tonic–clonic, and complex partial seizures. The early method of measuring valproic acid in patient specimens was by gas–liquid chromatography, but currently it is mostly measured by immunoassays. These immunoassays are relatively free from interferences.

2.6.4 Interferences in Immunoassays for Newer Anticonvulsants

Interferences in lamotrigine immunoassays have been reported. Morgan et al. observed an average 21% overestimation of lamotrigine values using Seradyn QMS lamotrigine assay compared to a reference chromatographic method [65]. Baldelli et al. evaluated QMS lamotrigine assay (Thermo Scientific) and observed an average 15.6% positive bias with QMS assay compared to a HPLC-UV method. However, the positive bias was concentration dependent, with higher bias (up to 49.5%) observed when lamotrigine concentrations were relatively low (<2 mg/L) [66]. No significant interferences in immunoassays for zonisamide and topiramate have been reported.

2.7 INTERFERENCES IN IMMUNOASSAYS FOR TCAs

TCAs share structural similarity with a number of other drugs. These drugs may interfere with serum or urine level measurement of TCAs using various immunoassays. Cyclobenzaprine is used as a skeletal muscle relaxant and is structurally related to amitriptyline, differing by only one double bond. Phenothiazine antipsychotics, carbamazepine, and oxcarbazepine are examples of three-ringed molecules that also share structural similarity to TCAs. In addition, the class of tetracyclic antidepressants including amoxapine, maprotiline, mianserin, and mirtazapine also has structural similarities with TCAs. Moreover, metabolites of TCAs also interfere with immunoassays.

In general, a tertiary amine TCA is metabolized to secondary amine, and the metabolite usually has almost 100% cross-reactivity with an antibody used for immunoassay for TCAs. Therefore, for monitoring tertiary amines, immunoassays in general indicate total concentration of the parent drug along with the active metabolite. As a result, manufacturers of such immunoassays recommend the use of these assays only for screening purposes in case of a suspected overdose involving TCAs but not for routine TDM of TCAs. For routine TDM, a chromatographic method must be used.

Urine screening for the presence of TCAs using immunoassays is also available. Melanson et al. compared serum TCAs using a reference HPLC method and urine TCA qualitative results obtained by using the Syva RapidTest and the Biosite Triage method and observed that serum concentrations of amitriptyline, desipramine, doxepin, imipramine, and nortriptyline ranging from subtherapeutic to toxic triggered a positive response to both urine immunoassays for TCAs, but neither immunoassay detected clomipramine even at levels greater than the therapeutic range. False-positive results were more common with Biosite assays if cyclobenzaprine was present in the urine. For virtually all positive urinary TCAs, it was not possible to identify whether the patient had subtherapeutic, therapeutic, or toxic serum TCA levels [67].

2.7.1 Interference of Phenothiazines and Their Metabolites in Immunoassays for TCAs

Schroeder et al. evaluated EMIT TCA assay in serum for application in identifying overdose using 87 patients. The authors reported that EMIT correctly identified 53 negative patients whose TCA levels were less than 300 ng/mL, but out of 34 remaining patients who showed TCA concentrations greater than 300 ng/mL, only 22 patients had confirmed TCA levels greater than 300 ng/mL using the reference HPLC method. Phenothiazine and phenothiazine metabolites were present in the remaining 12 unconfirmed specimens [68]. Ryder and Glick reported a case in which a patient who ingested thioridazine and flurazepam tested positive for TCAs by immunoassay. Investigation showed that the false-positive TCA result was due to thioridazine that was present in a therapeutic concentration of 125 ng/mL [69].

2.7.2 Interference of Carbamazepine in Immunoassays for TCAs

Carbamazepine is metabolized to carbamazepine-10,11-epoxide, an active metabolite. Both the parent drug and the epoxide metabolite interfere with immunoassays for TCAs due to structural similarities. Another structurally related drug, oxcarbazepine, also interferes with immunoassays for TCAs. The FPIA for TCAs for application on the AxSYM analyzer demonstrated significant cross-reactivities with both carbamazepine and its epoxide metabolite. In 30 patients receiving carbamazepine but no TCAs, the apparent TCAs ranged from 31.8 to 130 ng/mL (carbamazepine level ranged from 1.4 to 20.9 μg/mL), indicating that carbamazepine may falsely indicate the presence of TCAs in a patient never exposed to it [70]. Song and Chiu also reported interference of carbamazepine on the Abbott TDx and AxSYM TCA assay [71]. Chattergoon et al. reported that two patients with a history of ingestion of carbamazepine showed positive urinary screen for TCAs using an immunoassay. However, confirmatory HLPC assay showed a negative result [72].

2.7.3 Interference of Cyproheptadine and Quetiapine With Immunoassays for TCAs

Cyproheptadine is an antihistamine that is used for treating various allergic symptoms. Wians et al. reported a case of a 14-year-old girl who, after ingesting approximately 120 mg of cyproheptadine tablets, tested positive for TCAs in serum by EMIT assay. In vitro studies indicated that a cyproheptadine concentration of 400 ng/mL may cause false-positive TCA results, although serum obtained from a volunteer who was given a 12-mg dose of cyproheptadine for 3 days showed no TCAs when tested by the same EMIT assay [73].

Case Report: A 5-year-old child was taken to the emergency room due to unusual behavior after awakening from a nap. On examination, it was observed that she was agitated and her toxicology screen was positive for TCAs. The family was advised to bring all household medications to the emergency room; these included cyproheptadine tablets, hydroxyzine syrup, carbinoxamine syrup, nitroglycerine tablets, and multivitamin tablets. Her aunt reported that the child was playing with some tablets on the floor, and later those pills were identified as cyproheptadine using gas chromatography/mass spectrometric analysis. When the child's serum was further analyzed by a reference HPLC method, no peak was observed for nordoxepin, doxepin, desipramine, imipramine, nortriptyline, and amitriptyline, but a peak was observed with the same retention time as cyproheptadine standard, indicating that the false-positive serum TCA level detected by the EMIT assay was indeed due to interference of cyproheptadine metabolite [74].

Quetiapine, an atypical antipsychotic used for treating schizophrenia and acute episodes of bipolar disorders and also as an augmentor for treatment of depression, cross-reacts with the plasma TCA immunoassay and the false-positive response is dependent on quetiapine concentration. A 34-year-old man with a history of refractory-schizoaffective disorder and amphetamine dependence was admitted to the hospital for his depressive and psychotic symptoms. His only prescribed medication during admission was quetiapine 600 mg/day. On admission, his urine toxicology screen showed positive for TCA (at a level over the cutoff of 300 ng/mL) but negative for other illicit drugs, and the false-positive TCA was determined to be due to quetiapine interference with this homogeneous enzyme immunoassay for TCAs (Diagnostics Reagents) on the Hitachi 911 analyzer [75].

2.7.4 Interference of Miscellaneous Drugs With Immunoassays for TCAs

Hydroxyzine and cetirizine are antihistamines that may interfere with the FPIA assay for TCAs for application on the Abbott AxSYM analyzer. However, neither hydroxyzine nor cetirizine should interfere after recommended low-dose therapy, and such interference is expected with ingestion of higher dose or overdose [76]. Interference of diphenhydramine, an antihistamine in the EMIT assay for TCAs in serum, is interesting because unlike phenothiazine, diphenhydramine is an ethanolamine. A 21-year-old woman ingested 2 g of diphenhydramine. In vitro testing indicated that when drug-free serum was supplemented with up to 60 ng/mL of diphenhydramine, the response to TCA screen was negative, and specimens containing 690 ng/mL or greater amounts of diphenhydramine were clearly positive. Usually, ingestion of 100 mg of diphenhydramine leads to a serum diphenhydramine level of approximately 112 ng/mL 2 hours after ingestion. Therefore, positive screen for TCAs is more likely a result of diphenhydramine overdose [77]. Buflomedil in therapeutic concentration (between 1 and 4 μg/mL) does not interfere with TCAs assay, but overdosed patients (with serum buflomedil level >13 μg/mL) may show a false-positive response. A 16-year-old comatose boy with general seizure was admitted to the hospital. A blood specimen tested by the EMIT assay for TCAs (Behring, Cupertino, CA) was positive, and the result was later confirmed as a false positive due to the presence of buflomedil at a concentration of 28 μg/mL in the specimen [78].

2.8 CONCLUSIONS

Immunoassays are commonly used for TDM in clinical laboratories, and as expected, immunoassays are subject to interferences. Digoxin immunoassays are affected by interferences more commonly than any other immunoassays used for TDM. Digoxin immunoassays are affected by DLIS, Digibind, or DigiFab, as well as many other drugs, but newer assays based on monoclonal antibodies are less affected (except for Digibind and DigiFab interference). Carbamazepine-10,11-epoxide, the metabolite of carbamazepine, interferes with various carbamazepine immunoassays, and the magnitude of cross-reactivity may vary from 0% to 94%. Immunoassays for various immunosuppressants are significantly affected by metabolite cross-reactivities, and chromatographic methods should be used for routine TDM of various immunosuppressants. Immunoassays for TCAs should only be used for diagnosis of overdose (along with electrocardiograph and patient history) but should not be used for TDM due to cross-reactivity of metabolites as well as various structurally related drugs with immunoassays [79].

Interference from endogenous antibodies such as heterophilic antibodies is relatively common in the analysis of analytes with large molecular weights, such as human chorionic gonadotropin serum and various tumor markers. However, interference of heterophilic antibodies with immunoassays for TDM is relatively rare. The ACMIA assay, marketed by Siemens Diagnostics for analysis of cyclosporine and tacrolimus, is affected by endogenous antibodies including heterophilic antibody. There are only two isolated reports of heterophilic antibody interferences in serum digoxin measurement by an immunoassay. One report was published in 1996 showing interference of heterophilic antibody on a Roche enzyme immunoassay [80]. However, this assay had been improved and modified later. Another report indicated interference of heterophilic antibody in DGNA digoxin assay for application on the Dimension analyzer [81]. However, Siemens later marketed LOCI digoxin assay with more specificity toward digoxin. Therefore, interference of heterophilic antibody in newer digoxin immunoassays using more specific monoclonal antibody is unlikely.

Although immunoassays are widely used in clinical laboratories for routine TDM, it should be recognized that assays based on LC-MS/MS are usually robust and suffer from very little interference. Therefore, even when a robust immunoassay is available for TDM, such as vancomycin, a particular immunoassay may show significant bias when results are compared with those of LC-MS/MS-based methods. Oyaert et al. recently reported that although three vancomycin immunoassays (immunoassays for application on ARCHITECT i2000 analyzer, Vitros 5000 analyzer, and Dimension Vista 1500 analyzer) showed good agreement with values determined by the reference LC-MS/MS method (within 10%), another immunoassay for application on the Cobas 8000 analyzer showed a mean proportional difference exceeding 20% [82]. Therefore, isolated cases of interference may be observed even with a robust immunoassay for a specific drug. When the clinical picture does not correlate with the drug concentration, it is important to investigate the issue of interference and, if possible, to repeat the test using a different immunoassay for the same analyte. A discordant result indicates the possibility of interference, and to resolve such discordance, it is advisable to send the sample to a reference laboratory for further evaluation.

References

[1] Datta P, Larsen F. Specificity of digoxin immunoassays toward digoxin metabolites. Clin Chem 1994;40:1348–9.

[2] Datta P. Oxaprozin and 5-(*p*-hydroxyphenyl)-5-phenylhydantoin interference in phenytoin immunoassays. Clin Chem 1997;43:1468–9.

[3] Datta P, Dasgupta A. Bidirectional (positive/negative) interference in a digoxin immunoassay: importance of antibody specificity. Ther Drug Monit 1998;20:352–7.

[4] Fonseca-Wolheim FD. Hemoglobin interference in the bichromatic spectrophotometry of NAD(P)H at 340/380 nm. Eur J Clin Chem Clin Biochem 1993;31:595–601.

[5] Bertholf RL, Johannsen LM, Bazooband A, Mansouri V. False-positive acetaminophen results in a hyperbilirubinemic patient. Clin Chem 2003;49:695–8.

[6] Polson J, Wians FH, Orsulak P, Fuller D, et al. False positive acetaminophen concentrations in patients with liver injury. Clin Chim Acta 2008;391:24–30.

[7] Dasgupta A, Zaldi S, Johnson M, Chow L, et al. Use of fluorescence polarization immunoassay for salicylate to avoid positive/negative interference by bilirubin in the Trinder salicylate assay. Ann Clin Biochem 2003;40:684–8.

[8] Sonntag O, Glick MR. Serum-index und interferogram-ein neuer weg zur prufung und darstellung von interferengen durch serumchromogene. Lab Med 1989;13:77–82.

[9] Wenk RE. Mechanism of interference by hemolysis in immunoassays and requirements for sample quality. Clin Chem 1998;44:2554.

[10] Bornhorst JA, Roberts RF, Roberts WL. Assay-specific differences in lipemic interference in native and intralipid-supplemented samples. Clin Chem 2004;50:2197–201.

[11] Gruber KA, Whitaker JM, Buckalew VM. Endogenous digitalis-like substances in plasma of volume expanded dogs. Nature 1980;287:743–5.

[12] Craver JL, Valdes R. Anomalous serum digoxin concentration in uremia. Ann Intern Med 1983;98:483–4.

[13] Jortani SA, Valdes Jr. R. Digoxin and its related endogenous factors. Crit Rev Clin Lab Sci 1997;34:225–74.

[14] Ebara H, Suzuki S, Nagashima K, Kuroume T. Natriuretic activity of digoxin like immunoreactive substance extracted from cord blood. Life Sci 1988;42:303–9.

[15] Dasgupta A, Trejo O. Suppression of total digoxin concentration by digoxin-like immunoreactive substances in the MEIA digoxin assay: elimination of interference by monitoring free digoxin concentrations. Am J Clin Pathol 1999;111:406–10.

[16] Jones TE, Morris RG. Discordant results from "real-world" patient samples assayed for digoxin. Ann Pharmacother 2008;42:1797–803.

[17] McMillin GA, Qwen W, Lambert TL, De B, Frank FL, Bach PR, et al. Comparable effects of DIGIBIND and DigiFab in thirteen digoxin immunoassays. Clin Chem 2002;48:1580–4.

[18] Fyer F, Steimer W, Muller C, Zilker T. Free and total digoxin in serum during treatment of acute digoxin poisoning with Fab fragments; case study. Am J Crit Care 2010;19:387–91.

[19] Steimer W, Muller C, Eber B, Emmanuilidis K. Intoxication due to negative canrenone interference in digoxin drug monitoring. Lancet 1999;354:1176–7 [letter].

[20] Steimer W, Muller C, Eber B. Digoxin assays: frequent, substantial and potentially dangerous interference by spironolactone, canrenone and other steroids. Clin Chem 2002;48:507–16.

[21] DeFrance A, Armbruster D, Petty D, Cooper K, Dasgupta A. Abbott ARCHITECT clinical chemistry and immunoassay systems: digoxin assays are free of interferences from spironolactone, potassium canrenoate and their common metabolite canrenone. Ther Drug Monit 2011;33:128–31.

[22] Dasgupta A, Johnson MJ, Sengupta TK. Clinically insignificant negative interference of spironolactone, potassium canrenoate and their common metabolite canrenone in new dimension vista LOCI digoxin immunoassay. J Clin Lab Anal 2012;26:143–7.

[23] Yamada T, Suzuki K, Kanada Y, Kato R, et al. Interference between eplerenone and digoxin in fluorescence polarization immunoassay, and affinity column mediated immunoassay. Ther Drug Monit 2010;32:774–7.

[24] Chow L, Johnson M, Wells A, Dasgupta A. Effect of the traditional Chinese medicine Chan Su, Lu-Shen-Wan, Dan Shen and Asian ginseng on serum digoxin measurement by Tina-Quant (Roche) and Synchron LX system (Beckman) digoxin immunoassays. J Clin lab Anal 2003;17:22–7.

[25] Dasgupta A, Datta P. Rapid detection of oleander poisoning by using digoxin immunoassays: comparison of five assays. Ther Drug Monit 2004;26:658–63.

[26] Fink SL, Robey TE, Tarabar AF, Hodsdon ME. Rapid detection of convallatoxin using five digoxin immunoassays. Clin Toxicol (Phila) 2014;52:659–63.

[27] Steimer W. Performance and specificity of monoclonal immunoassays for cyclosporine monitoring: how specific is specific? Clin Chem 1999;45:371–81.

[28] Schutz E, Svinarov D, Shipkova M, Niedmann PD, et al. Cyclosporin whole blood immunoassays (AxSYM, CEDIA, and Emit): a critical overview of performance characteristics and comparison with HPLC. Clin Chem 1998;44:2158–64.

[29] Hamwi A, Veitl M, Manner G, Ruzicka K, et al. Evaluation of four automated methods for determination of whole blood cyclosporine concentrations. Am J Clin Pathol 1999;112:358–65.

[30] Terrell AR, Daly TM, Hock KG, Kilgore DC, et al. Evaluation of a no-pretreatment cyclosporin A assay on the Dade Behring Dimension RxL clinical analyzer. Clin Chem 2002;48:1059–65.

[31] Butch AW, Fukuchi AM. Analytical performance of the CEDIA® cyclosporine PLUS whole blood immunoassay. J Anal Toxicol 2004;28:204–10.

[32] Wallemacq P, Maine GT, Berg K, Rosiere T, et al. Multisite analytical evaluation of Abbott Architect cyclosporine assay. Ther Drug Monit 2010;32:145–51.

[33] Soldin SJ, Hardy RW, Wians FH, Balko JA, et al. Performance evaluation of the new ADVIA Centaur system cyclosporine assay (single-step extraction). Clin Chim Acta 2010;411:806–11.

[34] De Jonge H, Geerts I, Declercq P, de Loor H, et al. Apparent elevation of cyclosporine whole blood concentration in a renal allograft recipient. Ther Drug Monit 2010;32:529–31.

[35] Armedariz Y, Garcia S, Lopez R, Pou L, et al. Hematocrit influences immunoassay performance for the measurement of tacrolimus in whole blood. Ther Drug Monit 2005;27:766–9.

[36] Westley IS, Taylor PJ, Salm P, Morris RG. Cloned enzyme donor immunoassay tacrolimus assay compared with high-performance liquid chromatography-tandem mass spectrometry in liver and renal transplant recipients. Ther Drug Monit 2007;29:584–91.

[37] Bazin C, Guinedor A, Barau C, Gozalo C, et al. Evaluation of the Architect tacrolimus assay in kidney, liver and heart transplant recipients. J Pharm Biomed Appl 2010;53:997–1002.

[38] Altinier S, Varagnolo M, Zaninotto M, Boccagni P, et al. Heterophilic antibody interference in a non-endogenous molecule assay: an apparent elevation in the tacrolimus concentration. Clin Chim Acta 2009;402:193–5.

[39] Morris RG, Salm P, Taylor PJ, Wicks FA. Comparison of the reintroduced MEIA assay with HPLC-MS/MS for the determination of whole blood sirolimus from transplant recipients. Ther Drug Monit 2006;28:164–8.

[40] Schmidt RW, Lotz J, Schweigert R, Lackner K, et al. Multi-site analytical evaluation of a chemiluminescent magnetic microparticle immunoassay (CMIA) for sirolimus on the Abbott ARCHITECT analyzer. Clin Biochem 2009;42:1543–8.

[41] Holt DW, Mandelbrot DA, Tortorici MA, Korth-Bradley JM, et al. Long term evaluation of analytical methods used in sirolimus therapeutic drug monitoring. Clin Transplant 2014;28:243–51.

[42] Dasgupta A, Davis B, Chow L. Evaluation of QMS everolimus assay using Hitachi 917 analyzer: comparison with liquid chromatography/mass spectrometry. Ther Drug Monit 2011;33:149–54.

[43] Hoffer E, Kurnik D, Efrati R, Scherb I, et al. Comparison of everolimus QMS immunoassay on Architect ci4100 and liquid chromatography/mass spectrometry: lack of agreement in organ-transplanted patients. Ther Drug Monit 2015;37:214–19.

[44] Sallustio BC, Noll BD, Morris RG. Comparison of blood sirolimus, tacrolimus and everolimus concentrations measured by LC-MS/MS, HPLC-UV and immunoassay methods. Clin Biochem 2011;44:231–6.

[45] Hosotsubo H, Takahara S, Imamura R, Kyakuno M, et al. Analytical validation of the enzyme multiplied immunoassay technique for the determination of mycophenolic acid in plasma from renal transplant recipients compared with a high performance liquid chromatographic assay. Ther Drug Monit 2001;23:669–74.

[46] Decavele AS, Favoreel N, Heyden FV, Verstraete AG. Performance of the Roche total mycophenolic acid assay on the Cobas Integra 400, Cobas 6000 and comparison to LC-MS/MS in liver transplant patients. Clin Chem Lab Med 2011;49:1159–65.

[47] Westley IS, Ray JE, Morris RG. CEDIA mycophenolic acid assay compared with HPLC-UV in specimens from transplant recipients. Ther Drug Monit 2006;28:632–6.

[48] Shipkova M, Schutz E, Besenthal I, Fraunberger P, et al. Investigation of the crossreactivity of mycophenolic acid glucuronide metabolites and of mycophenolic acid mofetil in the Cedia MPA assay. Ther Drug Monit 2010;32:79–85.

[49] Dasgupta A, Tso G, Chow L. Comparison of mycophenolic acid concentrations determined by a new PETINIA assay on the dimension EXL analyzer and a HPLC-UV method. Clin Biochem 2013;46:685–7.

[50] Rainey PM, Rogers KE, Roberts WL. Metabolite and matrix interference in phenytoin immunoassays. Clin Chem 1996;42:1645–53.

[51] Matzke GR, Sawchuk RJ. Elevated serum phenytoin concentrations in uremic patient measured by enzyme multiplied immunoassay. Drug Intell Clin Pharm 1981;15:386–8.

[52] Tutor-Crespo MJ, Hermida J, Tutor JC. Phenytoin immunoassay measurements in serum from patients with renal insufficiency: comparison with high-performance liquid chromatography. J Clin Lab Anal 2007;21:119–23.

[53] Datta P, Scurlock D, Dasgupta A. Analytic performance evaluation of a new turbidimetric immunoassay for phenytoin on the ADVIA 1650 analyzer: effect of phenytoin metabolite and analogue. Ther Drug Monit 2005;27:305–8.

[54] Dasgupta A, Warner B, Datta P. Use of alkaline phosphatase to correct underestimation of fosphenytoin concentrations in serum measured by phenytoin immunoassays. Am J Clin Pathol 1999;111:557–62.

[55] Roberts WL, De BK, Coleman JP, Annesley TM. Falsely increased immunoassay measurements of total and unbound phenytoin in critically ill uremic patients receiving fosphenytoin. Clin Chem 1999;45:829–37.

[56] Annesley T, Kurzyniec S, Nordblom G, et al. Glucuronidation of prodrug reactive site: isolation and characterization of oxymethylglucuronide metabolite of fosphenytoin. Clin Chem 2001;46:910–18.

[57] Tutor-Crespo MJ, Hermida J, Tutor JC. Relative proportions of serum carbamazepine and its pharmacologically active 10,11-epoxide derivative: effect of polytherapy and renal insufficiency. Ups J Med Sci 2008;113:171–80.

[58] Warner T, Patsalos PN, Prevett M, Elyas AA, et al. Lamotrigine induced carbamazepine toxicity: an interaction with carbamazepine 10,11-epoxide. Epilepsy Res 1992;11:147–50.

[59] Fitzgerald BJ, Okas AJ. Elevation of carbamazepine 10,11-epoxide by quetiapine. Pharmacotherapy 2002;22:1500–3.

[60] Spina E, Pisani F, Perucca E. Clinically significant pharmacokinetic drug interactions with carbamazepine: an update. Clin Pharmacokinetic 1996;31:198–214.

[61] Hermida J, Tutor JC. How suitable are currently used Carbamazepine immunoassays for quantifying carbamazepine 10,11-epoxide in serum samples? Ther Drug Monit 2003;25:384–8.

[62] Parant F, Bossu H, Gagnieu MC, Lardet G, Moulsma M. Cross reactivity assessment of carbamazepine 10,11-epoxide, oxcarbazepine, and 10 hydroxy carbazepine in two automated carbamazepine immunoassay: PENTINA and EMIT 2000. Ther Drug Monit 2003;25:41–5.

[63] McMillin GA, Juenke JM, Johnson M, Dasgupta A. Discordant carbamazepine values between two immunoassays: carbamazepine values determined by ADVIA Centaur correlate better with those determined by PETINIA assay. J Clin Lab Anal 2011;25:212–16.

[64] Parant F, Moulsma M, Gagnieu MC, Lardet G. Hydroxyzine and metabolites as a source of interference in carbamazepine particle-enhanced turbidimetric inhibition immunoassay (PETINIA). Ther Drug Monit 2005;27:457–62.

[65] Morgan PE, Fischer DS, Evers R, Flanagan RJ. A rapid and simple assay for lamotrigine in serum/plasma by HPLC and comparison with an immunoassay. Biomed Chromatogr 2011;25:775–8.

[66] Baldelli S, Castoldi S, Charbe N, Cozi V, et al. Comparison of the QMS analyzer with HPLC-UV for the quantification of lamotrigine concentrations in human plasma samples. Ther Drug Monit 2015;37:689–94.

[67] Melanson SE, Lewandrowski EL, Griggs DA, Flood JG. Interpreting tricyclic antidepressant measurements in urine in an emergency department: comparison of two qualitative point of care tricyclic antidepressant drug immunoassays with quantitative serum chromatographic analysis. J Anal Toxicol 2007;31:270–5.

[68] Schroeder TJ, Tasset JJ, Otten EJ, Hedges JR. Evaluation of Syva EMIT toxicological serum tricyclic antidepressant assay. J Anal Toxicol 1986;10:221–4.

[69] Ryder KW, Glick MR. The effect of thioridazine on the Automatic Clinical Analyzer serum tricyclic anti-depressant screen. Am J Clin Pathol 1986;86:248–9.

[70] Dasgupta A, McNeese C, Wells A. Interference of carbamazepine and carbamazepine 10,11-epoxide in the fluorescence polarization immunoassay for tricyclic antidepressants: estimation of the true tricyclic antidepressant concentration in the presence of carbamazepine using a mathematical model. Am J Clin Pathol 2003;121:418–25.

[71] Song D, Chiu W. Analytical interference of carbamazepine on the Abbott TDx and Abbott AxSYM tricyclic antidepressant assay. Pathology 2009;41:688–9.

[72] Chattergoon DS, Verjee Z, Anderson M, Johnson D, et al. Carbamazepine interference with an immunoassay for tricyclic antidepressant. J Toxicol Clin Toxicol 1998;36:109–13.

[73] Wians Jr. FH, Norton JT, Wirebaugh SR. False-positive serum tricyclic antidepressant screen with cyproheptadine. Clin Chem 1993;39:1355–6.

[74] Yuan CM, Spandorfer PR, Miller SL, Henretig FM, et al. Evaluation of tricyclic antidepressant false positivity in a pediatric case of cyproheptadine (periactin) overdose. Ther Drug Monit 2003;25:299–304.

[75] Sloan KL, Haver VM, Saxon AJ. Quetiapine and false-positive urine drug testing for tricyclic antidepressants. Am J Psychiatry 2000;157:148–9 [letter to the editor].

[76] Dasgupta A, Wells A, Datta P. False positive serum tricyclic antidepressant concentrations using fluorescence polarization immunoassay due to the presence of hydroxyzine and cetirizine. Ther Drug Monit 2007;29:134–9.

[77] Sorisky A, Watson DC. Positive diphenhydramine interference in the EMIT-st assay for tricyclic antidepressants in serum. Clin Chem 1986;32:715.

[78] Mura P, Kintz P, Robert R, Papet Y. Buflomedil is a potent interfering substance in immunoassay of tricyclic antidepressants. J Anal Toxicol 1998;22:254 [letter to the editor].

[79] Dasgupta. Impact of interferences including metabolite crossreactivity on therapeutic drug monitoring results. Ther Drug Monit 2012;34:496–506.

[80] Liendo C, Ghali JK, Graves SW. A new interference in some digoxin assays: anti-murine heterophilic antibodies. Clin Pharmacol Ther 1996;60:593–8.

[81] Hermida-Cadahia EE, Calvo MM, Tutor JC. Interference of circulating endogenous antibodies on the Dimension DGNA digoxin immunoassay: elimination with a heterophilic antibody blocking reagent. Clin Biochem 2010;43:1475–7.

[82] Oyaert M, Peersman N, Kieffer D, Deiteren K, et al. Novel LC/MS/MS method for plasma vancomycin: comparison with immunoassays and clinical impact. Clin Chim Acta 2015;441:63–70.

CHAPTER 3

Application of Chromatography Combined With Mass Spectrometry in Therapeutic Drug Monitoring

Madhuri Manohar[1] and Mark A. Marzinke[1,2]

[1]Division of Clinical Pharmacology, Department of Medicine, Johns Hopkins University School of Medicine, Baltimore, MD, United States

[2]Division of Clinical Chemistry, Department of Pathology, Johns Hopkins University School of Medicine, Baltimore, MD, United States

CONTENTS

3.1 Introduction......45

3.2 Liquid Chromatography.....48

3.3 Mass Spectrometry...........51

3.3.1 Ion Source...............51

3.3.2 Mass Analyzers......53

3.3.2.1 Quadrupole Analyzers.............54

3.3.2.2 Ion Trap Analyzers..54

3.3.2.3 TOF Analyzers........55

3.3.3 Detectors................55

3.4 Preanalytical Stage........................55

3.5 Application of LC-MS/MS Methods in TDM......................56

3.5.1 Immunosuppressants.......56

3.5.2 Anticonvulsants......57

3.5.3 Antidepressants.....58

3.5.4 Antifungal Drugs....59

3.5.5 Others Drug Classes.............................61

3.6 LC-MS/MS Limitations..............61

3.7 Conclusions......63

References.............63

3.1 INTRODUCTION

Recent estimates suggest that 60–70% of medical decisions in a health care setting are based on laboratory testing [1,2]. An estimated 7–10 billion tests are ordered annually by physicians in the United States for clinical disease management [3], and they may be used for diagnosis, prognosis, treatment, and preventative interventions. In addition, laboratory testing plays a significant role in assessing and optimizing the therapeutic efficacy of drugs as well as mitigating potential drug-related toxicities. This approach is referred to as therapeutic drug monitoring (TDM). However, interindividual variability can substantially impact the body's effect on a drug (pharmacokinetics (PK)) as well as the drug's effect on the body (pharmacodynamics (PD)). Hepatic and renal status, regimen compliance, and genetic variations may all impact the PK and PD of a drug [4]. The implementation and application of TDM may allow for modifications in dosing regimens to provide the right dose for maximal therapeutic efficacy and reduced toxicity (Fig. 3.1). With a demand for TDM in clinical laboratories and optimal patient care, the need for the development and use of robust tools with enhanced analytical performance characteristics has become imperative [5,6].

In clinical and reference laboratories, an immunoassay approach has been the primary platform used for drug quantification for decades. With several commercially available kits, particularly U.S. Food and Drug Administration (FDA)-approved in vitro diagnostic assays, methods may be completely automated with reduced need for specialized personnel and easily interfaced with

W. Clarke & A. Dasgupta (Eds): Clinical Challenges in Therapeutic Drug Monitoring. DOI: http://dx.doi.org/10.1016/B978-0-12-802025-8.00003-9

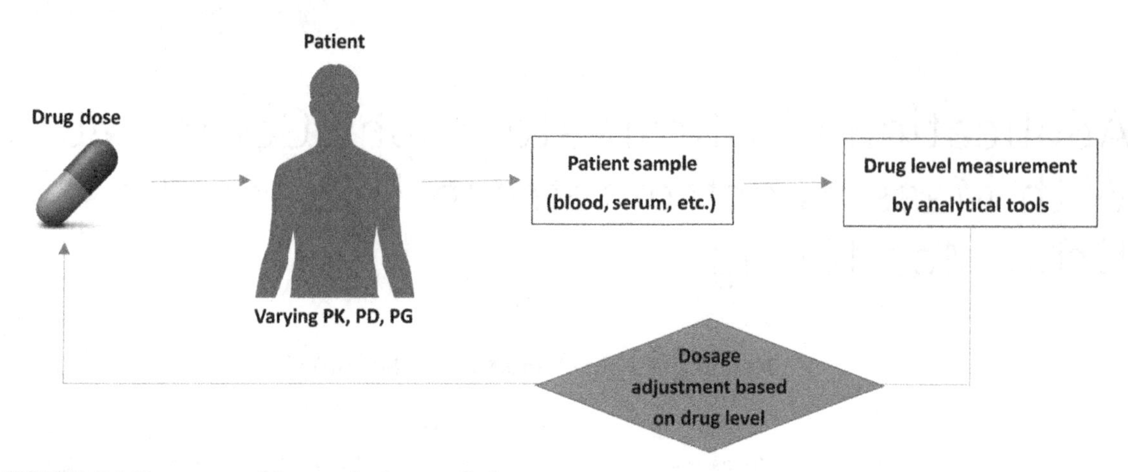

FIGURE 3.1 The process of therapeutic drug monitoring.
Following drug ingestion or administration to a patient, absorption, distribution, metabolism, elimination, as well as therapeutic efficacy may be variable. Collection of blood for drug analysis can be used to determine if the patient is within a target therapeutic window or has subtherapeutic or toxic drug concentrations. Based on drug concentrations (which may be timed or after the drug reaches steady-state concentrations), dose adjustments may be made to improve treatment efficacy.

laboratory information systems [7]. However, antibody-based methodologies may suffer from reduced specificity and sensitivity. Antibody cross-reactivity between the parent drug and structurally similar compounds may result in overestimation of drug concentrations, leading to misinterpretation of results and inappropriate clinical follow-up [8,9]. Furthermore, lack of multiplexed drug quantitation provides an additional challenge in the use of immunoassays for TDM [10]. However, despite these limitations, immunoassays are still commonly employed in many hospital laboratories on account of their high throughput and capacity for automation for several key analytes, including both aminoglycoside antibiotics [11] and immunosuppressants [12].

Conventional immunoassays may lack required sensitivity and specificity for accurate measurement of therapeutic drugs, and as a result, liquid chromatographic (LC) methods combined with mass spectrometry (MS; referred to as LC-MS and LC-MS/MS) were initially developed in clinical laboratories. Such analytical platforms have been implemented in both clinical and reference laboratory settings for nearly two decades [13]. The ability of LC to separate individual compounds from other drugs and metabolites present in the biological matrix, combined with selective MS techniques to provide further mass-based delineation, has resulted in its superior sensitivity and specificity over immunoassays. As a result, for certain drugs, such as immunosuppressants, LC-MS and LC-MS/MS methodologies are preferred over immunoassays for TDM [14]. In addition, the simultaneous measurement of multiple drugs within a single specimen source using an LC-MS-based platform has decreased the need for higher sample volumes for carrying out various individual tests, and it may even provide improvements on turnaround times (TATs) [10]. Traditionally, TDM measurements have been carried out in blood matrices such as plasma or serum [4]; however, drug quantification in tissue biopsies, dried blood spots, and oral fluid are becoming potential alternatives for TDM measurements. LC-MS platforms provide the advantage of drug quantification in such niche matrices [15].

Despite advantages in the clinical laboratory setting, LC-MS methods are associated with several limitations. Currently, LC-MS and LC-MS/MS assays are laboratory-developed tests (LDTs) and are not subjected to the FDA 501K notification or premarket approval processes. However, additional regulation of LDTs has been proposed, and the nuances regarding oversight are still being discussed in both private and public sectors [16]. LC-MS and LC-MS/MS assays may also exhibit interlaboratory variability that could potentially impact TDM measurements in patient populations that require prolonged follow-up from multiple health care settings [17]. Furthermore, the presence of certain interfering substances in specimen samples such as salts and phospholipids can cause matrix effects that may result in significant fluctuations during compound ionization [18,19]. Due to the high technical expertise required for its operation, there is also a need for highly skilled personnel, leading to challenges in automating the platform. The initial high installation costs, especially for implementation in small- and medium-scale laboratories, time investment for personnel training and method validation, and integration into laboratory workflow for maximum throughput and reduced TAT are some of the factors that require careful considerations before incorporating LC-MS- or LC-MS/MS-based platforms into the clinical laboratory [20].

Despite the aforementioned limitations, LC-MS- and LC-MS/MS-based assays are utilized in cases in which traditional immunoassays cannot be implemented. In addition, with combinatorial drug therapies being recommended as first-line treatments for antiretroviral agents [21], immunosuppressants [22], and anticonvulsants [23], the use of multiplexed LC-MS assays provides analytical measurements with diminished volume requirements. The much needed sensitivity and specificity of such assays for TDM have been a major motivation for its use in clinical laboratories. Further enhancements to improve assay sensitivity and selectivity while mitigating matrix effects and ion suppression can be achieved with more rigorous pre-LC sample cleanup.

This chapter provides an overview of the principles of LC and MS, the current technological advances and performance characteristics, their potential applications and limitations in a clinical laboratory for TDM, as well as considerations during implementation. Of note, because tandem MS is the most widely used LC-MS-based assay in clinical laboratories, such methods are referred to as LC-MS/MS throughout this chapter.

3.2 LIQUID CHROMATOGRAPHY

Chromatography was pioneered by Michael Tswett in the first decade of the 20th century to separate plant pigments [24]. Currently, the technique is utilized in a number of basic, translational and clinical settings. Chromatography may be dichotomized as interplay between two physio-chemical phases: a stationary phase (SP) and a mobile phase (MP). The SP typically consists of a fixed column with a specialized coating layer; the MP (in either a gaseous or a liquid form) flows through the packing of the SP. Differential interactions of compounds with the SP and the MP dictate their separation and elution from the column. A detector analyzes the different compounds as they exit the SP in a time-based manner, displaying the output in the form of a chromatogram. Compounds that exhibit stronger interactions with the SP are eluted at a later time in comparison to compounds with reduced SP interactions. Individual chromatographic peaks represent separated molecules at specific retention times, whereas the peak height and area correlate with the concentration of the particular compound present in the mixture [25,26].

Based on the type of MP used, chromatography is broadly classified as gas- or liquid-based. In gas chromatography (GC), the MP used is in the gaseous state and is applied onto a solid SP [25]. The use of GC comes with an inherent caveat that the compound must be volatile or must be chemically derivatized to enhance its volatility [27]. In addition, GC is operated at high temperatures, making it unsuitable for studying proteins and nucleic acids, which may denature at high temperatures. GC is employed in clinical settings primarily in toxicology screening and confirmatory analysis of drugs of abuse due to its high sensitivity [28].

On the other hand, LC uses liquid solvents as MPs. LC has gained prominence due to its intuitive approach in studying biological-based samples that are typically aqueous-based. LC operates at lower temperatures, thereby proving advantageous for separating compounds that are thermally unstable. The need for higher resolution and faster separation in LC led to the development of high-performance liquid chromatography (HPLC) and subsequently ultra high-performance liquid chromatography (UHPLC). Operated at pressures up to approximately 15,000 psi, UHPLC uses several pumps and injectors to combine the solute with the MP and load the complex onto a column containing the SP. Typically, HPLC is operated in two modes: normal phase and reversed phase [26,29]. Normal-phase liquid chromatography (NPLC) is the conventional form of HPLC (which will heretofore be used to describe both HPLC and UHPLC systems) wherein the SP is polar. Polar solutes exhibit stronger interaction with the SP, in comparison to a nonpolar MP, which is typically composed of organic solvents. Nonpolar solutes are

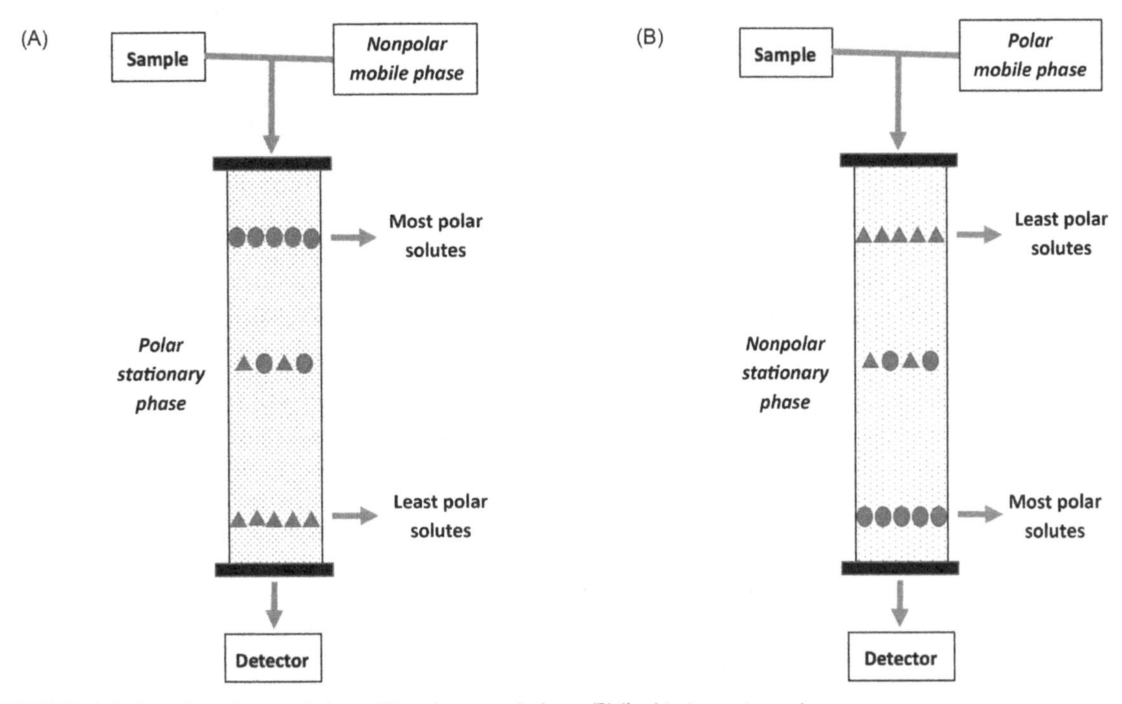

FIGURE 3.2 Overview of normal-phase (A) and reversed-phase (B) liquid chromatography.
(A) Polar solutes exhibit higher affinity toward the polar stationary phase; nonpolar solutes elute first. (B) Nonpolar solutes exhibit higher affinity toward the nonpolar stationary phase; polar solutes elute first.

eluted first, and based on the composition and ratio of polar and nonpolar solvents, polar (hydrophilic) compounds will elute later (Fig. 3.2A) [26,29]. In the early 1940s, Martin and Synge developed reversed-phase liquid chromatography (RPLC), which employs a nonpolar (hydrophobic) SP and a polar MP [30]. Using RPLC, small nonpolar molecules, including many drugs, will have an affinity for the SP, and separation and elution may be attained by flowing through aqueous (polar) MP solvents using an isocratic or gradient approach (Fig. 3.2B) [26,29].

In both NPLC and RPLC, separation is achieved by the appropriate combination of the SP and the MP. In NPLC, the SP comprises columns that are silica- or alumina-based with polar functional groups, such as diol, cyano, and amino groups. The interaction between the solute and the column is largely via hydrophilic interactions. In contrast, using an RPLC approach, the SP consists of a column that is supported with a layer of hydrophobic alkyl-based chains—most commonly octyl (C8) or octadecyl (C18). Consequent interactions with the SP are hydrophobic in nature, with an increased affinity for small, nonpolar molecules [31]. In most columns, the packing material constitutes approximately 60–65% of the space, whereas the remaining empty space allows for the flow of the MP and solutes and their efficient interactions with the SP [26]. Commercially, different column configurations of varying pore size, length, diameter, and particle size are available that can assist in the appropriate separation of solutes. Columns with smaller diameters (1.0 or 2.1 mm), shorter lengths (50 mm), and smaller particle diameter (1.7 to 3.5 μm) are preferred for better separation of small molecules [32]. In addition, the selection of an appropriate column configuration also dictates the flow rate that can be used on the LC platform. Lower flow rates (<0.3 mL/min) result in higher resolution as well as reduced usage of MPs [32].

Compound elution is achieved by either an isocratic or a gradient separation mode [27,29]. An isocratic elution employs a single MP system, which is typically a mixture of nonpolar and polar solvents. Due to the poor resolving capability and longer time requirement for separation, a multiple MP system is preferred. A gradient elution employs at least two MPs: polar and nonpolar solvent. MP may be referred to as weak or strong, depending on potential interactions with the SP. A weak MP is one that has opposite polarity as the SP [25]. For example, in RPLC with a nonpolar column, the aqueous-based MP is the weak MP and the organic-based MP is the strong MP. In a gradient-type elution, an online program is typically set up that alters the gradient, either suddenly or gradually, to modulate the ratio of the organic and aqueous MPs. This ensures that there is sufficient compound separation and faster and timely elution of various solutes [27]. During the preparation of MPs, additives such as mild acids (formic acid and acetic acid) are incorporated into the MPs at low amounts (0.05–5%), which aids in providing additional protons for interactions, thereby resulting in better peak shape of the output chromatogram [32].

Hydrophilic interaction chromatography (HILIC) is a variant of NPLC and is capable of better separation of polar compounds on the polar SP, in comparison to NPLC. Whereas the SP is polar as in NPLC, there is an introduction of aqueous solvents in the MP, as in the case of RPLC. Separation is brought about by the elution of nonpolar compounds by the organic MP, followed by polar compounds with the gradual introduction of the aqueous MP [33]. HILIC has gained prominence in metabolomics for the separation of various polar metabolites [34].

For the previously discussed LC techniques, several detector systems are available to analyze the separated compounds, including fluorescence detectors, light scattering detectors, ultraviolet (UV) diode-array detectors, and electrochemical detectors. In terms of sensitivity, these detectors possess some limitations that could result in narrow detection ranges and poor UV absorbance by certain compounds [25]. Also, in a clinical setting, HPLC-UV systems have exhibited erroneous false-positive results due to interfering substances, in some confirmed cases [35]. This has led to the incorporation of MS at the back end of an LC platform. In addition to the increased sensitivity and selectivity in detection, MS systems provide further mass-based and charge-based separation to the separated compounds [36,37]. Common MS-complexed chromatography systems, such as GC-MS, LC-MS, and LC-MS/MS, are discussed in more detail in the next section.

3.3 MASS SPECTROMETRY

Pioneered by J.J. Thomson in the early years of the 20th century, the use of MS has grown from being an important biophysical tool for studying atomic masses to a significant component of biomarker discovery, translational research, and clinical laboratory testing [38,39]. Currently, MS is widely applied to several fields, such as neonatal screening [40], isotope enrichment [41], peptide sequencing, small-molecule quantification [32], toxicology screening and subsequent confirmatory testing [42], as well as metabolomics and proteomics-centered translational research [43]. The advent of electrospray ionization [44,45] has further facilitated the interfacing of mass analyzers with LC platforms [46], enhancing its translational and clinical utility.

A mass spectrometer consists of three essential components: an ion source, a mass analyzer, and a detector (Fig. 3.3). The ion source ionizes the desired analyte, the mass analyzer selects the ionized analyte based on its mass to charge ratio (m/z), and the detector captures the selected ions, providing readable output [47,48]. The various components of a mass spectrometer are discussed in the following sections.

3.3.1 Ion Source

The ionization event involves the conversion of a chromatographically eluted compound or molecule into a charged state, which is a prerequisite for downstream MS analysis. The selection of an appropriate ion source is largely dependent on the physical and chemical properties of the analyte of interest, such as its physical state (gas, liquid, or solid), thermal stability, and molecular weight, and these factors also influence the degree of ionization required for analysis [49]. The process of ionization is broadly classified into hard and soft ionization techniques [50]; a hard ionization involves high potential energy imparted to the analyte resulting in extensive ion fragmentation, whereas a soft ionization involves lower energy levels and reduced in-source ion fragmentation [32]. The selection of ionization approach is dependent on the downstream application of the MS analysis. Typically, extensive fragmentation provides better clarity in studying molecular structures and functional groups of compounds, whereas reduced fragmentation via soft

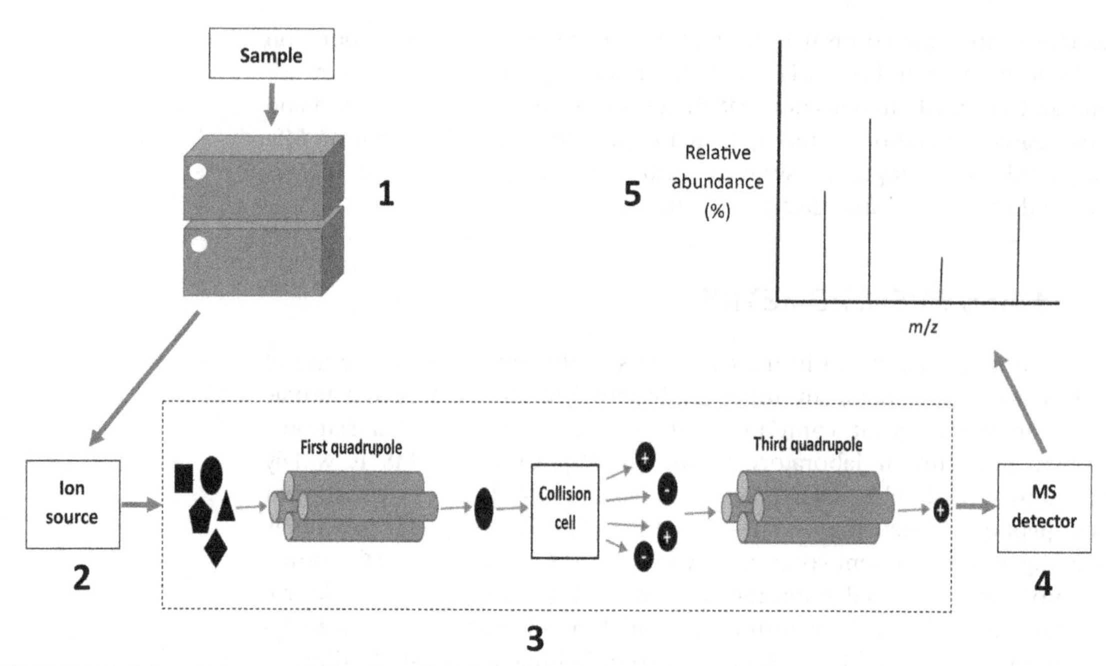

FIGURE 3.3 Schematic representation of an LC-MS/MS platform.
The extracted sample is introduced into the liquid chromatography (LC) system (1). The LC system separates different constituents of the sample; based on compound affinities for chromatographic columns, eluted analytes are diverted to waste or introduced into the mass spectrometer (2–4). In the mass spectrometer, the molecules are ionized at the ion source under high heat and voltage requirements (2), and they subsequently enter the tandem mass spectrometer (MS) (3). Precursor ion selection occurs in the first quadrupole, undergoes collision-induced dissociation in the collision cell (non-mass resolving quadrupole), and the product ions are selected in the third quadrupole. The MS detector (4) captures the selected product ions and displays an output in the form of a mass spectrum (5), which represents the relative abundance of the different ions of varying mass-to-charge (*m/z*) values.

ionization provides more information on the molecular formula of the compound [51].

The most common hard ionization techniques are electron ionization (EI) and chemical ionization (CI). EI and CI utilize electrons and chemical reagents, respectively, to bombard the analyte of interest, leading to the cationic production of the analyte [48]. Such MS ion sources are housed under vacuum conditions and thus are better suited to measure analytes in the gaseous phase. Consequently, they work well when complexed with an upfront GC platform and are termed GC-MS platforms. They can be very useful for ionizing low-molecular-weight compounds (<1000 Da); measured compounds should also be volatile or capable of volatilization via appropriate derivatization [50].

However, as previously mentioned, in a clinical setting, biological specimen sources are mostly aqueous-based. Typically separated via LC, the analyte of interest is in a dissolved liquid state. Ionization of a liquid solution via EI or CI may not be possible; furthermore, certain analytes may not be volatile or easily derivatized [48]. The development of ion sources capable of operating at atmospheric pressure allowed for the more straightforward pairing of an MS with an upstream LC system [44,45]. Electrospray ionization (ESI) is one of the most popular atmospheric pressure ionization processes employed in clinical and research laboratories due to its ease of use. ESI employs an electric field to volatilize and ionize the analyte into a spray of large charged droplets. These large droplets are further subjected to heat and inert gas in order to compress the larger droplets into smaller ones. The smaller droplets experience higher ion repulsion leading to their further reduction to single ion droplets [52]. Depending on the mode of operation (positive or negative), ions are produced either by proton addition ($M + H^+$) or proton subtraction ($M-H^+$), where M describes the analyte of interest [32]. ESI is a soft ionization technique resulting in reduced fragmentation of the compound at the ionization source [53].

Other ionization sources include atmospheric pressure chemical ionization (APCI) and atmospheric pressure photoionization (APPI). In contrast to ESI, in both APCI and APPI, ionization occurs after the transformation of the analyte from the liquid to gaseous state. This ionization technique facilitates reduced occurrence of matrix effects. In APCI, a corona discharge needle [47] producing a spray of electrons is used, whereas in APPI, photons/ultraviolet light [54] is used instead to bombard the analyte and release ions. Whereas ESI works well with most biological specimens, especially high-molecular-weight polar molecules, both APCI and APPI are commonly suited for the ionization of steroidal and lipid compounds [32].

Matrix-assisted laser desorption/ionization (MALDI) and surface-enhanced laser desorption/ionization (SELDI) are commonly employed to ionize peptides, which are either mixed with a suitable matrix and then coated onto a solid surface (in the case of MALDI) [47] or applied onto a premodified coated surface (in the case of SELDI) [55]. Typically, the matrix is made up of light-absorbing molecules (MALDI) or is chemically modified to improve binding affinity (SELDI) [56]. When interfaced with high-resolution, untargeted mass analyzers, such as a time-of-flight (TOF) instrument, MALDI has been used in screening and identification of not only drugs but also microorganisms [57].

3.3.2 Mass Analyzers

Following ionization, ions enter into the mass analyzer, wherein they are selected based on their *m*/*z*. Some of the common mass analyzers are discussed next.

3.3.2.1 Quadrupole Analyzers

As the name suggests, quadrupole analyzers consist of four metallic cylindrical rods placed circularly parallel to one another to create a central space for the passage of ions through it. With the application of varying voltages across these quadrupoles, differential electric fields are created, causing only certain ions to be selected by the detector, while others are deflected away from the detector [47]. Instruments can be operated in either full scan mode, in which all ions of varying *m/z* values are monitored, or in selective ion monitoring (SIM) or selective reaction monitoring (SRM) modes, wherein an ion of a specific *m/z* is monitored [58]. Triple quadrupole (QQQ) instruments are a popular variation of these analyzers due to their enhanced specificity.

Commonly used for TDM and small-molecule quantification, a QQQ consists of two mass-selecting quadrupoles, Q1 and Q3, which flank a non-mass quadrupole (Q2). Q1 and Q3 select for the parent (precursor) and product ion, respectively. Q2 is also referred to as a collision cell, where the precursor ion selected in Q1 may undergo collision-induced dissociation. With the use of direct current and radio frequency voltages, applied in pulses, different *m/z* values can be selected [58]. The selected ions are bombarded by an inert gas (nitrogen, argon, krypton, or xenon), resulting in fragmentation into product ions, which may be selected for at Q3. The selective identification of both parent and precursor ions specific to the compound of interest demonstrates the highly specific identification of an analyte of interest. The approach of drug identification or quantification based on precursor and product ion characteristics is referred to as tandem mass spectrometry (MS/MS), and analytes of interest may be monitored in SRM mode [53,59]. As previously described, this is a common downstream analyzer used in conjunction with an LC system, resulting in an LC-MS/MS platform.

3.3.2.2 Ion Trap Analyzers

Similar to quadrupole analyzers, ion trap analyzers also separate ions based on their *m/z* value. They trap ions for a fixed time in a central chamber, before sequentially releasing these ions based on their *m/z* values, thereby creating a mass spectrum [48]. Ion trap analyzers exhibit improved sensitivity due to higher diversion of the trapped ions to the detector in comparison to quadrupole analyzers [31]. However, they may exhibit lower accuracy for smaller mass ions due to limited ion trapping range [60]. A modified ion trap analyzer is the Orbitrap MS developed by Makarov [61,62], which traps ions in an orbital manner, resulting in improved resolution and mass accuracy [63]. Trap analyzers have been utilized in the identification of compounds present in trace levels in mixtures as well as in the qualitative identification of antiretroviral drugs [64] and identification and quantification of pain management drugs [65].

3.3.2.3 *TOF Analyzers*

TOF instruments separate ions based on the premise that lower mass ions travel faster, whereas heavier ions travel slowly. Thereby, separation is caused based on the velocity of the ions, which is dependent on the TOF of the ions from the ion source to the detector, rather than based on the m/z values [48]. TOF analyzers require high sensitive detectors downstream for optimal performance. MALDI-TOF, a common TOF analyzer, is used in peptide mass fingerprinting, for screening and identification of unknown proteins, rather than for quantitative analysis [43]. Both trap and TOF analyzers are high-resolution mass spectrometer platforms.

3.3.3 Detectors

As the ions exit the mass analyzer, they strike a detector that captures and amplifies the signal, providing a readable output called the mass spectrum. The mass spectrum represents a range of m/z values on the x-axis plotted against their relative abundance on the y-axis (Fig. 3.3) [58]. Some of the commonly used detectors are electron multipliers and Faraday cups [48].

3.4 PREANALYTICAL STAGE

Currently, in a clinical setting, TDM is carried out using patient specimens in the form of whole blood, plasma, and serum [4]. However, nontraditional niche matrices (tissue biopsy, oral fluids, dried blood spots, etc.) are gaining widespread relevance for TDM [15]. In all aforementioned matrices, patient specimens contain the desired drug for analysis as well as several interfering components, such as salts and phospholipids. These "additives" have a potential to prevent accurate drug quantification. Matrix effects are commonly observed in MS-based quantification approaches, resulting in over-recovery or suppression of the desired analyte [18,19]. Some of the common techniques employed in laboratories to mitigate the influence of matrix effects on the final output are the use of an appropriate sample cleanup procedure prior to LC injection and the use of an isotopically labeled internal standard.

The simplest forms of sample preparation for LC-MS/MS platforms are dilute-and-shoot and protein precipitation approaches. These techniques are extremely quick to perform and eliminate plasma proteins to a certain extent. However, techniques such as solid phase extraction (SPE) and liquid–liquid extraction (LLE) are more efficient in eliminating potential interfering compounds that affect the degree of ionization of the desired analyte. In addition, SPE and LLE may also concentrate the analyte of interest, thereby resulting in enhanced quantification limits [66,67]. Implementation of SPE and LLE in a clinical setting may not always be feasible because of the extensive manual operations involved, which could impact TAT and cause other blocks in workflow. Although there has been development in online and automated SPE, fully integrated systems in the clinical laboratory are currently unavailable [67].

Historically, LC-MS/MS methods have used structural analogs of the analyte as an internal standard for quantification. Structural analogs are easier to produce and procure, and they are less expensive. Their use in clinical assays still persists, particularly for multiplexed assays [68]. However, an optimal approach to characterize the extent of matrix effects on drug concentrations is through the use of an isotopically labeled form of the analyte as the internal standard [69]. Isotopically labeled internal standards mimic the compound of interest. As a result, any potential variation caused by matrix effects is replicated in both the analyte and the internal standard, resulting in reduced skewing of data [32,70]. Also, it has been suggested that the use of the same internal standard for quantification of a specific drug across laboratories may reduce error rates between laboratories [17]. Although isotopically labeled internal standards can reduce matrix effects and increase standardization, they are costly, and in some cases the desired internal standard for an analyte may not be available [19].

3.5 APPLICATION OF LC-MS/MS METHODS IN TDM

LC-MS/MS methods are becoming increasingly more common for the TDM of several drug classes, such as immunosuppressants, anticonvulsants, antidepressants, antifungal drugs, antiviral drugs, and antibiotics [20,71]. The high selectivity and specificity to delineate different drug compounds, the ability to measure two or more drugs simultaneously in a single analytical run, and reduced drug class or metabolite cross-reactivity are some of the key factors that have favored the implementation of LC-MS/MS methods in clinical laboratories [72,73]. Currently, TDM and clinical decisions are based on drug measurements in common matrices such as plasma, serum, and whole blood [4]. However, several LC-MS/MS methods are being developed for drug quantification in niche matrices such as tissue biopsies, dried blood spots, and oral fluids [15,74,75] to provide improved clinical relevance or ease of collection for TDM, respectively. Some of the significant contributions of LC-MS/MS methods to different drug classes are briefly reviewed here. These topics are discussed extensively in other chapters of this book.

3.5.1 Immunosuppressants

Immunosuppressants are used in the treatment and control of several autoimmune conditions, such as rheumatoid arthritis [76,77], psoriasis [78], and inflammatory bowel disease [79], and in solid organ transplantation [80–82]. Corticosteroids and antimetabolites are less frequently monitored in bodily fluids, whereas cyclosporine A (CyA), tacrolimus (Tac), sirolimus (Sir), everolimus (Eve), and mycophenolic acid (MPA) are routinely measured to optimize therapy [83]. Tac and Sir have very narrow therapeutic ranges of 5–20 and 5–15 ng/mL, respectively [84]. Toxic concentrations may lead to nephrotoxicity, hepatotoxicity, elevated blood pressure, increased susceptibility to viral infections, and hypertension [85,86]. Drug–drug interactions among immunosuppressants also exist, including between MPA and CyA: CyA can modulate MPA metabolism, thereby impacting drug pharmacokinetics [87,88]. Therefore, TDM has become imperative for these aforementioned drugs due to their therapeutic window, significant drug–drug interactions, and toxic side effects [89].

The lack of specificity exhibited by immunoassays for the quantification of CyA [8,90], MPA [9], and Tac [91] and their respective metabolites has been extensively reviewed. HPLC-UV has shown better specificity to quantify the parent drugs and their respective metabolites [86]. However, the sensitivity that can be achieved by HPLC-UV compares closely to the lower limits of therapeutic concentrations of Eve and Tac (<20 ng/mL). As a result, although HPLC-UV is beneficial and has been used for the measurement for CyA and MPA, it lacks the necessary sensitivity for Tac and Eve quantification [22]. This has led to the use of MS instead of UV as a detector, which was first investigated for Tac quantification in the blood and urine of liver transplant patients with the use of a CI-based LC-MS method [92]. Since then, LC-MS and LC-MS/MS platforms have become more common for TDM measurements for immunosuppressants [93,94]. Table 3.1 reviews some recently developed LC-MS methods for the quantification of several immunosuppressants in various matrices.

3.5.2 Anticonvulsants

Also known as antiepileptic drugs (AEDs), anticonvulsants are primarily used as a prophylactic measure for the management of epilepsy [83]. TDM for AEDs gained prominence in the mid-1960s, and by the late 1970s, it was routinely used in clinical management [98]. Many AEDs possess narrow therapeutic index; toxic concentrations of valproate and phenytoin have resulted in seizures, proving counterproductive for their prophylactic use [20]. Typically administered with one or more AEDs and other medications, the need for TDM arose because of potential drug–drug interactions exhibited by both first-generation and newer AEDs, which has been extensively confirmed [99,100]. For example, valproate significantly increases the concentrations of lamotrigine and phenytoin when administered concomitantly [20,83]. Apart from other AEDs, they may also impact concentrations of oral contraceptives and oral anticoagulants [83,99].

Table 3.1 Selective LC-MS Methods Recently Developed for Measurements of Immunosuppressants in Various Matrices

Compound(s)	Matrix	Unique Features	Reference
Cyclosporine A Tacrolimus Sirolimus Everolimus Mycophenolic acid	Whole blood (except plasma for MPA)	■ Simultaneous measurement of all five drugs: measurement of MPA in parallel with the other four drugs due to same LC conditions ■ Online SPE ■ Use of deuterated internal standard for all analytes—decreased matrix effects ■ Good sensitivity (CyA, 2–1250 ng/mL; Tac, 0.5–42.2 ng/mL; Sir, 0.62–49.2 ng/mL; Eve, 0.53–40.8 ng/mL; MPA, 0.01–7.5 μg/mL) ■ Analytical run time: 3.5 min	[95]
Tacrolimus	Peripheral blood mononuclear cells (PBMCs)	■ Automated online SPE system ■ Detection of tacrolimus in PBMCs—ideal for TDM to study graft rejection ■ Good sensitivity (Tac, 0.391–100 ng/mL) ■ Analytical run time: 5.0 min	[96] [97]

CyA, cyclosporine A; Eve, everolimus; MPA, mycophenolic acid; Sir, sirolimus; Tac, tacrolimus.

Although immunoassay-based platforms have long been used for TDM of AEDs, LC-MS/MS methods have gained wider implementation in clinical laboratories, largely due to their multiplexed capabilities and increased selectivity [101]. Multiplexed LC-MS/MS assays are useful in monitoring patients who are on cotherapies with first-generation AEDs (carbamazepine, phenytoin, and valproic acid) and second-generation AEDs (levetiracetam, gabapentin, and lamotrigine) [102]. The high selectivity of the LC-MS/MS methods can also allow for the simultaneous quantification of a parent drug and its biologically active or inert metabolite. Some of the recently published LC-MS/MS methods are reviewed in Table 3.2.

3.5.3 Antidepressants

Tricyclic antidepressants (TCAs) are first-generation antidepressants used clinically since the 1950s [83]. TDM was not initially emphasized for TCAs due to the poor correlation between plasma concentrations and therapeutic outcomes [104]. However, studies in the 1980s demonstrated that large

Table 3.2 Selective LC-MS Methods Recently Developed (2010–2015) for Measurements of Anticonvulsants

Compound(s)	Matrix	Unique Features	Reference
GBP, LEV, VPA, LTG, CBZ-epoxide, ZNS, OXC, TPM, CBZ, PHT	Plasma	▪ Simultaneous measurement of 9 AEDs + 1 active metabolite (CBZ-epoxide) ▪ VPA, TPM, ZNS—detected in negative mode (internal standard, d_6-VPA) ▪ Remaining drugs—detected in positive mode (internal standard, d_{10}-PHT) ▪ Analytical run time: 12.0 min ▪ Limited sample volume: 10 μL	[102]
Rufinamide	Dried blood spots	▪ MS—TurboIon spray, operated in positive mode, MRM ▪ Analytical run time: 4.0 min ▪ Analytical measuring range: 0.008–0.8 mg/L (0.48–47.60 mg/L in DBS) ▪ Limited sample volume: 3.3–3.4 μL	[103]

CBZ, carbamazepine; CBZ-epoxide, carbamazepine-10,11-epoxide; GBP, gabapentin; LEV, levetiracetam; LTG, lamotrigine; OXC, oxcarbazepine; PHT, phenytoin; TPM, topiramate; VPA, valproic acid; ZNS, zonisamide.

interindividual metabolic variability existed among different patient populations on account of age, race, and alcohol intake, which could result in cardiac and central nervous system-related toxicities [83]. TCAs are currently used in pain management, and they are largely monitored to assess regimen adherence [105]. Depression is largely treated using newer drug classes, including selective serotonin reuptake inhibitors, serotonin–norepinephrine reuptake inhibitors, and monoamine oxidase inhibitors [20]. Table 3.3 reviews some of the multiplexed LC-MS methods for the quantification of several antidepressants and their metabolites.

3.5.4 Antifungal Drugs

Antifungal drugs are primarily used to combat infections in immunocompromised individuals and post-transplantation. Three classes of antifungals currently in clinical use are polyenes, triazoles, and echinocandins [107]. Polyenes are broad-spectrum compounds and, hence, TDM is not necessary. Triazoles, a major class of antifungals, are exhibiting declining efficacy due to

Table 3.3 Selective LC-MS Methods Recently Developed (2005–2015) for Measurements of Antidepressants

Compound(s)	Matrix	Unique Features	Reference
48 Compounds	Serum	■ Simultaneous measurement of 48 TDM-relevant antidepressants, antipsychotics, and their metabolites ■ Sample preparation—protein precipitation ■ Drugs classified into three subgroups with differing sensitivities of measurement to carry out stepwise dilution during sample preparation Low, 1–100 μg/L; medium, 10–1000 μg/L; high, 100–10,000 μg/L ■ MS—ESI operated in positive mode with MRM ■ Analytical run time: 8.0 min ■ Sample volume: 100 μL	[73]
14 Compounds + metabolites	Plasma	■ Online SPE system—reduces significant sample preparation time ■ MS—ESI operated in positive mode with MRM ■ Analytical run time: 20.0 min ■ Sample volume: 50 μL ■ Analytical measuring range: 10–1000 μg/L	[106]

the development of drug-resistant fungal strains [108]. This has led to the development and use of echinocandins, a new class of antifungals. Whereas TDM is recommended for azoles due to potential drug–drug interactions, there is no definitive recommendation for echinocandins [107]. Also, the correlation between blood concentrations and treatment is still under investigation [109]. Combinatorial therapies with azoles and echinocandins are becoming preferred treatment options, resulting in the need for TDM to study potential drug–drug interactions and PK variability [20]. Some of the well-established drug–drug interactions for antifungals include those with immunosuppressants. CyA and Tac are primarily metabolized by CYP3A enzymes, whereas ketoconazole inhibits CYP3A. Consequently, coadministration involves reduced CyA dosages to prevent toxicity [110]. Coadministration of omeprazole, a drug commonly used to treat gastroesophageal reflex disease, results in reduced circulating concentrations of posaconazole [111]. Also, such drug interactions have been reported to more severely impact critically ill patients compared to healthy individuals [112]. Consequently, TDM has become imperative, and Table 3.4 lists some of the recently developed multiplexed LC-MS methods that can be used for TDM.

Table 3.4 Selective LC-MS Methods Recently Developed (2010–2015) for Measurements of Antifungal Agents

Compound(s)	Matrix	Unique Features	Reference
Echinocandins (ANF, CSF, MCF) + triazoles (ISC, PSC, VRC)	Plasma, PBMC, PMN, RBC	■ Simultaneous measurement of echinocandins and triazole agents in various compartments of peripheral blood ■ Two internal standards—for monitoring echinocandins and triazoles ■ MS–ESI operated in positive mode with SRM ■ Sample volume: 30 μL ■ Analytical run time: 8.0 min	[109]
Triazoles (VRC, PSC, FLZ, ITZ)	Serum	■ Sample preparation—involves protein precipitation (less time-consuming) ■ MS–ESI operated in positive mode with SRM ■ Sample volume: 10 μL (FLZ, VRC), 100 μL (PSC, ITZ) ■ Analytical run time: 3.6 min	[112]

ANF, anidulafungin; CSF, caspofungin; FLZ, fluconazole; ISC, isavuconazole; ITZ, itraconazole; MCF, micafungin; PBMC, peripheral blood mononuclear cells; PMN, polymorphonuclear leukocytes; PSC, posaconazole; RBC, red blood cells; VRC, voriconazole.

3.5.5 Others Drug Classes

In the case of antibiotics, TDM is vital for vancomycin and three specific aminoglycosides (amikacin, gentamycin, and tobramycin) due to the nephrotoxicity and ototoxicity exhibited by these drugs and their narrow therapeutic drug windows [20,113]. Clinically, TDM is carried out by immunoassays on account of high throughput and automation. However, with antibiotic drug resistance becoming commonplace, patients are on multiple antibiotic regimens. In such cases, multiplexed LC-MS/MS methods are highly beneficial in comparison to immunoassays [113–115]. For antituberculosis drugs, TDM is typically not recommended because of the broad-spectrum nature of these drugs and prolonged treatment regimens [20]. Regarding antiretroviral drugs, although TDM is not currently in clinical practice, it is suggested that TDM can greatly benefit in studying patient adherence [21].

3.6 LC-MS/MS LIMITATIONS

Although LC-MS and LC-MS/MS methods have been effectively applied in the TDM of several drug compounds, several limitations exist that need to be addressed with careful consideration during implementation in a clinical laboratory.

First, these methods are LDTs. Unlike immunoassays, for which commercialization is more prevalent, in the case of these methods, the driver is typically the laboratory itself, overseeing the development, validation, and implementation of the assay [93]. Consequently, nonharmonization has resulted in significant interlaboratory variability [17]. The factors responsible for such heterogeneity range from the manner in which a sample is handled and processed upon receipt to the reagents and stock materials used for processing, the frequency of quality control checks carried out on the instrumentations, and the use of different extraction, chromatographic, and mass spectrometric parameters [17]. The use of solvents from different vendors, both for the preparation of stock solutions and for LC mobile phases, can impact the stability of the analyte and internal standard and can impact MS ionization [115,116].

Several validation guidelines have been published and endorsed by different regulatory agencies, such as the Food and Drug Administration (FDA) [117], the Clinical and Laboratory Standards Institute (CLSI) [118], and the European Medicines Agency (EMA) [119], for use by individual laboratories to carry out extensive method validation to assess and improve the robustness of their in-house assay. Overall, these guidelines emphasize several key analytical performance characteristics and quality control/quality assurance checks that assess the robustness of the method for clinical application. Validation parameters include the use of appropriate standards and quality controls for calibration curve generation, the assessment of individual peak shape and peak height in comparison to background noise, precision and accuracy, repeatability and reproducibility, the limits of detection and quantification and their therapeutic correlation, matrix effects, possibility of carryover, dilutional integrity, and overall acceptance and rejection criteria. However, from a global perspective, there is a need for a standardized document for LC-MS method validation in a clinical laboratory [16,120,121]. The CLSI C62-A document has attempted to address many of these issues for laboratories developing clinical LC-MS/MS methods. Regardless, the lack of standardization can potentially obscure the interpretation of laboratory results by physicians [122], which in turn can influence dosage adjustments in patients who require routine and long-term monitoring from multiple health centers. Until the issues of regulatory differences are addressed, individual laboratories can reduce this inherent variation by the use of commercialized FDA-approved LC-MS kits as a reference that can significantly enhance standardization [93,123]. Also, efforts to cross-validate assays across laboratories with the extensive participation in a proficiency testing panel may enable laboratories the ability to characterize and control procedural errors [17].

It is also important to note that analytes measured via LC-MS/MS-based methods may exhibit matrix effects, which may present as either ion suppression or ion enhancement [18,124], and can potentially affect the accuracy of the results. Such effects have been shown to be more common in ESI compared to APCI or APPI ionization sources [125]. Matrix effects are compound-dependent; however, ion modulation is more common in polar compounds [19]. Also, the use of pooled matrix for the preparation of standard and reference materials in laboratories also contributes to matrix effects, leading to further interlaboratory variation in results [19]. Some of the ways to mitigate matrix effects include the use of an appropriate sample cleanup procedure and the use of an isotopically labeled internal standard, as discussed previously.

From an implementation perspective, especially for small- and medium-scale laboratories, there is a high cost associated with its initial setup and operation [13,126]. In addition, LC-MS methods are not yet a "push-button" technology and are less automated in comparison to immunoassays. There is a requirement for highly skilled staff for its operation and maintenance [20]. Due to the higher physical involvement, LC-MS methods are more prone to human errors, which could further contribute to interlaboratory variability. Also, there is a significant time investment required for personnel training as well as for method validation. Rigorous sample preparation techniques, although beneficial in improving assay sensitivity and accuracy, result in additional physical labor and time.

3.7 CONCLUSIONS

In the past two decades, LC combined with MS has shown tremendous potential for its implementation in clinical laboratories in the field of TDM. Enhanced sensitivity, specificity, precision, and accuracy have been favorable factors for its use, whereas matrix effects, higher interlaboratory variability, and increased physical labor have been trade-offs. With the growing technological advancements in LC and MS instrumentations, there is a promising hope that superior analytical performance characteristics can be achieved in clinical laboratories without the need for compromise, thereby serving the field of patient health care with high quality and timely contributions.

References

[1] Forsman RW. Why is the laboratory an afterthought for managed care organizations? Clin Chem 1996;42(5):813–16.

[2] Silverstein M. An approach to medical errors and patient safety in laboratory services. A white paper prepared for the Quality Institute Meeting. Making the laboratory a partner in patient safety. Atlanta, GA. Suppl. 2. Division of Laboratory Systems, Centers for Disease Control and Prevention; April 2003.

[3] Steindel SJ, Rauch WJ, Simon MK, Handsfield J. National Inventory of Clinical Laboratory Testing Services (NICLTS). Development and test distribution for 1996. Arch Pathol Lab Med 2000;124(8):1201–8.

[4] Gross AS. Best practice in therapeutic drug monitoring. Br J Clin Pharmacol 2001;52:5S–10S.

[5] Pippenger CE. Commentary: therapeutic drug monitoring in the 1990s. Clin Chem 1989;35 (7):1348–51.

[6] Bowers LD. Analytical goals in therapeutic drug monitoring. Clin Chem 1998;44 (2):375–80.

[7] Taylor PJ, Tai CH, Franklin ME, Pillans PI. The current role of liquid chromatography-tandem mass spectrometry in therapeutic drug monitoring of immunosuppressant and antiretroviral drugs. Clin Biochem 2011;44(1):14–20.

[8] Steimer W. Performance and specificity of monoclonal immunoassays for cyclosporine monitoring: how specific is specific? Clin Chem 1999;45(3):371–81.

[9] Shipkova M, Schutz E, Armstrong VW, Niedmann PD, Wieland E, Oellerich M. Overestimation of mycophenolic acid by EMIT correlates with MPA metabolite. Transplant Proc 1999;31(1–2):1135–7.

[10] Yang Z, Wang S. Recent development in application of high performance liquid chromatography-tandem mass spectrometry in therapeutic drug monitoring of immunosuppressants. J Immunol Methods 2008;336(2):98–103.

[11] Jolley ME, Stroupe SD, Wang CHJ, Panas HN, Keegan CL, Schmidt RL, et al. Fluorescence polarization immunoassay I. Monitoring aminoglycoside antibiotics in serum and plasma. Clin Chem 1981;27(7):1190–7.

[12] Gounden V, Soldin SJ. Tacrolimus measurement: building a better immunoassay. Clin Chem 2014;60(4):575–6.

[13] Grebe SK, Singh RJ. LC-MS/MS in the clinical laboratory—where to from here? Clin Biochem Rev 2011;32(1):5–31.

[14] Hoofnagle AN, Wener MH. The fundamental flaws of immunoassays and potential solutions using tandem mass spectrometry. J Immunol Methods 2009;347(1–2):3–11.

[15] Ghareeb M, Akhlaghi F. Alternative matrices for therapeutic drug monitoring of immunosuppressive agents using LC-MS/MS. Bioanalysis 2015;7(8):1037–58.

[16] Marzinke MA, Clarke W. Laboratory developed tests in the clinical laboratory: challenges for implementation. Bioanalysis 2015;7(15):1817–20.

[17] Christians U, Vinks AA, Langman LJ, Clarke W, Wallemacq P, van Gelder T, et al. Impact of laboratory practices on interlaboratory variability in therapeutic drug monitoring of immunosuppressive drugs. Ther Drug Monit 2015;37(6):718–24.

[18] Annesley TM. Ion suppression in mass spectrometry. Clin Chem 2003;49(7):1041–4.

[19] Taylor PJ. Matrix effects: the Achilles heel of quantitative high-performance liquid chromatography-electrospray-tandem mass spectrometry. Clin Biochem 2005;38 (4):328–34.

[20] Adaway JE, Keevil BG. Therapeutic drug monitoring and LC-MS/MS. J Chromatogr B Analyt Technol Biomed Life Sci 2012;883–884:33–49.

[21] Anonymous Panel on Antiretroviral Guidelines for Adults and Adolescents. Guidelines for the use of antiretroviral agents in HIV-1-infected adults and adolescents. Department of Health and Human Services; April 8, 2015.

[22] Deters M, Kaever V, Kirchner GI. Liquid chromatography/mass spectrometry for therapeutic drug monitoring of immunosuppressants. Analytica Chimica Acta 2003;492:133–45.

[23] Brodie MJ. Antiepileptic drug therapy the story so far. Seizure 2010;19(10):650–5.

[24] Guiochon GA. Chromatography, today and tomorrow. Anal Chim Acta 1993;283:309–19.

[25] Hage DS. Chromatography and electrophoresis. In: Clarke W, editor. Contemporary practice in clinical chemistry. 2nd ed. Washington, DC: AACC Press; 2011. p. 101–19.

[26] Loura LMS. Principles of liquid chromatography. In: Fernando R, editor. Liquid chromatography: principles, technology and applications. New York, USA: Nova Science Publishers, Inc.; 2013. p. 1–36.

[27] Snyder LR. HPLC: past and present. Anal Chem 2000;72(11):412A–20A.

[28] Maurer HH. Systematic toxicological analysis of drugs and their metabolites by gas chromatography-mass spectrometry. J Chromatogr 1992;580:3–41.

[29] Bird IM. High performance liquid chromatography: principles and clinical applications. BMJ 1989;299:783–7.

[30] Martin AJP, Synge RLM. A new form of chromatography employing two liquid phases. Biochem J 1941;35:1358–68.

[31] McMaster MC. LC/MS: a practical user's guide. Hoboken, NJ: John Wiley & Sons, Inc.; 2005.

[32] Pitt JJ. Principles and applications of liquid chromatography-mass spectrometry in clinical biochemistry. Clin Biochem Rev 2009;30(1):19–34.

[33] Buszewski B, Noga S. Hydrophilic interaction liquid chromatography (HILIC)—a powerful separation technique. Anal Bioanal Chem 2012;402(1):231–47.

[34] Periat A, Boccard J, Veuthey JL, Rudaz S, Guillarme D. Systematic comparison of sensitivity between hydrophilic interaction liquid chromatography and reversed phase liquid chromatography coupled with mass spectrometry. J Chromatogr A 2013;1312:49–57.

[35] Lertxundi U, Manrique MH, Echaburu SD, Martinez M. A false positive for clozapine using high-pressure liquid chromatography with ultraviolet detection. Acta Neuropsychiatr 2012;24(4):245–6.

[36] Vogeser M, Seger C. A decade of HPLC-MS/MS in the routine clinical laboratory—goals for further developments. Clin Biochem 2008;41(9):649–62.

[37] Niessen WMA, Tinke AP. Liquid chromatography-mass spectrometry: general principles and instrumentation. J Chromatogr A 1995;703:37–57.

[38] Griffiths J. A brief history of mass spectrometry. Anal Chem 2008;80:5678–83.

[39] Yates III JR. A century of mass spectrometry: from atoms to proteomes. Nat Methods 2011;8(8):633–7.

[40] Millington DS, Kodo N, Norwood DL, Roe CR. Tandem mass spectrometry: a new method for acylcarnitine profiling with potential for neonatal screening for inborn errors of metabolism. J Inherit Metab Dis 1990;13:321–4.

[41] van Eijk HMH, Rooyakkers DR, Soeters PB, Deutz NEP. Determination of amino acid isotope enrichment using liquid chromatography–mass spectrometry. Anal Biochem 1999;271:8–17.

[42] Herrin GL, McCurdy HH, Wall WH. Investigation of an LC-MS-MS (QTrap) method for the rapid screening and identification of drugs in postmortem toxicology whole blood samples. J Anal Toxicol 2005;29(7):599–606.

[43] Hortin GL. The MALDI-TOF mass spectrometric view of the plasma proteome and peptidome. Clin Chem 2006;52(7):1223–37.

[44] Fenn JB, Mann M, Meng CK, Wong SF, Whitehouse CM. Electrospray ionization for mass spectrometry of large biomolecules. Science 1989;246:64–71.

[45] Fenn JB. Electrospray wings for molecular elephants (Nobel lecture). Angew Chem Int Ed Engl 2003;42(33):3871–94.

[46] Pullen F. The fascinating history of the development of LC-MS; a personal perspective. Chromatogr Today 2010;February/March:4–6.

[47] Glish GL, Vachet RW. The basics of mass spectrometry in the twenty-first century. Nat Rev Drug Discov 2003;2(2):140–50.

[48] Straseski JA, Clarke W. Mass spectrometry. In: Clarke W, editor. Contemporary practice in clinical chemistry. 2nd ed. Washington, DC: AACC Press; 2011. p. 131–40.

[49] Lavagnini I, Magno F, Seraglia R, Traldi P. Quantitative applications of mass spectrometry. Hoboken, NJ: John Wiley & Sons Inc.; 2006.

[50] Vogeser M, Parhofer KG. Liquid chromatography tandem-mass spectrometry (LC-MS/MS)—technique and applications in endocrinology. Exp Clin Endocrinol Diabetes 2007;115 (9):559–70.

[51] Clement RE, Karasek FW. Gas chromatography/mass spectrometry/computer instrumentation. In: Karasek FW, Hutzinger O, Safe S, editors. Mass spectrometry in environmental sciences. New York, NY: Plenum Press; 1985. p. 21–48.

[52] Kebarle P, Verkerk UH. Electrospray: from ions in solution to ions in the gas phase, what we know now. Mass Spectrom Rev 2009;28(6):898–917.

[53] Dooley KC. Tandem mass spectrometry in the clinical chemistry laboratory. Clin Biochem 2003;36(6):471–81.

[54] Raffaelli A, Saba A. Atmospheric pressure photoionization mass spectrometry. Mass Spectrom Rev 2003;22(5):318–31.

[55] Liu C. The application of SELDI-TOF-MS in clinical diagnosis of cancers. J Biomed Biotechnol 2011;2011:245821.

[56] Hillencamp F, Karas M. Matrix-assisted laser desorption/ionisation, an experience. Int J Mass Spectrom 2000;200:71–7.

[57] Dingle TC, Butler-Wu SM. MALDI-TOF mass spectrometry for microorganism identification. Clin Lab Med 2013;33(3):589–609.

[58] de Hoffman E. Tandem mass spectrometry: a primer. J Mass Spectrom 1996;31:129–37.

[59] Yost RA, Fetterolf DD. Tandem mass spectrometry (MS/MS) instrumentation. Mass Spectrom Rev 1983;2:1–45.

[60] Aebersold R, Mann M. Mass spectrometry-based proteomics. Nature 2003;422 (6928):198–207.

[61] Makarov A. Electrostatic axially harmonic orbital trapping: a high-performance technique of mass analysis. Anal Chem 2000;72(6):1156–62.

[62] Zubarev RA, Makarov A. Orbitrap mass spectrometry. Anal Chem 2013;85(11):5288–96.

[63] Perry RH, Cooks RG, Noll RJ. Orbitrap mass spectrometry: instrumentation, ion motion and applications. Mass Spectrom Rev 2008;27(6):661–99.

[64] Marzinke MA, Breaud A, Parsons TL, Cohen MS, Piwowar-Manning E, Eshleman SH, et al. The development and validation of a method using high-resolution mass spectrometry (HRMS) for the qualitative detection of antiretroviral agents in human blood. Clin Chim Acta 2014;433:157–68.

[65] Cao Z, Kaleta E, Wang P. Simultaneous quantitation of 78 drugs and metabolites in urine with a dilute-and-Shoot LC-MS-MS Assay. J Anal Toxicol 2015;39(5):335–46.

[66] Van Eeckhaut A, Lanckmans K, Sarre S, Smolders I, Michotte Y. Validation of bioanalytical LC-MS/MS assays: evaluation of matrix effects. J Chromatogr B Analyt Technol Biomed Life Sci 2009;877(23):2198–207.

[67] Vogeser M, Kirchhoff F. Progress in automation of LC-MS in laboratory medicine. Clin Biochem 2011;44(1):4–13.

[68] Seger C, Tentschert K, Stoggl W, Griesmacher A, Ramsay SL. A rapid HPLC-MS/MS method for the simultaneous quantification of cyclosporine A, tacrolimus, sirolimus and everolimus in human blood samples. Nat Protoc 2009;4(4):526–34.

[69] Taylor PJ, Brown SR, Cooper DP, Salm P, Morris MR, Pillans PI, et al. Evaluation of 3 internal standards for the measurement of cyclosporin by HPLC–mass spectrometry. Clin Chem 2005;51(10):1890–3.

[70] Stockvis E, Rosing H, Beijnen JH. Stable isotopically labeled internal standards in quantitative bioanalysis using liquid chromatography/mass spectrometry: necessity or not? Rapid Commun Mass Spectrom 2005;19(3):401–7.

[71] Saint-Marcoux F, Sauvage FL, Marquet P. Current role of LC-MS in therapeutic drug monitoring. Anal Bioanal Chem 2007;388:1327–49.

[72] Premaud A, Rousseau A, Picard N, Marquet P. Determination of mycophenolic acid plasma levels in renal transplant recipients co-administered sirolimus: comparison of an enzyme multiplied immunoassay technique (EMIT) and liquid chromatography-tandem mass spectrometry. Ther Drug Monit 2006;28(2):274–7.

[73] Kirchherr H, Kuhn-Velten WN. Quantitative determination of forty-eight antidepressants and antipsychotics in human serum by HPLC tandem mass spectrometry: a multi-level, single-sample approach. J Chromatogr B Analyt Technol Biomed Life Sci 2006;843 (1):100–13.

[74] Cheung CY, van der Heijden J, Hoogtanders K, Christiaans M, Liu YL, Chan YH, et al. Dried blood spot measurement: application in tacrolimus monitoring using limited sampling strategy and abbreviated AUC estimation. Transpl Int 2008;21(2):140–5.

[75] Manicke NE, Abu-Rabie P, Spooner N, Ouyang Z, Cooks RG. Quantitative analysis of therapeutic drugs in dried blood spot samples by paper spray mass spectrometry: an avenue to therapeutic drug monitoring. J Am Soc Mass Spectrom 2011;22(9):1501–7.

[76] Appel GB, Radhakrishnan J, Ginzler EM. Use of mycophenolate mofetil in autoimmune and renal diseases. Transplantation 2005;80(2 Suppl.):S265–71.

[77] Doan T, Massarotti E. Rheumatoid arthritis: an overview of new and emerging therapies. J Clin Pharmacol 2005;45(7):751–62.

[78] Heydendael VM, Spuls PI, Opmeer BC, de Borgie CA, Reitsma JB, Goldschmidt WF, et al. Methotrexate versus cyclosporine in moderate-to-severe chronic plaque psoriasis. N Engl J Med 2003;349(7):658–65.

[79] Faubion Jr WA, Loftus Jr EV, Harmsen WS, Zinsmeister AR, Sandborn WJ. The natural history of corticosteroid therapy for inflammatory bowel disease: a population-based study. Gastroenterology 2001;121:255–60.

[80] Morris PJ. Transplantation—a medical miracle of the 20th century. N Engl J Med 2004;351 (26):2678–80.

[81] Halloran PF. Immunosuppressive drugs for kidney transplantation. N Engl J Med 2004;351 (26):2715–29.

[82] Calne RY, White DJ, Thiru S, Evans DB, McMaster P, Dunn DC, et al. Cyclosporin A in patients receiving renal allografts from cadaver donors. Lancet 1978;2(8104-5):1323–7.

[83] Hammett-Stabler CA, Dasgupta A, editors. Therapeutic drug monitoring data—a concise guide. 3rd ed. Washington, DC: AACC Press; 2007.

[84] Millner L, Rodriguez C, Jortani SA. A clinical approach to solving discrepancies in therapeutic drug monitoring results for patients on sirolimus or tacrolimus: towards personalized medicine, immunosuppression and pharmacogenomics. Clin Chim Acta 2015;450:15–18.

[85] Rezzani R. Cyclosporine A and adverse effects on organs: histochemical studies. Prog Histochem Cytochem 2004;39(2):85–128.

[86] Yang Z, Peng Y, Wang S. Immunosuppressants: pharmacokinetics, methods of monitoring and role of high performance liquid chromatography/mass spectrometry. Clin Appl Immunol Rev 2005;5:405–30.

[87] Cremers S, Schoemaker R, Scholten E, den Hartigh J, Konig-Quartel J, van Kan E, et al. Characterizing the role of enterohepatic recycling in the interactions between mycophenolate mofetil and calcineurin inhibitors in renal transplant patients by pharmacokinetic modelling. Br J Clin Pharmacol 2005;60(3):249–56.

[88] Kim T, Jancel T, Kumar P, Freeman AF. Drug-drug interaction between isavuconazole and tacrolimus: a case report indicating the need for tacrolimus drug-level monitoring. J Clin Pharm Ther 2015;40(5):609–11.

[89] Oellerich M, Armstrong VW. The role of therapeutic drug monitoring in individualizing immunosuppressive drug therapy: recent developments. Ther Drug Monit 2006;28(6):720–5.

[90] Soldin SJ, Steele BW, Witte DL, Wang E, Elin RJ, College of American Pathologists Study. Lack of specificity of cyclosporine immunoassays. Results of a college of American pathologists study. Arch Pathol Lab Med 2003;127(1):19–22.

[91] Ghoshal AK, Soldin SJ. IMx tacrolimus II assay: is it reliable at low blood concentrations? A comparison with tandem MS/MS. Clin Biochem 2002;35(5):389–92.

[92] Christians U, Braun F, Schmidt M, Kosian N, Schiebel HM, Ernst L, et al. Specific and sensitive measurement of FK506 and its metabolites in blood and urine of liver-graft recipients. Clin Chem 1992;38(10):2025–32.

[93] Annesley TM, McKeown DA, Holt DW, Mussell C, Champarnaud E, Harter L, et al. Standardization of LC-MS for therapeutic drug monitoring of tacrolimus. Clin Chem 2013;59(11):1630–7.

[94] Aucella F, Lauriola V, Vecchione G, Tiscia GL, Grandone E. Liquid chromatography-tandem mass spectrometry method as the golden standard for therapeutic drug monitoring in renal transplant. J Pharm Biomed Anal 2013;86:123–6.

[95] Buchwald A, Winkler K, Epting T. Validation of an LC-MS/MS method to determine five immunosuppressants with deuterated internal standards including MPA. BMC Clin Pharmacol 2012;12:2.

[96] Capron A, Lerut J, Latinne D, Rahier J, Haufroid V, Wallemacq P. Correlation of tacrolimus levels in peripheral blood mononuclear cells with histological staging of rejection after liver transplantation: preliminary results of a prospective study. Transpl Int 2012;25(1):41–7.

[97] Pensi D, De Nicolo A, Pinon M, Calvo PL, Nonnato A, Brunati A, et al. An UPLC-MS/MS method coupled with automated on-line SPE for quantification of tacrolimus in peripheral blood mononuclear cells. J Pharm Biomed Anal 2015;107:512–17.

[98] Pippenger CE. Therapeutic drug monitoring assay development to improve efficacy and safety. Epilepsy Res 2006;68(1):60–3.

[99] Perucca E. Clinically relevant drug interactions with antiepileptic drugs. Br J Clin Pharmacol 2006;61(3):246–55.

[100] Johannessen SI, Landmark CJ. Antiepileptic drug interactions—principles and clinical implications. Curr Neuropharmacol 2010;8(3):254–67.

[101] Juenke JM, Miller KA, Ford MA, McMillin GA, Johnson-Davis KL. A comparison of two FDA approved lamotrigine immunoassays with ultra-high performance liquid chromatography tandem mass spectrometry. Clin Chim Acta 2011;412(19–20):1879–82.

[102] Kim KB, Seo KA, Kim SE, Bae SK, Kim DH, Shin JG. Simple and accurate quantitative analysis of ten antiepileptic drugs in human plasma by liquid chromatography/tandem mass spectrometry. J Pharm Biomed Anal 2011;56(4):771–7.

[103] la Marca G, Malvagia S, Filippi L, Innocenti M, Rosati A, Falchi M, et al. Rapid assay of rufinamide in dried blood spots by a new liquid chromatography-tandem mass spectrometric method. J Pharm Biomed Anal 2011;54(1):192–7.

[104] Linder MW, Keck Jr. PE. Standards of laboratory practice: antidepressant drug monitoring. National Academy of Clinical Biochemistry. Clin Chem 1998;44(5):1073–84.

[105] Poklis JL, Wolf CE, Goldstein A, Wolfe ML, Poklis A. Detection and quantification of tricyclic antidepressants and other psychoactive drugs in urine by HPLC/MS/MS for pain management compliance testing. J Clin Lab Anal 2012;26(4):286–94.

[106] de Castro A, Ramirez Fernandez Mdel M, Laloup M, Samyn N, De Boeck G, Wood M, et al. High-throughput on-line solid-phase extraction-liquid chromatography-tandem mass spectrometry method for the simultaneous analysis of 14 antidepressants and their metabolites in plasma. J Chromatogr A 2007;1160(1–2):3–12.

[107] Ashbee HR, Barnes RA, Johnson EM, Richardson MD, Gorton R, Hope WW. Therapeutic drug monitoring (TDM) of antifungal agents: guidelines from the British Society for Medical Mycology. J Antimicrob Chemother 2014;69(5):1162–76.

[108] Denning DW, Hope WW. Therapy for fungal diseases: opportunities and priorities. Trends Microbiol 2010;18(5):195–204.

[109] Farowski F, Cornely OA, Vehreschild JJ, Hartmann P, Bauer T, Steinbach A, et al. Quantitation of azoles and echinocandins in compartments of peripheral blood by liquid chromatography-tandem mass spectrometry. Antimicrob Agents Chemother 2010;54(5):1815–19.

[110] Wilkinson GR. Drug metabolism and variability among patients in drug response. N Engl J Med 2005;352(21):2211–21.

[111] Li W, Zeng S, Yu LS, Zhou Q. Pharmacokinetic drug interaction profile of omeprazole with adverse consequences and clinical risk management. Ther Clin Risk Manag 2013;9:259–71.

[112] Alffenaar JW, Wessels AM, van Hateren K, Greijdanus B, Kosterink JG, Uges DR. Method for therapeutic drug monitoring of azole antifungal drugs in human serum using LC/MS/MS. J Chromatogr B Analyt Technol Biomed Life Sci 2010;878(1):39–44.

[113] Cazorla-Reyes R, Romero-Gonzalez R, Frenich AG, Rodriguez Maresca MA, Martinez Vidal JL. Simultaneous analysis of antibiotics in biological samples by ultra high performance liquid chromatography-tandem mass spectrometry. J Pharm Biomed Anal 2014;89:203–12.

[114] Oertel R, Neumeister V, Kirch W. Hydrophilic interaction chromatography combined with tandem-mass spectrometry to determine six aminoglycosides in serum. J Chromatogr A 2004;1058(1–2):197–201.

[115] Annesley TM. Methanol-associated matrix effects in electrospray ionization tandem mass spectrometry. Clin Chem 2007;53(10):1827–34.

[116] Napoli KL. Organic solvents compromise performance of internal standard (ascomycin) in proficiency testing of mass spectrometry-based assays for tacrolimus. Clin Chem 2006;52(4):765–6.

[117] US Department of Health and Human Services UF. Center for Drug Evaluation and Research, Center for Veterinary Medicine. Guidance for industry: bioanalytical method validation. FDA, editor. Rockville, MD; 2001.

[118] Clinical and Laboratory Standards Institute (CLSI). Liquid chromatography-mass spectrometry methods; approved guidelines. Wayne, PA: CLSI document C62-A. Clinical and Laboratory Standards Institute; 2014.

[119] Committee for Medicinal Products for Human Use (CHMP), European Medicines Agency. Guideline on bioanalytical method validation. London, UK; 2011.

[120] Clarke W, Rhea JM, Molinaro R. Challenges in implementing clinical liquid chromatography-tandem mass spectrometry methods—the light at the end of the tunnel. J Mass Spectrom 2013;48(7):755–67.

[121] Li S. Standardization of LC-MS/MS in clinical laboratory. J Chromatogr Sep Tech 2015;6 (2):1000e128.

[122] Wians Jr. FH. Clinical laboratory tests: which, why, and what do the results mean? Lab Med 2009;40(2):105–13.

[123] Carter GD, Jones JC. Use of a common standard improves the performance of liquid chromatography-tandem mass spectrometry methods for serum 25-hydroxyvitamin-D. Ann Clin Biochem 2009;46(Pt 1):79–81.

[124] Seger C. Usage and limitations of liquid chromatography-tandem mass spectrometry (LC-MS/MS) in clinical routine laboratories. Wien Med Wochenschr 2012;162 (21–22):499–504.

[125] Vogeser M, Seger C. Pitfalls associated with the use of liquid chromatography-tandem mass spectrometry in the clinical laboratory. Clin Chem 2010;56(8):1234–44.

[126] Kushnir MM, Rockwood AL, Bergquist J. LC-MS/MS in clinical laboratories. Bioanalysis 2013;5(1):5–6.

CHAPTER 4

Monitoring Free Drug Concentration: Clinical Usefulness and Analytical Challenges

Amitava Dasgupta
Department of Pathology and Laboratory Medicine, University of Texas Health Science Center at Houston, Houston, TX, United States

CONTENTS

4.1 Introduction 71
4.2 Drug–Protein Binding 72
4.3 Drugs Requiring Free Drug Monitoring .. 73
4.4 Conditions in Which Monitoring Free Anticonvulsants is Necessary 75
4.4.1 Clinical Utility of Monitoring Free Phenytoin Concentrations 76
4.4.2 Clinical Utility of Monitoring Free Valproic Acid Concentration 78
4.4.3 Clinical Utility of Monitoring Free Carbamazepine Concentrations 80
4.5 Mechanisms of Elevated Free Anticonvulsant Levels in Various Pathophysiological Conditions 80
4.5.1 Mechanism of Elevated Free Anticonvulsant Concentrations in Uremia 80
4.5.2 Mechanism of Elevated Free Anticonvulsant Concentrations in Liver Diseases 81

4.1 INTRODUCTION

Therapeutic drug monitoring (TDM) is defined as the management of a patient's drug regimen based on the serum, plasma, or whole blood concentration of a drug. The International Association for Therapeutic Drug Monitoring and Clinical Toxicology has adopted the following definition: "Therapeutic drug monitoring (TDM) is defined as the measurement made in the laboratory of a parameter that, with appropriate interpretation, will directly influence prescribing procedures. Commonly, the measurement is in a biological matrix of a prescribed xenobiotic, but it may also be of an endogenous compound prescribed as a replacement therapy in an individual who is physiologically or pathologically deficient in that compound" [1]. TDM is valuable when the drug in question has a narrow therapeutic index and toxicity may be encountered at a concentration slightly above the upper end of the therapeutic range. There are approximately 6000 prescription and nonprescription (over-the-counter) drugs available for clinical use in the United States, but most of these drugs have wide therapeutic index and do not require routine TDM. Approximately 20–26 drugs are routinely monitored in clinical laboratories, whereas there are an additional 25–30 drugs that may benefit from TDM. However, due to technical difficulty in measuring these drugs that are not routinely monitored (immunoassays may not be commercially available for these drugs; therefore, chromatographic methods such as liquid chromatography combined with tandem mass spectrometry must be used), only large medical centers, academic medical centers, and reference laboratories offer TDM for these drugs.

Although TDM is more common for drugs used in treating a chronic illness such as epilepsy and cardiac dysfunction, certain antibiotics, such as vancomycin and aminoglycosides, are routinely monitored in treating an acute

W. Clarke & A. Dasgupta (Eds): Clinical Challenges in Therapeutic Drug Monitoring. DOI: http://dx.doi.org/10.1016/B978-0-12-802025-8.00004-0

4.5.3 Mechanism of Elevated Free Anticonvulsant Concentrations in AIDS 82
4.5.4 Mechanism of Elevated Free Anticonvulsant Concentrations in Pregnancy....83
4.6 Elevated Free Anticonvulsant Due to Drug–Drug Interactions 83
4.6.1 Displacement of One Anticonvulsant by Another Anticonvulsant from Protein Binding Sites 83
4.6.2 Displacement of Anticonvulsant from Protein Binding Site by Various Drugs 85
4.7 Free Drug Monitoring of Drugs Bound to AGP....86
4.8 Monitoring Free Concentrations of Immunosuppressants..87
4.8.1 Monitoring Free Mycophenolic Acid....87
4.9 Special Situation: Monitoring Free Digoxin....90
4.10 Monitoring Free Concentrations of Protease Inhibitors 91
4.11 Saliva and Tears for Determination of Free Drug Concentration....92
4.12 Methods for Monitoring Free Drug Concentration....92
4.13 Conclusions 94
References 94

life-threatening infection due to the inherent toxicity of these drugs, mainly nephrotoxicity and ototoxicity. The pharmacological response of a patient to a given dosage regime is determined by several factors, including patient compliance, bioavailability, serum (or whole blood) drug level, rate of elimination, and the access of the drug to the receptor site as well as the receptor sensitivity. TDM is very useful when there is a good correlation between serum or whole blood drug concentration and the clinical response. Treatment failure due to noncompliance can also be determined by routine TDM. Gurwitz et al. commented on the clinical utility of TDM in order to avoid adverse drug events among elderly persons [2].

Many drugs are bound to serum proteins, and it is important to remember that it is the free (unbound) drug which is responsible for pharmacological activity of the drug.

Typically, the total drug concentration (bound drug + free drug) is measured for the purpose of TDM. However, for certain strongly protein-bound drugs, monitoring the free drug concentration may be necessary for particular patient populations.

4.2 DRUG–PROTEIN BINDING

The protein binding of a drug can be low, moderate, or high, whereas for some drugs such as lithium, the drug is not bound to serum protein at all. The major drug-binding protein in serum is albumin, followed by α_1 acid glycoprotein (AGP) and lipoproteins. Drugs exist in peripheral circulation as free (unbound) and bound to protein forms following the principle of reversible equilibrium and the law of mass action. Only free drug is capable of crossing the plasma membrane and binding with the receptor for pharmacological action [3]. Moreover, for central nervous system drugs, mostly free fraction can cross the blood–brain barrier [4].

In general, there is equilibrium between free drug and protein-bound drug:

$$[D] + [P] = [DP]$$

$$K = [DP]/[D][P]$$

where [D] is the drug concentration, [P] is the binding protein concentration, [DP] is the drug–protein complex, and K is the association constant (liters/mole). The greater the affinity of the protein for the drug, the higher the K value. The free fraction of a drug represents the relationship between bound and free drug concentration and is often referred as "α":

$$\alpha = \text{Free drug concentration}/\text{Total drug concentration(Bound + Free)}$$

Albumin is the major protein in circulation that binds various endogenous molecules, such as long-chain fatty acids and steroids, as well as many drugs. Albumin is also a transporter of various ions, such as copper, zinc, and cadmium. Moreover, albumin is a carrier of toxic waste in the body, such as bilirubin, delivering it to the liver for hepatic excretion. Albumin is primarily synthesized by the liver (~13.9 g/day), has a molecular weight of 66,348 Da, and is composed of three homologous domains numbered I, II, and III. Each domain contains two subdomains, A and B, that possess common structural motifs. Two major regions responsible for ligand binding to human serum albumin are known as Sudlow's site I and site II, and these sites are located in subdomain IIA and IIIA. Albumin is encoded by a single gene located on the long arm of chromosome 4 (4q13.3) [5].

Free fraction (α) of a drug typically does not vary with total drug concentration because protein binding sites usually exceed the number of drug molecules present. For example, the normal reference range of molar concentration of albumin is 527–784 μmol/L (3.5–5.2 g/dL), whereas the reference range of phenytoin, which is strongly bound to serum protein, is 40–79 μmol/L (10–20 μg/mL). Therefore, the normal molar concentration of binding protein albumin significantly exceeds the expected phenytoin concentration in serum. Similarly, another drug-binding protein, AGP, has a normal concentration of 12.5–30 μmol/L (50–120 mg/dL), assuming molecular weight of AGP as 40,000, which also exceeds the expected concentration of lidocaine (5.4–26 μmol/L; 1.5–6.0 μg/mL) in serum. However, saturation of binding sites may be possible if AGP concentration is at the lower end of normal while lidocaine concentration is at the higher end of normal.

For certain drugs, the number of protein binding sites in albumin may approach or be less than the number of drug molecules at the upper end of therapeutic range. For example, the free fraction of valproic acid is subject to more variation than other highly protein-bound antiepileptic drugs, such as phenytoin and carbamazepine [6,7]. Normal albumin concentration in serum is 527–784 μmol/L, whereas the therapeutic range of valproic acid is 50–100 μg/mL or 347–693 μmol/L. Because 1 mol of albumin can bind 1 mol of valproic acid, at the upper end of the therapeutic range, the concentration of valproic acid in micromoles per liter may exceed the albumin concentration micromoles per liter of in serum if the albumin concentration is at the lower end of the therapeutic range. Moreover, people with uremia and liver disease may have an albumin concentration lower than the reference range, and saturation of the albumin binding site may occur at a much lower concentration of valproic acid in these patients.

4.3 DRUGS REQUIRING FREE DRUG MONITORING

It is usually considered that if the protein binding of a drug is less than 80%, it is not a candidate for free drug monitoring. An exception is free digoxin monitoring (digoxin is only 25% protein bound), which is very useful in patients overdosed with digoxin and being treated with Digibind or DigiFab, the FAB fragment of anti-digoxin antibody. Protein binding of some commonly monitored therapeutic drugs is given in Table 4.1.

In today's practice, free drug monitoring is most common with the classical anticonvulsants phenytoin, carbamazepine, and valproic acid because these drugs are strongly bound to serum albumin. Soldin reported that in his personal experience, clinicians request the free drug levels of free phenytoin the most [8]. The first comprehensive report demonstrating the clinical utility of free drug monitoring was published in 1973. In a population of 30 epileptic patients, the authors found a better correlation between toxicity and free phenytoin concentrations compared to toxicity and total phenytoin concentrations [9]. Kits are available commercially for monitoring free phenytoin and free valproic acid concentrations. Moreover, the College of American Pathologists (the accrediting agency for clinical laboratories) also provides free anticonvulsant levels in their external survey specimens. Therapeutic ranges of total and free phenytoin, valproic acid, and carbamazepine are given in Table 4.2.

In addition to monitoring free concentrations of these anticonvulsants, the clinical utility of monitoring free mycophenolic acid has been demonstrated for certain patient populations (patients with uremia, liver disease,

Table 4.1 Protein Binding of Commonly Monitored Therapeutic Drugs

Drug	Protein Binding (%)	Binding Protein
Amikacin	<10	
Kanamycin	<10	
Tobramycin	0–10	
Ethosuximide	0	
Procainamide	10–15	Albumin
Theophylline	40	Albumin
Phenobarbital	40	Albumin
Phenytoin	90	Albumin
Carbamazepine	80	Albumin
Valproic acid	90–95	Albumin
Primidone	15	Albumin
Digoxin	25	Albumin
Quinidine	80	α_1 Acid glycoprotein
Lidocaine	60–80	α_1 Acid glycoprotein
Cyclosporine	98	Lipoproteins
Tacrolimus	97	Lipoprotein

Table 4.2 Therapeutic Range of Total and Free Phenytoin, Valproic Acid, and Carbamazepine[a]

Drug	Therapeutic Range (Total Drug)	Therapeutic Range (Free Drug)[a]	Free Fraction (%)
Phenytoin	10–20 μg/mL (40–79 μmol/L)	1.0–2.0 μg/mL (4.0–8.0 μmol/L)	8–14
Valproic acid	50–100 μg/mL (346–693 μmol/L)	4.3–10.8 μg/mL (30–75 μmol/L)	5–15
Carbamazepine	4–12 μg/mL (17–51 μmol/L)	1.4–3.1 μg/mL (6.0–13.0 μmol/L)	20–35

[a]*Free ranges of anticonvulsant drugs are adopted from Chan K, Beran RG. Value of therapeutic drug level monitoring and unbound (free) levels. Seizure 2008;17:572–5.*

hypoalbuminemia, and hyperbilirubinemia). Moreover, there are also some indications for monitoring free concentrations of certain protease inhibitors.

4.4 CONDITIONS IN WHICH MONITORING FREE ANTICONVULSANTS IS NECESSARY

Phenytoin along with carbamazepine is considered as the drug of choice for partial and generalized tonic–clonic seizure. Valproic acid, a branch-chain carboxylic acid, is considered a primary drug in the treatment of absence, generalized tonic–clonic, and myoclonic seizures. These drugs are voltage-dependent blockers of sodium channels. This mechanism selectively dampens pathological activation of sodium channels without interacting with normal functions of sodium channels. Significant interindividual variation can be observed in the free fraction of phenytoin, carbamazepine, and valproic acid, especially in the presence of uremia and liver disease. Drug–drug interactions can also lead to elevated free drug concentration. When protein binding is changed, the total concentration no longer reflects the pharmacologically active free drug in the plasma. Measuring free drug concentration for antiepileptic drugs eliminates a potential source of interpretative errors in TDM using traditional total drug concentrations. In general, free concentrations of classical anticonvulsants (phenytoin, carbamazepine, and valproic acid) should be monitored in the following circumstances:

- Uremic patients
- Patients with chronic liver disease
- Patients with hypercholesterolemia (probably due to elevated free fatty acids)

- Patients with hypoalbuminemia (critically ill patients, burn patients, elderly, pregnancy, AIDS, etc.)
- Suspected drug–drug interactions in which one strongly protein-bound drug can displace another strongly protein-bound anticonvulsant

Patient populations for which monitoring free anticonvulsants may be useful are listed in Table 4.3.

4.4.1 Clinical Utility of Monitoring Free Phenytoin Concentrations

Phenytoin is 90% bound to serum proteins, mainly albumin, and does not show any concentration-dependent binding within the therapeutic range. However, many published papers have indicated that free phenytoin concentration correlates better with therapeutic efficacy as well as toxicity compared to total phenytoin. Kilpatrick et al. reported that unbound phenytoin

Table 4.3 Special Patient Populations for Which Free Anticonvulsant Monitoring of Free Phenytoin, Valproic Acid, and Carbamazepine May Be Useful

Pathophysiological Condition	Comments
Uremia	■ In uremia, free levels are significantly elevated partly due to hypoalbuminemia (a common condition associated with uremia) and partly due to accumulation of various uremic compounds that may displace strongly protein-bound anticonvulsants from albumin binding site ■ Structure of albumin may also be altered in uremia, causing binding defect ■ In general, protein binding of phenytoin and valproic acid is more affected than that of carbamazepine in uremic patients
Chronic liver disease	■ Liver is the site of albumin synthesis, and hypoalbuminemia is a common condition associated with chronic liver disease, causing impaired drug–protein binding ■ Hyperbilirubinemia is usually observed in chronic liver disease, which may also cause elevated free anticonvulsant levels
Critically ill/trauma patients	■ Critically ill patients usually have hypoalbuminemia that may cause elevated free anticonvulsant levels. However, in severely ill head trauma patients, free phenytoin may be elevated despite the absence of hypoalbuminemia and renal and hepatic failure
Burn patients	■ Elevated free anticonvulsant levels may be related to hypoalbuminemia
Pregnant women	■ At constant dosages, plasma concentrations of anticonvulsants such as phenytoin, valproic acid, and carbamazepine tend to decrease during pregnancy and then return to normal within the first or second month after delivery. Disproportionate decreases are also observed in free drug level
AIDS patients	■ Elevated free phenytoin may be related to hypoalbuminemia as well as displacement of phenytoin by other strongly protein-bound drugs
Elderly	■ Elevated free anticonvulsants may be related to hypoalbuminemia as well as drug–drug interaction
Patients with hypercholesterolemia	■ May be related to elevated concentrations of free fatty acids

concentration reflected the clinical status of a patient equally or better than the total phenytoin concentration [10]. Booker and Darcey reported that free phenytoin concentration correlated better with toxicity, and the authors observed no toxicity at free phenytoin concentrations of 1.5 μg/mL or less [9]. In patients with greatly decreased albumin levels, free phenytoin is the better indicator of clinical outcome (therapeutic range for free: 0.8–2.1 μg/mL) [11]. Dutkiewicz et al. demonstrated that in hypercholesterolemia and in mixed hyperlipidemia, the serum level of free phenytoin was elevated [12]. The effect was probably related to displacement of phenytoin by free fatty acids [13]. In eclampsia, free phenytoin concentrations are usually abnormally high, although total phenytoin concentrations may be within therapeutic range. Unfortunately, neither total nor free phenytoin concentrations are good predictors of seizure control [14]. The binding of phenytoin to serum albumin can be altered significantly in uremia. The lower protein binding capacity of phenytoin in uremia can be related to hypoalbuminemia, structural modification of albumin, and accumulation of uremic compound in blood that displaces phenytoin from protein binding sites [15–17].

Monitoring free phenytoin concentration is clinically important in patients with hypoalbuminemia in order to avoid drug toxicity. Lindow and Wijdicks described severe phenytoin toxicity associated with hypoalbuminemia in critically ill patients that was confirmed by direct measurement of free phenytoin [18]. Free phenytoin concentration should be measured or at least theoretically calculated if free phenytoin measurement is unavailable to avoid misinterpretation of total phenytoin levels and consequent inappropriate dosage adjustment in critically ill patients [19]. Zielmann et al. reported that in 76% of 38 trauma patients, the free phenytoin fraction was increased to as high as 24% compared to 10% free levels observed in otherwise healthy subjects. The major causes of elevated free phenytoin were hypoalbuminemia, uremia, and hepatic disease. The authors recommended monitoring of free phenytoin in such patients [20].

Wolf et al. observed that the mean free to total phenytoin ratio was 0.13 (range, 0.06–0.42) in critically ill children and commented that total phenytoin is unreliable in directing phenytoin therapy in these children. The authors recommended routine measurement of free phenytoin in critically ill children [21]. Sadeghi et al. divided 40 head trauma patients into two groups: Group A consisted of 20 unconscious patients with severe head injury under mechanical ventilation, and group B consisted of 20 conscious self-ventilated patients. The authors observed that Pearson correlation analysis and Bland–Altman test showed weak to moderate correlation ($r = 0.528$) and poor agreement between free and total phenytoin concentrations in patients with severe head trauma and higher Acute Physiology and Chronic Health Evaluation II (APACHE II) scores in group A patients. In contrast, good correlation (r = 0.817) and moderate correlation between free and total phenytoin concentrations were found for group B patients who also showed lower APACHE II scores. The authors concluded that total phenytoin monitoring is not suitable for critically ill patients with severe head injury even in the absence of hypoalbuminemia and renal and hepatic failure. The authors recommended free phenytoin monitoring [22].

Thakral et al. reported a case in which a 19-year-old man with cryptogenic simple partial secondary generalized epilepsy developed blurred vision and xanthopsia after receiving phenytoin for status epilepticus. The free phenytoin concentration was found to be toxic. Phenytoin was withheld, and the patient experienced partial recovery. The authors concluded that phenytoin toxicity as revealed by an elevated free phenytoin concentration contributed to this acute visual dysfunction [23]. Burt et al. studied total and free phenytoin concentrations in 139 patients. Free phenytoin concentrations were 6.8–35.3% of total phenytoin concentrations (expected range, 8–12%). Clinical indications responsible for variations were hypoalbuminemia, drug interactions, uremia, pregnancy, and age. The authors concluded that monitoring total phenytoin is not as reliable as free phenytoin as a clinical indicator for therapeutic concentrations, and they recommended that therapeutic monitoring of phenytoin should be done only for the free concentration [24]. Iwamoto et al. also reiterated the need for free phenytoin monitoring in patients receiving phenytoin monotherapy because free phenytoin fraction was significantly influenced by aging, mean creatinine clearance, and serum albumin concentrations in the patient population they studied [25]. Hong et al. recommended monitoring free phenytoin concentrations in all patients with hypoalbuminemia and also commented that free phenytoin level measured directly was superior to calculated level using the Sheiner–Tozer equation [26].

4.4.2 Clinical Utility of Monitoring Free Valproic Acid Concentration

Valproic acid (therapeutic range of 50–100 μg/mL) is extensively bound to serum proteins, mainly albumin [27]. Fluctuations in protein binding occur within the therapeutic range due to saturable binding phenomenon leading to a wide variation in free fraction from 10% to 50% [28]. In addition, unbound valproic acid concentration may also vary during dosing interval in patients already stabilized on valproic acid [29]. Several studies have reported problems associated with predicting a therapeutic response of valproic acid from total serum concentrations [30,31]. Gidal et al. reported a case in which markedly elevated plasma free valproic acid in a hypoalbuminemic patient contributed to neurotoxicity. The total valproic acid concentration was 103 μg/mL, but the free valproic acid concentration was 26.8 μg/mL. This unexpected elevation was due to the low albumin level (3.3 g/dL) of the patient [32]. Haroldson et al. reported a case demonstrating the importance of monitoring free valproic acid in a heart transplant recipient with hypoalbuminemia. When the valproic acid dose was adjusted based on free valproic acid concentration rather than total valproic acid concentration, the patient improved and was eventually discharged from the hospital [33]. Lenn and Robertson demonstrated that the concentration of free valproic acid has clinical significance in the management of seizure as well as for avoiding undesirable side effects. The authors recommended using free valproic acid concentration for routine patient management [34]. Diurnal fluctuations in free and total plasma concentrations of valproic acid at steady state have been reported [35,36]. Ahmad et al. reported that total valproic acid concentrations show higher interindividual variation and tend to underestimate the effect of poor compliance, but the use of free valproic acid concentration offers an advantage in TDM [37]. Although it is assumed that unbound valproic acid concentration mirrors cerebrospinal fluid valproic acid concentration, Rapeport et al. reported a lack of correlation between free valproic acid levels and pharmacological effects [38].

Case Report: A 60-year-old male with Lennox syndrome and severe mental retardation had been treated with valproic acid for the past 26 years, and his serum valproic acid level was consistently around 100 μg/mL. Although his valproic acid level did not change on the day he presented to the clinic, his white blood cell count was 3300/mm^3 with 32% Pelger–Huët cells and a platelet count of 42,000/mm^3. He had never had Pelger–Huët anomalies in the past; however, his serum albumin levels had been declining gradually during the past month. When free valproic acid concentration was measured, it showed a significantly elevated level of 28.6 μg/mL (normal, 5–15 μg/mL). His total valproic acid was 103.7 μg/mL, indicating that his free valproic acid fraction was 27.6% (normal free fraction, 5–18%) and albumin concentration was 2.5 g/dL. After reducing valproic acid dosage from 1200 to 400 mg/day, his free valproic acid level was reduced to the normal range of 4.8 μg/mL and his albumin concentration in serum was 2.7 g/dL. At that time, Pelger–Huët cells completely disappeared from his blood, and his thrombocytopenia also improved after reducing valproic acid dosage. The authors concluded that the Pelger–Huët anomaly was induced by the increased free valproic acid under the condition of hypoalbuminemia [39].

Valproic acid is strongly protein bound and is considered not to be removable by extracorporeal means. However, in the case of severe overdose with valproic acid in which free concentration is high due to disproportionate protein binding, extracorporeal means such as hemodialysis and hemoperfusion can be used to treat valproic acid poisoning [40]. Khan et al. reported the effectiveness of low-efficiency dialysis with filtration in the management of acute valproic acid toxicity. This dialysis technique also prevented rebound phenomenon that can occur as the excess drug is released from its protein-bound stores [41].

4.4.3 Clinical Utility of Monitoring Free Carbamazepine Concentrations

Carbamazepine is effective in the treatment of primary or secondary generalized tonic–clonic epilepsy, all varieties of partial seizure, and myoclonic epilepsy. The plasma protein binding of carbamazepine is 70–80%. The primary and active metabolite, 10,11-epoxide, is only 50% bound to serum proteins. There seems to be less variability in the protein binding of carbamazepine compared to phenytoin and valproic acid in various pathophysiological conditions [42]. Froscher et al. showed that in patients with carbamazepine monotherapy, there was no closer relationship between free concentration and pharmacological effects compared to total concentration and pharmacological effects [43]. Lesser et al. found a broad overlapping of unbound carbamazepine causing toxicity and no toxicity [44]. Because carbamazepine-10,11-epoxide has a greater percentage of free fraction and is almost equipotent to carbamazepine, epoxide probably contributes significantly to the pharmacological effects of carbamazepine. Therapeutic monitoring of epoxide along with carbamazepine may be useful, especially in patients taking valproic acid or lamotrigine [45,46].

4.5 MECHANISMS OF ELEVATED FREE ANTICONVULSANT LEVELS IN VARIOUS PATHOPHYSIOLOGICAL CONDITIONS

Free anticonvulsant levels, most commonly free phenytoin and free valproic acid, are elevated in various pathophysiological conditions. Hypoalbuminemia is a major factor for elevated free anticonvulsant levels, but other factors may also play an important role in impaired protein binding of phenytoin and valproic acid and, to a lesser extent, carbamazepine in various disease states.

4.5.1 Mechanism of Elevated Free Anticonvulsant Concentrations in Uremia

In uremia, the free fraction of valproic acid can be as high as 20–30% compared to 8.45% as observed in healthy volunteers. The free fraction of phenytoin can be as high as 30%, whereas in normal volunteers the free fraction is usually 10%. Uremia also modifies the disposition of a highly metabolized drug by changes in plasma protein binding or hepatic metabolism [47]. Bruni et al.

studied four uremic patients receiving hemodialysis and reported that mean free fraction was 0.31 (31% free fraction) prior to dialysis, but it increased to 0.64 (64% free fraction) after hemodialysis in three patients [48].

High free drug concentrations in uremia are related to hypoalbuminemia, as well as the presence of endogenous uremic compounds that can displace strongly protein-bound drugs from protein binding sites. Monaghan et al. studied in detail the relationship between serum creatinine, blood urea nitrogen, albumin, and the unbound fraction of phenytoin in patients undergoing renal transplant. The authors concluded that estimation of free fraction of phenytoin in patients with a history of uremia and hypoalbuminemia should not be based on the measurement of serum creatinine and albumin [49].

Substantial evidence has accumulated indicating that certain compounds are retained in uremia that bind to serum proteins, thus reducing the number of binding sites available for drugs. Hippuric acid and indoxyl sulfate, the two compounds that are present in elevated concentrations in uremia, can cause displacement of strongly protein-bound drugs [50]. Takamura et al. identified 3-carboxy-4-methyl-5-propyl-2-furanpropionate as the major uremic toxin that causes impaired protein binding of furosemide. Oleate also plays a role [51]. Other uremic compounds, such as guanidine, methyl guanidine, and guanidinosuccinic acid, do not cause any displacement of drug from protein binding. Another study indicated that several endogenous compounds with low molecular weights (<500 Da) play significant roles in displacement of strongly protein-bound drugs, but mid-molecular-weight uremic toxins do not displace drugs [52]. In a review of drug–protein binding, Otagiri commented that reduced protein binding of drugs in uremia can be explained by a mechanism that involves a combination of direct displacement by free fatty acids and a cascade of effects of free fatty acids and unbound uremic toxins [53].

4.5.2 Mechanism of Elevated Free Anticonvulsant Concentrations in Liver Diseases

Patients with hepatic disease usually have hypoalbuminemia. Because albumin is the major binding protein for phenytoin, valproic acid, and carbamazepine, elevated free anticonvulsant concentrations are expected in patients with liver disease receiving these drugs. Elevated free phenytoin concentration occurs in patients with hepatic disease mainly due to hypoalbuminemia [54]. In hepatic failure, the hepatic clearance of unbound phenytoin may also be reduced as a result of hepatic tissue destruction and a reduction in hepatic enzyme activities responsible for metabolism of phenytoin. When this occurs, a reduction of phenytoin dose is necessary to maintain unbound phenytoin concentration below toxic level. Prabhakar and Bhatia reported that free phenytoin levels are elevated in patients with hepatic encephalopathy [55].

Hepatic disease can also alter pharmacokinetic parameters of valproic acid. Klotz et al. reported that alcoholic cirrhosis and viral hepatitis decreased valproic acid protein binding from 88.7% to 70.3% and 78.1%, respectively, with a significant increase in volume of distribution. Elimination half-life was also prolonged [56]. An increase in the unbound concentration of carbamazepine has been reported in patients with hepatic disease [57].

4.5.3 Mechanism of Elevated Free Anticonvulsant Concentrations in AIDS

Seizure is a common manifestation of central nervous system disease in patients with HIV infection. The incidence was approximately 10% in a population of hospitalized patients with advanced stage of disease [58]. Phenytoin is widely prescribed in the treatment of tonic–clonic seizures and other forms of epilepsy. Burger et al. investigated serum concentrations of phenytoin in 21 patients with AIDS [59]. The total phenytoin concentrations were significantly lower in patients with AIDS than in the control population, although phenytoin doses were significantly higher in patients with AIDS. Calculation of Michaelis–Menten parameters demonstrated that V_{max} values were similar in both patients with AIDS and the control group, but a nonsignificant trend of lower K_m values was observed in patients with HIV. The authors demonstrated that unbound phenytoin concentrations were significantly higher in patients with HIV infection and concluded that the lower protein binding of phenytoin in patients with AIDS could be related to hypoalbuminemia. Because unbound phenytoin is the pharmacologically active fraction, the authors recommended monitoring unbound phenytoin concentrations for patients with HIV infection receiving phenytoin [59].

In vitro experiments have confirmed the findings of Burger et al. [59]. Concentrations of free phenytoin and free valproic acid were significantly elevated in serum pools prepared from patients with AIDS and supplemented with phenytoin or valproic acid compared to serum pools prepared from normal subjects and also supplemented with the same amount of phenytoin or valproic acid. Hypoalbuminemia alone did not explain the elevation of free phenytoin or free valproic acid. Drug–drug interactions probably play a major role because an average patient with AIDS receives more than 10 medicines per day [60]. Toler et al. also described severe phenytoin toxicity as a result of decreased protein binding of phenytoin in a patient with AIDS, leading to an elevated free phenytoin concentration of 4.9 μg/mL. The toxicity was due to a high concentration of free phenytoin [61].

4.5.4 Mechanism of Elevated Free Anticonvulsant Concentrations in Pregnancy

In pregnancy, plasma volume usually increases and hypoalbuminemia may also be observed. The pharmacokinetics of many anticonvulsants undergo important changes in pregnancy due to modification in body weight, altered plasma composition, hemodynamic alteration, hormonal influence, and the contribution of the fetoplacental unit to drug distribution and disposition. Pregnancy thus affects absorption of drugs, binding to plasma protein, distribution, metabolism, and elimination [62]. At standard dosages, plasma concentrations of anticonvulsants such as phenytoin, valproic acid, carbamazepine, phenobarbital, and primidone tend to decrease during pregnancy and then return to normal within the first or second month after delivery. Marked decreases in total phenytoin concentrations (~40% of pre-pregnancy level) have been reported, whereas free phenytoin level decreased to a much lesser extent [63]. Reports on the decline in total and free carbamazepine during pregnancy are conflicting. One study reported a 42% decline in total carbamazepine concentration and a 22% decrease in free carbamazepine concentration from pregnancy to delivery in 22 patients [64]. For valproate, no significant change in free concentration was observed, despite reduction in total valproic acid concentration. For highly protein-bound drugs such as phenytoin and valproic acid, total plasma concentrations may be misleading during pregnancy, underestimating the pharmacological effects of the drug [65].

4.6 ELEVATED FREE ANTICONVULSANT DUE TO DRUG–DRUG INTERACTIONS

Several strongly protein-bound drugs can displace phenytoin and valproic acid from the albumin binding site, causing an elevated level of free drug. Elevated free anticonvulsant levels due to displacement of one strongly protein-bound anticonvulsant by another strongly protein-bound drug can be classified under two broad categories:

- Displacement of one strongly protein-bound anticonvulsant by another strongly protein-bound anticonvulsant, such as displacement of phenytoin from albumin binding site by valproic acid
- Displacement of one strongly protein-bound anticonvulsant by another strongly protein-bound drug, such as displacement of phenytoin from albumin binding site by ibuprofen

4.6.1 Displacement of One Anticonvulsant by Another Anticonvulsant from Protein Binding Sites

Displacement of phenytoin from albumin binding site by valproic acid has been reported. Tsanaclis et al. studied plasma protein binding of phenytoin in nine epileptic patients before and during addition of sodium valproate to the drug therapy. The mean free fraction of phenytoin increased from 13.5% to 18.2%, but the total phenytoin concentrations were reduced. The authors concluded that valproic acid displaces phenytoin from plasma protein binding sites but does not inhibit its metabolism [66]. Carvalho et al. described a case of a 12-year-old patient with refractory epilepsy syndrome who showed phenytoin toxicity following a concomitant therapy with phenytoin, valproic acid, and lamotrigine. The authors concluded that phenytoin toxicity was due to drug interaction between phenytoin and valproic acid, where valproic acid displaced phenytoin from protein binding site and also inhibited metabolism of phenytoin. Consequently, phenytoin dosage was reduced and phenytoin toxicity was resolved [67]. Pospisil and Perlik demonstrated in vivo significant decreases in phenytoin protein binding due to the presence of valproic acid or primidone [68]. Using 258 data pairs of total and free phenytoin analyzed from 155 cancer patients, Joerger et al. reported that comedication with valproic acid increased free phenytoin fraction by 52.5%, whereas comedication with carbamazepine increased free phenytoin fraction by 38.5%. The authors concluded that free phenytoin should be monitored in cancer patients receiving phenytoin along with valproic acid or carbamazepine [69].

Case Report: A 66-year-old Caucasian man with known epilepsy and ischemic heart disease was admitted to the hospital with vomiting, ataxia, and nystagmus presumably due to anticonvulsant drug toxicity. He was receiving valproic acid, phenytoin, carbamazepine, and levetiracetam. His phenytoin level on admission was 37 μmol/L (9.3 μg/mL), which was slightly subtherapeutic, and his valproic acid level was 499 μmol/L (72.0 μg/mL) and carbamazepine level was 27 μmol/L (6.4 μg/mL), both in therapeutic range. These results contradicted initial suspicion of anticonvulsant drug toxicity. The serum level of levetiracetam was not available. At that point, the neurologist requested measurement of free anticonvulsant levels. The patient's free phenytoin level of 5 μmol/L (1.3 μg/mL) and free carbamazepine level of 8.2 μmol/L (1.9 μg/mL) were both within therapeutic range, but his free valproic acid level was elevated to 93 μmol/L (13.4 μg/mL; normal free valproic acid level, 4.3–10.8 μg/mL). Chan and Beran determined that the patient was experiencing valproic acid toxicity. His valproic acid dose was omitted for 1 day and then reduced to 500 mg twice daily from initial dosage of 1 g twice daily. His symptoms were resolved. Eventually, phenytoin and levetiracetam were discontinued, but the patient was still receiving carbamazepine, although his dosage was simplified to 400 mg in the morning and 600 mg at night (initial dosage, 400 mg in the morning, 200 mg midday, and 400 mg at night) when he was discharged from the hospital on day 6. At that time, his total valproic acid level was 351 μmol/L (50.6 μg/mL) and free valproic acid level was 31 μmol/L (4.5 μg/mL), both therapeutic. His total (30 μmol/L; 7.1 μg/mL) and free (8.2 μmol/L; 1.9 μg/mL) carbamazepine levels were also within therapeutic range. The patient did not experience any anticonvulsant drug toxicity during a follow-up period of 8 months. The authors concluded that the monitoring of free anticonvulsant levels has a valuable role in clinical management of patients with epilepsy [70].

Xiong et al. described interaction between valproic acid and oxcarbazepine (an antimanic anticonvulsant) leading to toxic free valproic acid concentration in one patient. The patient developed valproic acid toxicity, and her platelet count dropped from 155,000 to 80,000/μL. At that time, her total and free valproic acid levels were 115.6 and 47.8 μg/mL, respectively, indicating that free fraction was elevated to 41.3%. Oxcarbazepine was discontinued, and her free valproic acid concentration was reduced to 26.8 μg/mL, whereas the total valproic acid concentration was 108.5 μg/mL (24.7% free). Unsure about valproic acid oxcarbazepine interaction, oxcarbazepine was reintroduced and the patient developed valproic acid toxicity. Again, symptoms resolved after discontinuation of oxcarbazepine, with platelet count, total, as well as free valproic acid levels returned to reference ranges. Oxcarbazepine is 40% bound to serum protein, and increased free valproic acid level with unchanged total valproic acid level may be related to decreased free valproic acid clearance due to reduced protein-mediated valproic acid transport into hepatocytes for clearance. The authors commented that monitoring free valproic acid is useful to identify such toxicity secondary to drug–drug interaction [71].

4.6.2 Displacement of Anticonvulsant from Protein Binding Site by Various Drugs

Several nonsteroidal antiinflammatory drugs, such as salicylate, ibuprofen, tolmetin, naproxen, mefenamic acid, and fenoprofen, can displace phenytoin, valproic acid, and carbamazepine from protein binding sites [72]. Sandyk reported a case in which phenytoin toxicity was induced by ibuprofen. This is due to displacement of phenytoin from protein binding by strongly protein-bound ibuprofen [73]. Blum et al. reported that tenidap sodium 120 mg/day at steady state increased the percentage of protein binding of phenytoin in plasma by 25%. The authors concluded that because tenidap increases the percentage of unbound phenytoin in plasma, when monitoring plasma phenytoin concentration, free phenytoin concentrations should also be monitored [74]. Reduced interaction between phenytoin and valproic acid with nonsteroidal antiinflammatory drugs in uremia has been described. The reduced interaction may be due to the presence of uremic compounds that block interaction between these anticonvulsants and antiinflammatory drugs [75]. Unexpected suppression of free phenytoin concentration by salicylate in uremic sera has also been reported [76].

Penicillins such as oxacillin and dicloxacillin can displace phenytoin from its binding sites. In vivo, the total phenytoin concentration in serum decreased during penicillin administration while the free phenytoin concentrations were increased [77]. However, phenytoin–oxacillin interaction is not significant at a lower dose of oxacillin usually prescribed in oral therapy, and this interaction is significant only at higher oxacillin doses, especially in patients with hypoalbuminemia [78]. In vitro and in vivo displacement of phenytoin by antibiotics ceftriaxone, nafcillin, and sulfamethoxazole also have been reported [79].

4.7 FREE DRUG MONITORING OF DRUGS BOUND TO AGP

Although albumin is the major drug-binding protein present in the serum, AGP, an acute phase reactant protein, binds basic drugs in serum. Lidocaine, which is bound to AGP in blood, was initially discovered as a local anesthetic, but later its antiarrhythmic properties were reported. Routledge et al. reported wide interindividual variation in free lidocaine concentration [80]. The percentage of unbound lidocaine was decreased in patients with uremia compared to controls (20.8% in uremic patients vs 30.8% in controls), as well as in renal transplant recipients. The cause of increased protein binding of lidocaine in these patients was significantly due to increased AGP concentration (134.9 mg/dL in patients vs 66.3 mg/dL in controls) because this protein is an acute phase reactant [80]. Shand commented that in situations in which AGP concentration is increased, particularly with myocardial infarction, the usual therapeutic range of total lidocaine may not apply and monitoring free concentration is more appropriate [81]. Displacement of lidocaine from protein binding by disopyramide may result in an elevated free lidocaine concentration because disopyramide has a stronger binding affinity for AGP [82]. Quinidine is also bound to AGP, and genetic polymorphism of the gene encoding this protein may influence binding of quinidine to AGP [83]. However, with many newer cardioactive drugs available for therapy, quinidine is infrequently used in clinical medicine today.

As another example, the protein binding of short-acting narcotic analgesic alfentanil is affected by disease. This drug is mainly bound to AGP, the concentration of which can be significantly increased in patients with renal failure, myocardial infarction, or rheumatoid arthritis and also in intensive care unit patients. Interestingly, protein binding of alfentanil was only increased in patients with myocardial infarction. In patients with liver cirrhosis, concentrations of both albumin and AGP were reduced, resulting in decreased protein binding of alfentanil. Disopyramide was able to displace alfentanil from protein binding leading to increased free fraction, whereas other strongly AGP-bound drugs, such as quinidine, lidocaine, and bupivacaine, had no effect [84].

4.8 MONITORING FREE CONCENTRATIONS OF IMMUNOSUPPRESSANTS

Immunosuppressants such as cyclosporine, tacrolimus, sirolimus, everolimus, and mycophenolic acid are strongly bound to serum proteins. Whereas cyclosporine, tacrolimus, sirolimus, and everolimus are measured in whole blood (due to a high distribution of these drugs in various cellular components of blood but mostly erythrocytes), mycophenolic acid is mostly distributed in serum (>99%). Therefore, mycophenolic acid is the only immunosuppressant that is monitored in serum or plasma. Although there are several publications indicating the clinical utility of monitoring free cyclosporine and tacrolimus, due to technical difficulty (equilibrium dialysis using stainless-steel equipment is needed), free cyclosporine and tacrolimus are not monitored in clinical laboratories. In contrast, free mycophenolic acid can be easily monitored in protein-free ultrafiltrate of serum similar to monitoring free phenytoin, valproic acid, or carbamazepine, and the clinical utility of monitoring free mycophenolic acid is well reported in the literature.

4.8.1 Monitoring Free Mycophenolic Acid

Mycophenolate mofetil is a prodrug that is converted into mycophenolic acid after oral administration. Another commercially available mycophenolic acid preparation is mycophenolate sodium. The plasma protein binding of mycophenolic acid increases with time after liver transplant from 92% to 98%, thus causing intraindividual variation in liver transplant recipients [85]. Mycophenolic acid is primarily bound to albumin, and binding to AGP in serum is minimal. Because mycophenolic acid is strongly protein bound, pharmacological activity of mycophenolic acid is due to its free fraction only. It has been demonstrated that only free mycophenolic acid is capable of inhibiting inosine monophosphate dehydrogenase. Moreover, only free mycophenolic acid can be metabolized to inactive mycophenolic acid glucuronide (major metabolite) and active mycophenolic acid acyl glucuronide (minor metabolite).

Traditionally, total mycophenolic acid level is monitored with an assumption that approximately 97–99% would be protein bound. In general, in transplant recipients with normal albumin concentration as well as normal renal and liver function, the free fraction of mycophenolic acid is 2% or 3%, and free mycophenolic acid concentration can be predicted from total mycophenolic acid concentration. Ensom et al. reported that average free fraction of mycophenolic acid in stable lung transplant recipients was 2.9% (range, 2.0–3.4%). Mean albumin concentration in these patients was 3.7 g/dL [86]. However, many conditions may lead to significantly increased free mycophenolic acid concentrations, and for these patients, monitoring only total mycophenolic acid concentration may provide misleading information regarding exposure of mycophenolic acid. Major causes of decreased protein binding of mycophenolic acid include the following:

- Hypoalbuminemia
- Uremia
- Liver disease (Hypoalbuminemia is a common feature of liver disease.)
- Elevated mycophenolic acid glucuronide
- Hyperbilirubinemia

Various conditions that may increase free mycophenolic acid are summarized in Table 4.4.

Hypoalbuminemia is the major cause of elevated free mycophenolic acid because albumin is the major mycophenolic acid binding protein in plasma. Many pathological conditions, such as uremia and liver disease, may also cause hypoalbuminemia. For these patients, traditionally monitored total mycophenolic acid concentration may be within the therapeutic window but free mycophenolic acid level may be elevated, causing toxicity. For these patients, monitoring free mycophenolic acid should be more beneficial. In a study of 42 renal transplant recipients who also received mycophenolic acid, Atcheson et al. observed a significant relationship between low plasma albumin concentrations and elevated free mycophenolic acid fraction (>3%) using Spearman correlation. Receiver operating characteristic curve analysis demonstrated that the cutoff value for albumin was 3.1 g/dL. In general, patients with albumin concentration less than 3.1 g/dL showed an elevated

Table 4.4 Various Clinical Conditions That May Increase Free Mycophenolic Acid Concentrations

- Significant hypoalbuminemia (albumin <3.1 g/dL) may cause increased concentration of free mycophenolic acid
- Time of transplant (usually free fraction is higher immediately after transplant and then free fraction is reduced with time, probably due to improved graft function and improved albumin concentration)
- Increased concentration of mycophenolic acid glucuronide metabolite may increase free mycophenolic acid concentration due to displacement of mycophenolic acid from albumin binding
- Significant renal dysfunction may increase free mycophenolic acid concentration due to hypoalbuminemia as well as accumulation of mycophenolic acid glucuronide metabolite
- Liver disease may also increase free mycophenolic acid concentration
- One strongly protein-bound drug can displace mycophenolic acid from the protein binding site, causing increased free mycophenolic acid concentration. For example, salicylate can displace mycophenolic acid from protein binding

percentage of free mycophenolic acid (>3%). At that cutoff, albumin was found to be a good predictor of elevated free mycophenolic acid concentration, with sensitivity and specificity of 0.75 and 0.80, respectively. The authors concluded that clinicians should consider monitoring free mycophenolic acid concentrations in hypoalbuminemic patients with plasma albumin levels less than 3.1 g/dL [87].

In uremic patients, free fraction of mycophenolic acid may be elevated due to hypoalbuminemia as well as accumulation of mycophenolic acid glucuronide metabolite in plasma. Mycophenolic acid glucuronide is capable of displacing mycophenolic acid from protein binding site because mycophenolic acid glucuronide is also strongly bound to protein (~82%). Elevated free fraction in these patients leads to higher clearance and lower area under the curve (AUC) for total mycophenolic acid. In patients with chronic renal failure, both total and free mycophenolic acid should be measured in order to avoid mycophenolic acid toxicity induced by increased free mycophenolic acid concentration. Kaplan et al. studied 8 renal transplant recipients (1 patient with both kidney and pancreas transplant) with chronic renal insufficiency and 15 renal transplant patients with preserved renal function and observed that average free mycophenolic acid fractions were more than double in renally compromised patients compared to patients with normal renal function ($5.8 \pm 2.7\%$ vs $2.5 \pm 0.4\%$). Such differences were both clinically and statistically significant. The authors concluded that mycophenolic acid protein binding was decreased and free mycophenolic acid concentrations were increased in chronic renal failure patients [88]. Weber et al. observed that renal impairment had no effect on total mycophenolic acid $AUC_{0-12\,h}$ values, but free fraction in children (median, 1.65%; range, 0.4–13.8%) who received renal transplantation was significantly modulated by renal function and serum albumin concentration. Therefore, patients with renal insufficiency and/or low serum albumin showed higher free fraction of mycophenolic acid as well as free mycophenolic acid $AUC_{0-12\,h}$. In addition, patients with renal insufficiency also showed significantly higher mycophenolic acid glucuronide $AUC_{0-12\,h}$ because mycophenolic acid glucuronide is cleared by the kidney [89]. High unbound mycophenolic acid concentration was also encountered in a hematopoietic cell transplant patient with sepsis and renal and hepatic dysfunction [90].

Case Report: A 58-year-old man with end-stage renal failure secondary to polycystic kidney disease received renal transplant and subsequently developed a highly elevated free mycophenolic acid fraction associated with severe toxicity over a period of 5 days. During this period, although his serum creatinine was reduced from 0.85 mmol/L (0.96 mg/dL) to 0.5 mmol/L (0.56 mg/dL), he became markedly jaundiced as bilirubin increased from 17 μmol/L (1.0 mg/dL) to 161 μmol/L (9.4 mg/dL). His serum albumin level was also reduced from 3.6 g/dL to less than 2 g/dL. Although his liver enzymes were normal before transplant, they were elevated after transplant (alkaline phosphatase, 137 U/L; γ-glutamyltransferase, 301 U/L). On day 5, his 2-h cyclosporine level (837 ng/mL) and total mycophenolic acid $AUC_{0-6\ h}$ (12.6 mg h/L) were low, but total mycophenolic acid glucuronide $AUC_{0-6\ h}$ was elevated (1317 mg h/L). Mycophenolic acid dose was not changed, but cyclosporine was substituted by tacrolimus. The patient subsequently experienced severe nausea, vomiting, hematemesis, and pancytopenia (nadir white cell count, 1.6×10^9/L; platelet count, 32×10^9/L; hemoglobin, 7.3 mg/dL). This severe toxicity was resolved after cessation of mycophenolic acid therapy. Retrospective analysis revealed that free mycophenolic acid $AUC_{0-6\ h}$ was significantly elevated (2.3 mg h/L), as was the free fraction, which was highly elevated to 18.3%, thus explaining the observed toxicity despite a total mycophenolic acid level that was relatively low. This case illustrates severe mycophenolic acid toxicity caused by increased free fraction despite the total mycophenolic acid level being relatively low. Moreover, high free mycophenolic acid concentration may be associated with hypoalbuminemia and hyperbilirubinemia [91].

4.9 SPECIAL SITUATION: MONITORING FREE DIGOXIN

Digoxin is a cardioactive drug that is only 25% bound to serum proteins, mainly albumin. Therefore, monitoring free digoxin is not indicated for patients with uremia, liver disease, or any other pathophysiological conditions that may cause hypoalbuminemia. Endogenous digoxin-like immunoreactive substances (DLIS) cross-react with polyclonal antibody-based digoxin immunoassays, and such interferences may be eliminated by taking advantage of the strong protein binding of DLIS (>95%) and only 25% protein binding of digoxin. However, newer digoxin immunoassays using specific monoclonal antibodies against digoxin are virtually free from DLIS interference. Spironolactone is a diuretic that also cross-reacts with older polyclonal antibody-based digoxin immunoassays, and again taking advantage of greater than 90% protein binding of spironolactone and 25% protein binding of digoxin, such interferences can be eliminated by using free digoxin monitoring. However, newer monoclonal antibody-based digoxin immunoassays are free from such interferences. For example, digoxin immunoassays on both ARCHITECT chemistry and immunoassay analyzers are free from interferences of spironolactone, potassium canrenoate, and their common metabolite, canrenone [92].

Anti-digoxin FAB fragments (Digibind and the more recently introduced DigiFab) have been used successfully for many years in the management of severe digoxin and digitoxin toxicity. Monitoring total digoxin concentration during Digibind or DigiFab therapy may produce confusing digoxin concentrations due to interactions of Digibind or DigiFab with different antibodies used in various digoxin immunoassays in different ways. McMillin et al. studied the effect of Digibind and DigiFab on 13 different digoxin immunoassays. Increasing concentrations of Digibind or DigiFab reduced the levels of total digoxin in all immunoassays except fluorescence polarization immunoassay on the TDx analyzer. However, no interference was observed when free digoxin concentrations in the protein-free ultrafiltrates were measured. The authors concluded that ultrafiltration remains the best strategy for accurate determination of free digoxin concentrations in the presence of FAB products [93]. In addition, it is expected that free digoxin level should be reduced significantly after initiation of Digibind or DigiFab therapy. Therefore, free digoxin level also correlates with success of therapy.

4.10 MONITORING FREE CONCENTRATIONS OF PROTEASE INHIBITORS

Currently, many antiretroviral drugs are used in the treatment of patients with AIDS undergoing highly active antiretroviral therapy. Although TDM of several antiretroviral agents has been suggested, it is not standard of care in this patient population. Current evidence indicates that TDM of certain protease inhibitors may be beneficial in patient management. Protease inhibitors with the exception of indinavir are strongly protein bound (>90%) mainly to AGP. The pharmacological effect of antiretroviral drugs is dependent on the unbound concentration of drugs capable of entering cells that harbor HIV [94]. Fayet et al. described a modified ultrafiltration method for monitoring of free concentrations of several antiretroviral agents and commented that monitoring free concentrations of lopinavir, saquinavir, and efavirenz may increase its clinical usefulness due to high variability in the free fraction [95].

Protein binding of indinavir varied between 54% and 70% in eight men, with a mean protein binding of 61%. In addition, the variability of protein binding was concentration dependent [96]. The mean free plasma unbound amprenavir concentration was 8.6% (range, 4.4–20%) in one study. Moreover, lopinavir was able to displace amprenavir from protein binding in vitro, but another strongly protein-bound protease inhibitor, ritonavir, had no effect [97]. Boffito et al. studied lopinavir protein binding in vivo through a 12-h dosing interval and measured free lopinavir concentrations using high-performance liquid chromatography combined with tandem mass spectrometry (HPLC-MS/MS). The mean unbound lopinavir concentration was 0.92% when measured using ultrafiltration but 1.32% using equilibrium dialysis. The unbound percentage of lopinavir was also found to be higher after 2 h than at baseline [98]. However, TDM of free concentrations of antiretroviral drugs is in the preliminary stage of development. Currently, only a few medical centers perform TDM of antiretroviral agents.

4.11 SALIVA AND TEARS FOR DETERMINATION OF FREE DRUG CONCENTRATION

Saliva is an ultrafiltrate of blood, and salivary drug levels are typically reflective of free drug levels, although there are exceptions. Drugs, which are not ionizable within the salivary pH range, are candidates for salivary TDM. Salivary flow rate varies significantly both between individuals and under different conditions; the use of stimulated saliva has advantage over resting saliva. The salivary flow rate, pH, and sampling condition and other pathophysiological factors may influence the concentration of a particular drug in saliva. However, under well-controlled and standardized conditions, saliva can be used as an alternative matrix for monitoring of carbamazepine, phenytoin, primidone, and ethosuximide.

Za'abi et al. concluded that monitoring of salivary phenytoin and carbamazepine proved to be a realistic alternative to plasma free level monitoring because excellent correlations were found between salivary levels and serum unbound levels of both phenytoin and carbamazepine [99]. Nakajima et al. compared tear valproic acid concentrations with total and free valproic acid concentrations in serum and concluded that the tear valproic acid concentrations correlated well with free valproic acid concentrations in serum [100]. Patsalos and Berry commented that overall there is compelling evidence that salivary TDM can be usefully applied to optimize therapy with carbamazepine, clobazam, ethosuximide, gabapentin, lacosamide, lamotrigine, levetiracetam, oxcarbazepine, phenobarbital, phenytoin, primidone, topiramate, and zonisamide. Salivary monitoring of valproic acid is probably not helpful. Another advantage of salivary monitoring is that for strongly protein-bound antiepileptic drugs, it provides levels of free drugs, which have pharmacological activities [101]. Another application of salivary TDM is determining compliance of patients with antiretroviral agents in resource-limited settings in which thin-layer chromatography is used for drug determination [102].

4.12 METHODS FOR MONITORING FREE DRUG CONCENTRATION

Although equilibrium dialysis is the gold standard for separation of bound drug from free drug, ultrafiltration using the Centrifree micropartition system is the most common technique for preparation of protein-free ultrafiltrate for monitoring free drug concentration in clinical laboratories. Usually, 0.8–1.0 mL of serum is centrifuged for 15–20 min to prepare the ultrafiltrates. Then free drug concentration can be measured in the protein-free ultrafiltrates using appropriate immunoassay or preferably by using a chromatographic method such as liquid chromatography combined with tandem mass spectrometry (LC-MS/MS). The time of centrifuging to prepare ultrafiltrate is crucial for measuring free drug concentrations. Liu et al. demonstrated that there is a significant difference between measured free valproic acid concentration in ultrafiltrates prepared by centrifuging specimens for 5 versus 10 or 20 min. The measured free concentrations were low if the specimen was centrifuged for 5 min. Therefore, the authors recommended centrifugation of specimens for at least 15 min [103]. McMillan et al. reported that ultrafiltrate volumes were directly proportional to the centrifugation time (15–30 min) and were inversely proportional to albumin concentrations of serum. Although ultrafiltrate volume was significantly increased with increasing centrifugation time, free phenytoin values did not change significantly, indicating that equilibrium was maintained between the ultrafiltrate and serum retained in the ultrafiltration device [104]. Dong et al. reported that the ratio of ultrafiltrate volume to specimen volume may affect free drug level monitored in the protein-free ultrafiltrate. Using carbamazepine as the model drug, the authors showed that depending on ultrafiltrate volume (40–400 μL), protein binding changed from 40% to 70%. The authors commented that hollow fiber centrifugal ultrafiltration is a superior technique to traditional ultrafiltration for the determination of free drug concentration [105]. In another report, Dong et al. indicated that when the V_u (volume of ultrafiltrate) to V_s (volume of sample) ratio was less than 0.4, the free drug level determined using the ultrafiltrate provided reliable results. However, when the ratio exceeded 0.4, the free drug level was overestimated [106].

Free drug concentrations in the ultrafiltrate can be measured using immunoassays for phenytoin and valproic acid because specially designed assays are commercially available for monitoring of such free drug levels. However, widely used fluorescence polarization immunoassay for free phenytoin, carbamazepine, and valproic acid for application on the TDx analyzer (Abbott Laboratories) is no longer commercially available. Digoxin immunoassay for total digoxin monitoring can be applied for monitoring free drug concentration because digoxin is poorly protein bound and almost 75% of the drug exists in free form. For other less common drugs, more sophisticated methods such as HPLC-MS/MS may be required to measure the free drug level in the protein-free ultrafiltrate. In fact, even for monitoring free phenytoin (where immunoassays are commercially available), the isotope dilution electrospray ionization tandem mass spectrometric method may provide better precision and accuracy compared to immunoassay [107].

Similarly, for determination of free mycophenolic acid concentration, chromatographic methods are most appropriate. Nevertheless, Rebollo et al. modified the enzyme-multiplied immunoassay technique (EMIT) mycophenolic acid immunoassay for application on the Viva-E analyzer for determination of free mycophenolic acid. The authors used Centrifree micropartition system for preparation of protein-free ultrafiltrate and then determined free mycophenolic acid concentration in the ultrafiltrate. However, the analyzer was programmed so 30 μL of ultrafiltrate was used (for determination of total mycophenolic acid using EMIT only a 3-μL specimen is needed). Appropriate adjustments of reagent volumes were also made. The limit of quantitation for free mycophenolic acid was 5 ng/mL, which was comparable to the limit of quantitation achieved by LC-MS/MS [108]. Nevertheless, HPLC-MS/MS offers more analytical specificity for determination of free as well as total mycophenolic acid in ultrafiltrate or plasma. In addition, this chromatographic method is capable of simultaneous measurement of mycophenolic acid glucuronide [109]. However, LC-MS/MS is the only method available for determination of free drug concentration in the protein-free ultrafiltrate where immunoassay for that drug is not available, such as for the protease inhibitor lopinavir [110].

4.13 CONCLUSIONS

TDM of strongly protein-bound antiepileptic drugs such as phenytoin, valproic acid, and carbamazepine is useful for patients with uremia, liver disease, and any pathophysiological condition that may cause hypoalbuminemia. Drug–drug interactions may also increase free fractions of antiepileptic drugs without significantly altering total drug concentrations. Monitoring free concentrations of immunosuppressant drugs such as mycophenolic acid also has clinical value. Monitoring free concentration of certain protease inhibitors may be useful, but further studies are needed for establishing guidelines.

References

[1] Watson I, Potter J, Yatscoff R, Fraser A, et al. Therapeutic drug monitoring [Editorial]. Drug Monit 1997;19:125.

[2] Gurwitz JH, Field TS, Harrold LR, Rothschild J, et al. Incidence and preventability of adverse drug events among older persons in the ambulatory settings. JAMA 2003;289:1107–16.

[3] Chan S, Gerson B. Free drug monitoring. Clinics Lab Med 1987;7:279–87.

[4] Maurer TS, Debartolo DB, Tess DA, Scott DO. Relationship between exposure and nonspecific binding of thirty three central nervous system drugs in mice. Drug Metab Dispos 2005;33:175–81.

[5] Merlot AM, Kallinowski DS, Richardson DR. Unraveling the mysteries of serum-albumin—more than just a serum protein. Front Physiol 2014;5:299.

[6] Klotz U, Antonin KH. Pharmacokinetics and bio-availability of sodium valproate. Clin Pharmacol Ther 1977;21:736–43.

[7] Meinardi H, Vander Kleijn E, Meijer JWA. Absorption and distribution of anti-epileptic drugs. Epilepsia 1982;23:23–6.

[8] Soldin SJ. Free drug measurements when and why? An overview. Arch Pathol Lab Med 1999;123:822–3.

[9] Booker HE, Darcey B. Serum concentrations of free diphenylhydantoin and their relationship to clinical intoxication. Epilepsia 1973;2:177–84.

[10] Kilpatrick CJ, Wanwimolruk S, Wing LMH. Plasma concentrations of unbound phenytoin in the management of epilepsy. Br J Clin Pharmacol 1984;17:539–46.

[11] Fedler C, Stewart MJ. Plasma total phenytoin: a possible misleading test in developing countries. Ther Drug Monit 1999;21:155–60.

[12] Dutkiewicz G, Wojcicki J, Garwronska-Szklarz B. The influence of hyperlipidemia on pharmacokinetics of free phenytoin. Neurol Neurochir Pol 1995;29:203–11.

[13] Dasgupta A, Crossey MJ. Elevated free fatty acid concentrations in lipemic sera reduce protein binding of valproic acid significantly more than phenytoin. Am J Med Sci 1997;313:75–9.

[14] Naidu S, Moodley J, Botha J, et al. The efficacy of phenytoin in relation to serum levels in severe pre-eclampsia and eclampsia. Br J Obstet Gynaecol 1992;99:881–6.

[15] Sjoholm I, Kober A, Odar-Cederlof I, Borga O. Protein binding of drugs in uremia and normal serum: the role of endogenous binding inhibitors. Biochem Pharmacol 1976;25:1205–13.

[16] McNamara PI, Lalka D, Gibaldi M. Endogenous accumulation products and serum protein binding in uremia. J Lab Clin Med 1981;98:730–40.

[17] Reidenberg MM, Drayer DE. Alteration of drug protein binding in renal disease. Clin Pharmacokinet 1984;9(Suppl. I):18–26.

[18] Lindow J, Wijdicks EF. Phenytoin toxicity associated with hypoalbuminemia in critically ill patients. Chest 1994;105:602–4.

[19] von Winckelmann SL, Spriet I, Willems L. Therapeutic drug monitoring of phenytoin in critically ill patients. Pharmacotherapy 2008;28:1391–400.

[20] Zielmann S, Mielck F, Kahl R, et al. A rational basis for the measurement of free phenytoin concentrations in critically ill trauma patients. Ther Drug Monit 1994;16:139–44.

[21] Wolf GK, McClain CD, Zurakowski D, Dodson B, et al. Total phenytoin concentrations do not accurately predict free phenytoin concentrations in critically ill children. Pediatr Crit Care Med 2006;7:434–9.

[22] Sadeghi K, Hadi F, Ahmadi A, Hamishehkar H, et al. Total phenytoin concentration is not well correlated with active free drug in critically ill head trauma patients. J Res Pharm Pract 2013;2:105–9.

[23] Thakral A, Shenoy R, Deleu D. Acute visual dysfunction following phenytoin-induced toxicity. Acta Neurol Belg 2003;103:218–20.

[24] Burt M, Anderson D, Kloss J, Apple F. Evidence based implementation of free phenytoin therapeutic drug monitoring. Clin Chem 2000;46:1132–5.

[25] Iwamoto T, Kagawa Y, Natio Y, Kuzuhara S, Okuda M. Clinical evaluation of plasma free phenytoin measurement and factors influencing its protein binding. Biopharm Drug Dispos 2005;27:77–84.

[26] Hong JM, Choi YC, Kim WJ. Differences between the measured and calculated free serum phenytoin concentrations in epileptic patients. Yonsei Med 2009;50:517–20.

[27] Urien S, Albengres E, Tillement JP. Serum protein binding of valproic acid in healthy subjects and in patients with liver disease. Int J Clin Pharmacol 1981;19:319–25.

[28] Bowdle TA, Patel IH, Levy RH, Wilensky AJ. Valproic acid dosage and plasma protein binding and clearance. Clin Pharmacol Ther 1980;28:486–92.

[29] Marty JJ, Kilpatrick CJ, Moulds RFW. Intra-dose variation in plasma protein binding of sodium valproate in epileptic patients. Br J Clin Pharmacol 1982;14:399–404.

[30] Gugler R, Von Unruh GE. Clinical pharmacokinetics of valproic acid. Clin Pharmacokinet 1980;5:67–83.

[31] Chadwick DW. Concentration-effect relationship of valproic acid. Clin Pharmacokinet 1985;10:155–63.

[32] Gidal BE, Collins DM, Beinlich BR. Apparent valproic acid neurotoxicity in a hypoalbuminemic patient. Ann Pharmacother 1993;27:32–5.

[33] Haroldson JA, Kramer LE, Wolff DL, Lake KD. Elevated free fractions of valproic acid in a heart transplant patient with hypoalbuminemia. Ann Pharmacother 2000;34:183–7.

[34] Lenn NJ, Robertson M. Clinical utility of unbound antiepileptic drug blood levels in the management of epilepsy. Neurology 1992;42:988–90.

[35] Bauer LA, Davis R, Wilensky A, Raisys VA, Levy RH. Diurnal variation in valproic acid clearance. Clin Pharmacol Ther 1984;35:505–9.

[36] Bauer LA, Davis R, Wilensky A, Raisys VA, Levy RH. Valproic acid clearance: unbound fraction and diurnal variation in young and elderly patients. Clin Pharmacol Ther 1985;37:697–700.

[37] Ahmad AM, Douglas Boudinot F, Barr WH, Reed RC, Garnett WR. The use of Monte Carlo stimulation to study the effect of poor compliance on the steady state concentrations of valproic acid following administration of enteric-coated and extended release divalprox sodium formulation. Biopharm Drug Dispos 2005;26:417–25.

[38] Rapeport WG, Mendelow AD, French G, et al. Plasma protein binding and CSF concentration of valproic acid in man following acute oral dosing. Br J Clin Pharmacol 1983;8:362–71.

[39] Suzuki K, Hiramoto A, Okumura T. A case on reversible Pelger-Huet anomaly depending on serum free fraction of valproic acid. Brain Dev 2015;37(3):344–6.

[40] Al Aly Z, Yalamanchili P, Gonzalez E. Extracorporeal management of valproic acid toxicity: a case report and review of literature. Semin Dial 2005;18:62–6.

[41] Khan E, Huggan P, Celi L, MacGinley R, et al. Sustained low-efficient dialysis with filtration (SLEED-f) in the management of acute sodium valproate intoxication. Hemodial Int 2008;12:211–14.

[42] Bertilsson L, Tomson T. Clinical pharmacokinetics and pharmacological effects of carbamazepine and carbamazepine 10,11-epoxide. Clin Pharmacokinet 1986;11:177–98.

[43] Froscher W, Burr W, Penin H, Vohl J, et al. Free level monitoring of carbamazepine and valproic acid: clinical significance. Clin Neuropharmacol 1985;8:362–71.

[44] Lesser RP, Pippenger CE, Luders H, Dinners DS. High dose monotherapy in treatment of intractable seizure. Neurology 1984;34:707–11.

[45] Al-Qudah AA, Hwang PA, Giesbrecht E, Soldin SJ. Contribution of 10,11-epoxide to neurotoxicity in epileptic children on polytherapy. Jordan Med J 1991;25:171–7.

[46] Potter JM, Donnelly A. Carbamazepine 10,11-epoxide in therapeutic drug monitoring. Ther Drug Monit 1998;20:652–7.

[47] Yuan R, Venitz J. Effect of chronic renal failure on the disposition of highly hepatically metabolized drugs. Int J Clin Pharmacol 2000;38:245–53.

[48] Bruni J, Wang LH, Marbury TC, Lee CS, et al. Protein binding of valproic acid in uremic patients. Neurology 1980;557–9.

[49] Monaghan MS, Marx MA, Olsen KM, Turner PD, Bergman KL. Correlation and prediction of phenytoin using standard laboratory parameters in patients after renal transplantation. Ther Drug Monit 2001;23:263–7.

[50] Gulyassy PF, Jarrard E, Stanfel L. Roles of hippurate and indoxyl sulfate in the impaired ligand binding by azotemic plasma. Adv Exp Med Biol 1987;223:55–8.

[51] Takamura N, Maruyama T, Otagiri M. Effects of uremic toxins and fatty acids on serum protein binding of furosemide: possible mechanism of the binding defect in uremia. Clin Chem 1997;43:2274–80.

[52] Dasgupta A, Malik S. Fast atom bombardment mass spectrometric determination of the molecular weight range of uremic compounds that displace phenytoin from protein binding: absence of midmolecular uremic toxins. Am J Nephrol 1994;14:162–8.

[53] Otagiri M. A molecular functional study on the interactions of drugs with plasma proteins [Review]. Drug Metab Pharmacokinet 2005;20:309–23.

[54] Reidenberg MM, Affirme M. Influence of disease on binding of drugs to plasma proteins. Ann NY Acad Sci 1973;226:115–26.

[55] Prabhakar S, Bhatia R. Management of agitation and convulsions in hepatic encephalopathy. Indian J Gastroenterol 2003;22(Suppl. 2):S54–8.

[56] Klotz U, Rapp T, Muller WA. Disposition of VPA in patients with liver disease. Eur J Clin Pharmacol 1978;13:55–60.

[57] Hooper W, Dubetz D, Bochner F, et al. Plasma protein binding of carbamazepine. Clin Pharmacol Ther 1975;17:433–40.

[58] Wong MC, Suite NDA, Labar DR. Seizures in human immunodeficiency virus infection. Arch Neurol 1990;47:640–2.

[59] Burger D, Meenhorst PL, Mulder JW, et al. Therapeutic drug monitoring of phenytoin in patients with the acquired immunodeficiency syndrome. Ther Drug Monit 1994;16:616–20.

[60] Dasgupta A, McLemore J. Elevated free phenytoin and free valproic acid concentrations in sera of patients infected with human immunodeficiency virus. Ther Drug Monit 1998;20:63–7.

[61] Toler SM, Wilkerson MA, Porter WH, Smith AJ, Chandler MH. Severe phenytoin intoxication as a result of altered protein binding in AIDS. DICP: Ann Pharmacother 1990;24:698–700.

[62] Pennell PB. Antiepileptic drug pharmacokinetics during pregnancy and lactation. Neurology 2003;61(Suppl. 2):S35–42.

[63] Tomson T, Lindbom U, Ekqvist B, et al. Epilepsy and pregnancy: a prospective study on seizure control in relation to free and total concentrations of carbamazepine and phenytoin. Epilepsia 1994;35:122–30.

[64] Yerby MS, Friel PN, McCormick K. Antiepileptic drug disposition during pregnancy. Neurology 1992;42(Suppl. 5):12–16.

[65] Tomson T. Gender aspect of pharmacokinetics of new and old AEDs; pregnancy and breast feeding. Ther Drug Monit 2005;27:718–21.

[66] Tsanaclis LM, Allen J, Perucca E, Routledge PA, Richens A. Effect of valproate on free plasma phenytoin concentrations. Br J Clin Pharmacol 1984;18:17–20.

[67] Carvalho IV, Carnevale RC, Visacri MB, Mazzola PG, et al. Drug interaction between phenytoin and valproic acid in a child with refractory epilepsy: a case report. J Pharm Pract 2014;27:214–16.

[68] Pospisil J, Perlik F. Binding parameters of phenytoin during monotherapy and polytherapy. Int J Clin Pharmacol Ther Toxicol 1992;30:24–8.

[69] Joerger M, Huitema AD, Boogerd W, van der Sande JJ, et al. Interactions of serum albumin, valproic acid and carbamazepine with the pharmacokinetics of phenytoin in cancer patients. Basic Clin Pharmacol Toxicol 2006;99:133–40.

[70] Chan K, Beran RG. Value of therapeutic drug level monitoring and unbound (free) levels. Seizure 2008;17:572–5.

[71] Xiong GL, Ferranti J, Leamon MH. Toxic interaction between valproate and oxcarbazepine: a case detected by free valproate level. J Clin Psychopharmacol 2008;28:472–3.

[72] Dasgupta A, Emerson L. Interaction of valproic acid with nonsteroidal anti-inflammatory drugs mefenamic acid and fenoprofen in normal and uremic sera: lack of interaction in uremic sera due to the presence of endogenous factors. Ther Drug Monit 1996;18:654–9.

[73] Sandyk R. Phenytoin toxicity induced by interaction with ibuprofen. Afr Med J 1982;62:592.

[74] Blum RA, Schentag JJ, Gardner MJ, Wilner KD. The effect of tenidap sodium on the disposition and plasma protein binding of phenytoin in healthy male volunteers. Br J Clin Pharmacol 1995;39(Suppl. I):35S–8S.

[75] Biddle D, Wells A, Dasgupta A. Unexpected suppression of free phenytoin concentration by salicylate in uremic sera due to the presence of inhibitors: MALDI mass spectrometric determination of molecular weight range of inhibitors. Life Sci 2000;66L:143–51.

[76] Dasgupta A, Thompson WC. Carbamazepine-salicylate interaction in normal and uremic sera: reduced interaction in uremic sera. Ther Drug Monit 1995;17:199–202.

[77] Arimori K, Nanko M, Otagiri M, Uekama K. Effect of penicillins on binding of phenytoin to plasma proteins in vitro and in vivo. Biochem Drug Dispos 1984;5:219–27.

[78] Dasgupta A, Sperelakis A, Mason A, Dean R. Phenytoin-oxacillin interactions in normal and uremic sera. Pharmacotherapy 1997;17:375–8.

[79] Dasgupta A, Dennen DA, Dean R, McLawhon RW. Displacement of phenytoin from serum protein carriers by antibiotics: studies with ceftriaxone, nafcillin and sulfamethoxazole. Clin Chem 1991;37:98–100.

[80] Routledge PA, Barchowsky A, Bjornsson TD, Kitchell BB, Shand DG. Lidocaine plasma protein binding. Clin Pharmacol Ther 1980;27:347–51.

[81] Shand DG. Alpha 1-acid glycoprotein and plasma lidocaine binding. Clin Pharmacokinet [Review] 1984;9(Suppl. 1):27–31.

[82] Bonde J, Jenen NM, Burgaard P, Angelo HR, et al. Displacement of lidocaine from human plasma proteins by disopyramide. Pharmacol Toxicol 1987;60:151–5.

[83] Li JH, Xu JQ, Cao XM, Ni L, et al. Influence of the ORM1 phenotypes on serum unbound concentration and protein binding of quinidine. Clin Chim Acta 2002;317:85–92.

[84] Belpaire FM, Bogaert MG. Binding of alfentanil to human alpha-1 glycoprotein, albumin and serum. Int J Clin Pharmacol 1991;29(3):96–102.

[85] Pisupati J, Jain A, Burckart G, Hamad I, et al. Intraindividual and interindividual variation in the pharmacokinetics of mycophenolic acid in liver transplant patients. J Clin Pharmacol 2005;45:34–41.

[86] Ensom MH, Partovi N, Decarie D, Dumont RJ, et al. Pharmacokinetics and protein binding of mycophenolic acid in stable lung transplant recipients. Ther Drug Monit 2002;24:310–14.

[87] Atcheson BA, Taylor PJ, Kirkpatrick CM, Duffull SB, et al. Free mycophenolic acid should be monitored in renal transplant recipients with hypoalbuminemia. Ther Drug Monit 2004;26:284–6.

[88] Kaplan B, Meier-Kriesche HU, Friedman G, Mulgaonkar S, et al. The effect of renal insufficiency on mycophenolic acid protein binding. J Clin Pharmacol 1999;39:715–20.

[89] Weber LT, Shipkova M, Lamersdorf T, Niedmann PD, et al. Pharmacokinetics of mycophenolic acid (MPA) and determination of MPA free fraction in pediatric and adult renal transplant recipients: German study group on mycophenolate mofetil therapy in pediatric renal transplant recipients. J Am Soc Nephrol 1998;9:1511–20.

[90] Jacobson P, Long J, Rogosheske J, Brunstein C, Eweisdorf D. High unbound mycophenolic acid concentrations in a hematopoietic cell transplantation patient with sepsis and renal and hepatic dysfunction. Biol Blood Marrow Transplant 2005;11:977–8.

[91] Mudge DW, Atcheson BA, Taylor PJ, Pillans PI, et al. Severe toxicity associated with a markedly elevated mycophenolic acid free fraction in a renal transplant recipient. Ther Drug Monit 2004;26:453–5.

[92] DeFrance A, Armbruster D, Petty D, Kelley C, et al. Abbott ARCHITECT clinical chemistry and immunoassay systems—digoxin assays are free of interferences from spironolactone, potassium canrenoate and their common metabolite canrenone. Ther Drug Monit 2011;33:128–31.

[93] McMillin GA, Owen WE, Lambert TL, De BK, et al. Comparable effects of DIGIBIND and DigiFab in thirteen digoxin immunoassays. Clin Chem 2002;48:1580–4.

[94] Boffito M, Black DJ, Blaschke TF, Rowland M, et al. Protein binding in antiretroviral therapies [Review]. AIDS Res Hum Retroviruses 2003;19(9):825–35.

[95] Fayet A, Beguin A, de Tejada BM, Colombo S, et al. Determination of unbound antiretroviral drug concentrations by a modified ultrafiltration method reveals high variability in free fraction. Ther Drug Monit 2008;30:511–22.

[96] Anderson PL, Brundage RC, Bushman L, Kakuda TN, Remmel RP, Fleccher CV. Indinavir plasma protein binding in HIV-1 infected adults. AIDS 2000;14:2293–7.

[97] Barrail A, Tiec CL, Paci-Bonaventure S, Furlan V, Vicent I, Taburet AM. Determination of amprenavir total and unbound concentrations in plasma by high performance liquid chromatography and ultrafiltration. Ther Drug Monit 2006;28:89–94.

[98] Boffito M, Hoggard PG, Lindup WE, Bonora S, et al. Lopinavir protein binding in vivo through 12-hour dosing interval. Ther Drug Monit 2004;26:35–9.

[99] Za'abi M, Deleu D, Batchelor C. Salivary free concentrations of anti-epileptic drugs: an evaluation in a routine clinical setting. Acta Neurol Belg 2003;103:19–23.

[100] Nakajima M, Yamato S, Shimada K, Sato S, et al. Assessment of drug concentrations in tears in therapeutic drug monitoring I: determination of valproic acid in tears by gas chromatography/mass spectrometry with EC/NCI mode. Ther Drug Monit 2000;22:716–22.

[101] Patsalos PN, Berry DJ. Therapeutic drug monitoring of antiepileptic drugs by use of saliva. Ther Drug Monit 2013;35:4–29.

[102] George L, Muro EP, Ndaro A, Dolmans W, et al. Nevirapine concentrations in saliva measured by thin layer chromatography and self-reported adherence in patients on antiretroviral therapy at Kilimanjaro Christian medical center. Ther Drug Monit 2014;36:366–70.

[103] Liu H, Montoya JL, Forman LJ, et al. Determination of free valproic acid: evaluation of Centrifree system and comparison between high performance liquid chromatography and enzyme immunoassay. Ther Drug Monit 1992;14:513–21.

[104] McMillan GA, Juenke J, Dasgupta A. Effect of ultrafiltrate volume on the determination of free phenytoin concentration. Ther Drug Monit 2005;27:630–3.

[105] Dong WC, Zhang ZQ, Jiang XH, Sun YG, et al. Effect of volume ratio to ultrafiltrate to sample solution on the analysis of free drug and measurement of free carbamazepine in clinical drug monitoring. Eur J Pharm Sci 2013;48:332–8.

[106] Dong WC, Zhang ZQ, Hou ZL, Jiang XH, et al. The influence of volume ratio of ultrafiltrate of sample on the analysis of non-protein binding drugs in human plasma. Analyst 2013;128:7369–75.

[107] Garg U, Peat J, Frazee III C, Nguyen T, et al. A simple isotope dilution electrospray ionization tandem mass spectrometry method for the determination of free phenytoin. Ther Drug Monit 2013;35:831–5.

[108] Rebollo N, Calvo MV, Martin-Suarez A, Dominguez-Gil A. Modification of the EMIT immunoassay for the measurement of unbound mycophenolic acid in plasma. Clin Biochem 2011;44:260–3.

[109] Figurski MJ, Korecka M, Fileds L, Waligorska T, et al. High performance liquid chromatography mass spectroscopy/mass spectroscopy method for simultaneous quantification of total and free fraction of mycophenolic acid and its glucuronide metabolites. Ther Drug Monit 2009;31:717–26.

[110] Illamola SM, Labat L, Benaboud S, Tubiana R, et al. Determination of total and unbound concentrations of lopinavir in plasma using liquid chromatography-tandem mass spectrometry and ultrafiltration methods. J Chromatogr B Analyt Technol Biomed Life Sci 2014;965:216–23.

CHAPTER 5

Therapeutic Drug Monitoring of Newer Antiepileptic Drugs

Gwendolyn A. McMillin[1,2] and Matthew D. Krasowski[3]

[1]Department of Pathology, School of Medicine, University of Utah, Salt Lake City, UT, United States

[2]ARUP Institute for Clinical and Experimental Pathology, ARUP Laboratories, Inc., Salt Lake City, UT, United States

[3]Department of Pathology, University of Iowa Hospitals and Clinics, Iowa City, IA, United States

CONTENTS

5.1 Introduction 101

5.2 Therapeutic Drug Monitoring of Antiepileptic Drugs 104

5.2.1 Therapeutic Ranges and Specimen Types for TDM of Newer Antiepileptic Drugs 104

5.2.2 Analytical Methods Used in TDM of Antiepileptic Drugs 108

5.2.3 Monitoring of Free Drug Fractions 110

5.2.4 Monitoring for Special Populations 110

5.2.5 Pharmacogenetics of Antiepileptic Drugs 111

5.3 Newer Generation of Antiepileptic Drugs 112

5.3.1 Clobazam 113

5.3.2 Eslicarbazepine Acetate............................114

5.3.3 Ezogabine (Retigabine) 114

5.3.4 Felbamate...........115

5.3.5 Gabapentin..........116

5.3.6 Lacosamide 116

5.3.7 Lamotrigine 116

5.3.8 Levetiracetam.....118

5.3.9 Oxcarbazepine....118

5.3.10 Perampanel........119

5.3.11 Pregabalin 119

5.3.12 Rufinamide..........120

5.1 INTRODUCTION

Drugs used to treat and prevent seizures (antiepileptic drugs (AEDs); also known as anticonvulsant drugs) have been among the most common drugs for which therapeutic drug monitoring (TDM) is performed. AEDs are a structurally diverse group of drugs that work by several molecular mechanisms but in general reduce frequency of seizures by enhancing inhibitory neurotransmission (eg, potentiating the effect of the neurotransmitter γ-aminobutyric acid (GABA) at $GABA_A$ receptors) or inhibiting excitatory processes (eg, by inhibiting ligand-gated glutamate receptors or blocking voltage-gated sodium channels) [1]. In addition to management of seizures, the newer agents may also be used (often "off-label") for other neurologic or psychiatric conditions, such as addiction, bipolar disorder, panic attacks, aggression, Parkinson's disease, and a variety of movement disorders (Table 5.1) [2–4]. Some AEDs now have primary indications for treating neuropathic pain, migraine and cluster headaches, and neuralgias. Compliance monitoring may be important for these nonseizure applications, but TDM of AEDs prescribed for clinical uses other than management of seizure disorders is not well defined and is not discussed here.

For treatment of seizure disorders, AED selection requires consideration of the etiology of the seizure and the type(s) of seizure [5,6]. Some seizure disorders are inherited (eg, Dravet spectrum disorders), but most are acquired (eg, head injury or brain malformations) or idiopathic. Characteristic presentations may be described based on cause (eg, febrile seizure), as a syndrome

W. Clarke & A. Dasgupta (Eds): Clinical Challenges in Therapeutic Drug Monitoring. DOI: http://dx.doi.org/10.1016/B978-0-12-802025-8.00005-2

5.3.13 *Stiripentol*...........120
5.3.14 *Tiagabine*.............121
5.3.15 *Topiramate*..........122
5.3.16 *Vigabatrin*............122
5.3.17 *Zonisamide*.........122

5.4 Conclusions........123

References.................123

Table 5.1 Antiepileptic Drug Names, Clinical Uses, and Saliva:Serum Concentration Ratio

Generic Drug Name(s)	Common U.S. Trade Name (Year Approved)	Primary Seizure Indication	Nonseizure Uses	Notes	Saliva:Serum Concentration Ratio
Clobazam	Onfi (2011)	Lennox–Gastaut	Anxiety, schizophrenia	*N*-desmethyl metabolite is active; substrate of CYP3A4 (parent) and CYP2C19 (metabolite)	~1
Eslicarbazepine acetate	Aptiom (2013)	Partial seizures	Bipolar disorder	Chiral prodrug (*S*-enantiomer); glucuronidated	Not determined
Ezogabine/ retigabine	Potiga (2010)	Partial seizures	Migraine headaches, neuropathic pain, tinnitus	*N*-acetyl metabolite has weak pharmacological activity; glucuronidated	Not determined
Felbamate	Felbatol (1993)	Lennox–Gastaut, or nonresponsive partial seizures	–	CYP3A4, CYP2E1 substrate; rare cases of aplastic anemia and liver failure	Not determined
Gabapentin	Neurontin (1994)	Partial seizures	Chronic pain, peripheral neuropathy, restless leg syndrome	Not extensively metabolized	<0.1
Lacosamide	Vimpat (2008)	Partial seizures	–	Chiral (*R*-enantiomer); CYP2C19 substrate	0.77–0.96
Lamotrigine	Lamictal (1994)	Generalized tonic–clonic and partial seizures, Lennox–Gastaut	Bipolar disorder, pervasive development disorders	Glucuronidated	0.4–1.19
Levetiracetam	Keppra (1999)	Generalized tonic–clonic, myoclonic and partial seizures	Migraine	Chiral (*S*-enantiomer); hydrolyzed by cytosolic enzymes	0.36–1.55
Oxcarbazepine	Trileptal (1999)	Partial seizures	–	Achiral prodrug; 10-hydroxycarbazepine metabolite occurs as *S*- and *R*-enantiomers in ratio of 4:1; subsequently glucuronidated	0.3–1.7
Perampanel	Fycompa (2012)	Partial seizures	–	CYP3A4 substrate; glucuronidated	Not determined

Continued...

Table 5.1 Antiepileptic Drug Names, Clinical Uses, and Saliva:Serum Concentration Ratio *Continued*

Generic Drug Name(s)	Common U.S. Trade Name (Year Approved)	Primary Seizure Indication	Nonseizure Uses	Notes	Saliva:Serum Concentration Ratio
Pregabalin	Lyrica (2007)	Partial seizures	Chronic pain, fibromyalgia, generalized anxiety, diabetic neuropathy, neuropathic pain associated with a spinal cord injury	Chiral (*S*-enantiomer); not extensively metabolized	Not determined
Rufinamide	Banzel (2008)	Lennox–Gastaut	–	Hydrolyzed by carboxyesterases	0.66
Stiripentol	Diacomit[a]	Severe myoclonic epilepsy in infancy	–	Supplied as a racemic mixture; extensive metabolism via multiple pathways	Not determined
Tiagabine	Gabitril (1998)	Partial seizures	Panic attacks, movement disorders	Chiral (*R*-enantiomer); extensive metabolism, CYP3A4 substrate	Not determined
Topiramate	Topamax (1996)	Generalized tonic–clonic and partial seizures, Lennox–Gastaut	Migraine, cluster headaches, bipolar disorder, obesity	Not extensively metabolized	0.63–1.13
Vigabatrin	Sabril (2009)	Partial seizures	–	Chiral (*S*-enantiomer is active); provided as racemic mixture; not metabolized	<0.1
Zonisamide	Zonegran (2005)	Partial seizures	Bipolar disorder, chronic pain, migraine, obesity	CYP3A4 substrate	Not determined

[a]*Can be legally prescribed in the United States for compassionate use as an orphan drug but is not FDA approved and marketed in the United States.*

(eg, Lennox–Gastaut), or as type of epilepsy (eg, temporal lobe epilepsy). Seizure types are often described as generalized (involving both hemispheres of the brain) or partial (focal). Partial seizures represent approximately 60% of all seizures types [5,6]. Partial seizures may be described as simple (no loss of consciousness) or complex (loss of consciousness), and they may be associated with autonomic, motor, and/or somatosensory symptoms. Partial seizures occur with or without generalization. Generalized seizures may be convulsive (eg, tonic–clonic) or nonconvulsive (eg, absence). Status epilepticus (continuous seizure) is life-threatening and represents a medical emergency. The most widely recognized classification of seizure types was defined by the International League Against Epilepsy in 1981 and revised in 2010 [5,6]. Most of the newer AEDs are indicated for partial seizures, with or without generalization (see Table 5.1). Many newer AEDs were originally approved as adjunct therapy for refractory seizures.

5.2 THERAPEUTIC DRUG MONITORING OF ANTIEPILEPTIC DRUGS

AED therapy has traditionally been managed by TDM [7,8]. The "first-generation" or "classical" AEDs were introduced prior to 1990, and many are still used today. These include carbamazepine, clonazepam, diazepam, ethosuximide, phenobarbital, phenytoin, primidone, and valproic acid. The first-generation AEDs in general have significant interindividual variability in their pharmacokinetics (absorption, distribution, metabolism, and excretion) and also possess a relatively high risk of toxicity due to narrow therapeutic indices. Although TDM for the first-generation AEDs is widely used clinically, only two randomized, controlled studies of TDM in AED therapy have been performed, and neither showed clear benefit [9,10]. However, these two studies and others demonstrate that pre- and post-analytic errors related to TDM (eg, incorrect timing of specimen acquisition and erroneous interpretation of data) are common, thereby reducing the clinical benefit of TDM [8]. TDM of AEDs can be quite challenging given that seizures occur irregularly and unpredictably, often with long periods in between episodes. In addition, toxicity from AEDs may resemble the underlying seizure disorder in clinical presentation.

The most basic assumption of TDM is that the drug concentration being measured (eg, in blood) correlates with the clinical effect produced by the drug at the target organ (in this case, the central nervous system). However, seizures can have variable presentation and intensity even within a single individual. This makes it difficult to define appropriate therapeutic ranges for any drug and any person. Treatment with multiple AEDs adds complexity because drug–drug interactions are common. Irreversibility of drug action or the development of tolerance to the effects of the drug also complicates the use of TDM. The presence of active metabolites can also present challenges in that TDM may need to involve the parent drug and one or more metabolites or only metabolite(s). Variables to consider when selecting and interpreting TDM approaches and results are illustrated in Fig. 5.1.

5.2.1 Therapeutic Ranges and Specimen Types for TDM of Newer Antiepileptic Drugs

AED TDM is most commonly performed on serum or plasma. Therapeutic ranges are well-defined for the first-generation AEDs [11]. Proposed therapeutic ranges

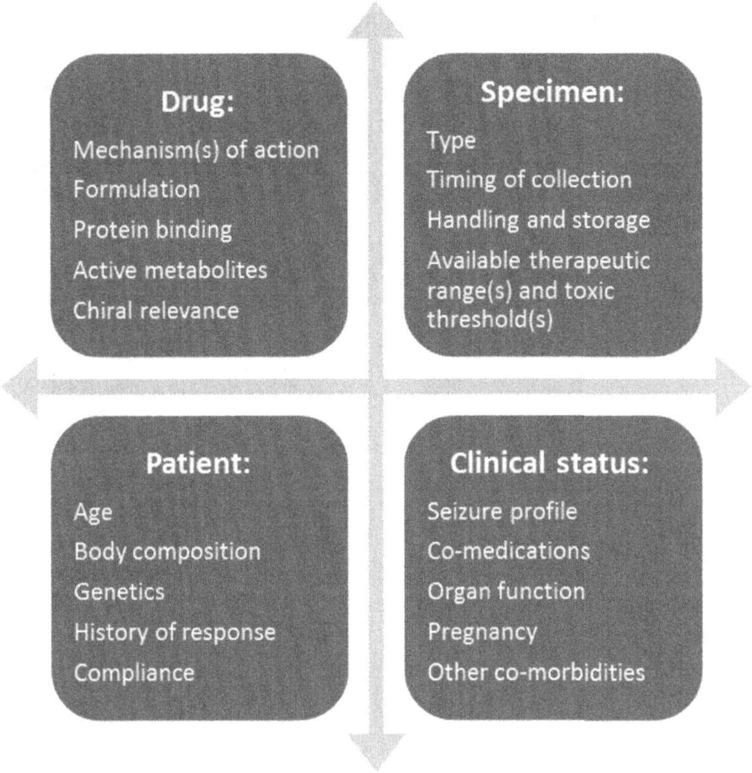

Drug:
Mechanism(s) of action
Formulation
Protein binding
Active metabolites
Chiral relevance

Specimen:
Type
Timing of collection
Handling and storage
Available therapeutic range(s) and toxic threshold(s)

Patient:
Age
Body composition
Genetics
History of response
Compliance

Clinical status:
Seizure profile
Co-medications
Organ function
Pregnancy
Other co-morbidities

FIGURE 5.1
Factors to consider when selecting and interpreting TDM for AEDs.

for the newer AEDs are included in Table 5.2. The newer-generation AEDs generally have wider therapeutic indices and less serious adverse effects than the first-generation AEDs [7]. However, establishing therapeutic ranges for the newer AEDs is challenging due to the wide range of serum/plasma concentrations associated with effective management of seizures [8]. Therapeutic ranges may vary based on seizure subtype or whether the AED is used alone or together with other AEDs. It is prudent to target clinical efficacy and not just a "standard" therapeutic range. Dosing requirements and therapeutic targets can change with age, clinical status, concomitant drug therapy, organ insufficiency, and pregnancy. Perucca has advocated for "individual therapeutic concentrations" for AEDs based on good seizure control [12]. Thus, the serum/plasma concentration target is determined when there is good seizure control. TDM can be adjusted as needed when changes in the patient occur that might alter AED pharmacokinetics [8,13,14].

TDM for AEDs may also be used early in therapy to ensure that steady-state concentrations have been achieved before efficacy is evaluated, particularly for drugs that have complicated pharmacokinetics [8,13]. With chronic maintenance therapy, TDM is useful to identify and avoid problems with

Table 5.2 Therapeutic Ranges and Methodologies for Measuring Concentrations of Antiepileptic Drugs

		ARK Diagnostics (Fremont, CA)[a]	Thermo Scientific[b] (International)	ChromSystems (Germany)[c]
Drug—Generic Name(s) or Drug Metabolite	**Therapeutic Range (mg/L)**	**Assay Range (mg/L)**	**Assay Range (mg/L)**	**MassTox TDM Kit**
Clobazam	0.03–0.3	—	—	—
N-Desmethylclobazam	0.3–3.0			
Eslicarbazepine acetate	As 10-OH-carbazepine[d]	—	—	As 10-OH-carbazepine[d]
Eslicarbazepine	3–35[e]	—	—	As 10-OH-carbazepine[d]
Ezogabine (retigabine)	Not established	—	—	—
Felbamate	30–60	—	—	2–100
Gabapentin	2–20	0.75–40	—	0.5–30
Lacosamide	5–15	—	—	0.2–12.5
Lamotrigine	3–14	0.84–40	0–40	0.2–30
Levetiracetam	12–46	2–100	—	1–100
Oxcarbazepine	As 10-OH-carbazepine[d]	—	—	0.1–10
10-OH-carbazepine	3–35[e]	—	—	0.5–50
Perampanel	Not established	—	—	—
Pregabalin	2.8–8.3	—	—	0.2–30
Rufinamide	3–30	—	—	0.5–60
Stiripentol	4–22	—	—	0.5–30
Tiagabine	0.02–0.2	—	—	0.01–0.8
Topiramate	5–20	1.5–54	0–32	0.5–30
Vigabatrin	0.8–36	—	—	0.6–50
Zonisamide	10–40	2–50	3–50	0.5–60

[a]*Based on 2012 package inserts for these immunoassays.*
[b]*Based on 2007 package inserts for these immunoassays.*
[c]*Based on ChromSystems 92921, antiepileptic drugs and metabolites in serum/plasma, using an API 4000 LC-MS/MS.*
[d]*10-OH-carbazepine is an abbreviation for 10-hydroxycarbazepine, the active metabolite of oxcarbazepine. This compound is chemically identical to licarbazepine, but the S- and R-isomers are not typically resolved analytically.*
[e]*Therapeutic range is based on 10-OH-carbazepine, which assumes equal activity and equipotency of the S- and R-isomers.*

drug–drug interactions, to manage changes in dose or drug formulation, and to evaluate compliance, particularly when seizure control is suboptimal. Children can be particularly difficult to manage with AEDs, in part due to nonadherence, which was shown to be 33% in one study [15]. Evaluation of toxicity or sudden death should consider drug concentrations as well [16].

In recent years, saliva (oral fluid) and dried blood spots have emerged as alternative specimens for many AEDs [17]. Saliva has the advantage of simplicity of collection, which may be especially beneficial for pediatric patients or for cases in which travel to a phlebotomy site is difficult. Many collection kits are now commercially available for saliva. Some collection kits include a buffer and preservatives to improve stability of the sample. One study has shown that salivary samples can be collected by the patient and mailed to the clinical laboratory without compromise of the specimen [18]. The ratio of saliva concentrations to serum concentrations is included in Table 5.1. Although it is tempting to use these ratios to estimate therapeutic ranges for AEDs in saliva, there are variables that need to be considered.

Drug distribution to saliva is mostly based on passive diffusion. Drugs that are highly bound to serum proteins appear at very low concentrations in saliva, which may present difficult analytical challenges in measuring very low concentrations of drug. Other variables that affect the concentration of drug and/or drug metabolites observed in saliva include drug ionization, lipid solubility, molecular weight, pH and pK_a, as well as salivary flow rate and metabolism. Drugs that are not ionizable within the range of pH seen in saliva appear to be the best candidates for TDM with saliva [19].

Dried blood spots (DBS) also have been proposed as an alternate specimen for AED TDM [15]. Similar to saliva, DBS may provide an advantage for remote collections or settings in which a phlebotomist is not available. The minimal sample volume required may be attractive when collecting from infants and small children. Another advantage of DBS over blood is that the specimens do not need to be processed to separate cells from plasma or serum, and they may be transported to a laboratory by routine post. Excellent stability of the AEDs has been demonstrated for several storage conditions that may be anticipated during transport, including 3 days at 40°C and 6 weeks at either ambient (25°C) or freezer (−80°C) temperatures [20]. A study of carbamazepine pharmacokinetics showed that DBS concentrations are comparable to plasma concentrations, suggesting that existing plasma or serum therapeutic ranges may be directly applicable to DBS [21]. It has been further suggested that plasma and DBS are comparable when the unbound fraction and erythrocyte-to-plasma ratios are constant, but clinical validation to translate DBS and plasma concentration of specific drugs is recommended [22].

Although DBS testing is not yet widely available, analytical methods have been published for all the first-generation AEDs [15,20], as well as several newer AEDs such as gabapentin [23], lamotrigine [15], levetiracetam [24], topiramate [25], and rufinamide [26]. Collection typically involves spotting blood from a finger stick onto Guthrie cards or filter paper (eg, Whatman 903) [22]. The blood is then dried onto the card before transport to the laboratory. In the laboratory, a standard punch (6-mm diameter) is typically extracted with organic solvents such as methanol and/or acetonitrile. The extract is further processed and analyzed by a chromatographic method as described later. Variations in the blood volume spotted and hematocrit have proven to be confounding variables in the recovery of some drugs. In one study, concentrations of AEDs varied very little over a range of blood volumes (20–50 μL) and hematocrits (30–55%) [20]. For drugs that are vulnerable to the clinically relevant variations in recovery based on hematocrit, correction methods have been proposed [22,27].

To evaluate distribution and elimination of AEDs centrally, cerebrospinal fluid (CSF) or interstitial fluid are sometimes tested [28,29]. It has been suggested that the efflux transporters that control transport of AEDs across the blood–brain barrier may be overexpressed in some patients with epilepsy. As such, cerebral microdialysis or collection by catheters may be pursued for surgery patients. Studies have demonstrated high intra- and interindividual variation in concentrations of AEDs in interstitial fluid collected from cortical regions. Furthermore, these concentrations are frequently different than CSF, suggesting that these two specimens represent different compartments.

5.2.2 Analytical Methods Used in TDM of Antiepileptic Drugs

Quantitative TDM of the newer AEDs is performed using both commercial methods and laboratory-developed tests [7]. The most common commercial methods are competitive homogeneous immunoassays that are performed with automated clinical chemistry analyzers such as those common to hospital laboratories [7]. All immunoassays incorporate either monoclonal or polyclonal antibodies that bind drugs and drug metabolites via a unique affinity for an epitope. Most immunoassay tests for AED TDM are very specific and reliable, with good performance characteristics [7,30]. Examples of commercial TDM immunoassay products for the newer AEDs are shown in Table 5.2. The noted interferences for gabapentin (pregabalin) and lamotrigine (trimethoprim) are explained by the structural similarities of the target and the interferent [30,31]. Immunoassays are not yet widely available to support TDM of AEDs using saliva, in part due to the very low concentrations of drug analytes that appear in this matrix. Immunoassays are also not currently available to support TDM of AEDs using DBS, CSF, or other specimen types. When immunoassays are not available, or when immunoassays do not perform adequately to support the need for testing, high-complexity methods that typically involve chromatography are required.

Chromatographic assays separate compounds of interest from components of the matrix, drug metabolites, known interferents, and other chemically similar structures [7,32]. Specificity of a chromatographic method will be dependent on the sample preparation, chromatographic conditions, and the detector. Sample preparation methods may include dilution, protein precipitation, filtration, chemical derivatization, and/or extraction. Gas chromatography (GC) and liquid chromatography (LC), including high-performance (HPLC) and ultra-performance configurations, are the most commonly utilized chromatographic methods. Capillary electrophoresis has recently been described as a means for separating drugs from the components of a biological specimen as well [33].

Although many detectors can be used to detect drugs, mass spectrometers (MS) are preferred due to high specificity. Tandem mass spectrometers (MS/MS) coupled to a liquid chromatograph (LC-MS/MS) is the most common high-complexity platform used to support TDM with serum/plasma today [32]. The most common platform used for DBS and oral fluid testing is also LC-MS/MS. Specificity of an LC-MS/MS method is based on chromatographic retention time, detection and quantification of the parent drug mass, and the transition of that mass to smaller masses (product ions) that are characteristic for the compound of interest. A commercially available multidrug LC-MS/MS reagent kit is described in Table 5.2. High-complexity methods are often designed to detect many drugs in the same assay to improve efficiency and decrease costs of testing. In comparison to LC-MS/MS, traditional LC methods often employ a detector that measures absorption of single or multiple wavelengths, such as ultraviolet/visible or diode array. These detectors are typically less costly to purchase and maintain than mass spectrometric detectors, but they may require longer run times and exhibit lower sensitivity and/or specificity than mass spectrometric methods [7]. Methods that employ MS/MS detection without chromatographic separation have also been described [34]. The advantages of a "direct injection" include much faster cycle time, less sample preparation, and lower sample requirements, but specificity of this approach may be compromised by isobaric interferences.

Some of the newer AEDs are chiral molecules (see Table 5.1). Enantioselective methods have been described that employ chiral chromatography, often coupled to some preanalytical chemical modification to one enantiomer (eg, using chiral derivatization or by adding chiral selector molecules) [35]. Selected chiral methods are cited in the discussion of individual drugs later. Enantiomeric separations may be important for select patients who are managed with AEDs known to exhibit enantiomer-specific effects. For example, only the *S*-enantiomer of vigabatrin is pharmacologically active, but vigabatrin is supplied as a racemic mixture. The *R*-enantiomer of vigabatrin is eliminated by the kidneys and has been reported to accumulate and potentially contribute to toxicity in patients with renal failure [36]. Sometimes enantiomers are equipotent and equally active, as is the case for the *S*- and *R*-enantiomers of 10-hydroxycarbazepine (metabolite of both eslicarbazepine acetate and oxcarbazepine), and enantioselective TDM is unnecessary. Enantiomeric separations are currently not required for supporting routine TDM of AEDs [35].

5.2.3 Monitoring of Free Drug Fractions

Some of the newer AEDs have a high proportion (>90%) of drug bound to serum proteins. For special populations, such as pregnant women, the elderly, and uremic patients, it may be appropriate to manage the TDM based on unbound or "free" fraction of drug [37]. To measure the free fraction of drug, the drug bound to protein should be physically separated from the drug that is not bound to protein. Measurement of free drug concentrations has been well characterized for the classical AED phenytoin, using ultrafiltrate and immunoassays that have been adapted for calibration at lower concentrations [37]. Free drug concentrations have also been shown to be useful for other drugs, such as carbamazepine and valproic acid [38]. Based on the percentage of protein binding, perampanel, stiripentol, and tiagabine represent new AEDs that may be good candidates for free drug analysis [39].

5.2.4 Monitoring for Special Populations

Many clinical scenarios, such as renal failure and hepatic dysfunction, are known to affect pharmacokinetics of AEDs and may therefore require special consideration of dose optimization with TDM. In cases of renal failure, drugs that are eliminated primarily by the kidneys are not highly protein bound, and small volumes of distribution will accumulate. Examples of such drugs are gabapentin, pregabalin, vigabatrin, levetiracetam, and topiramate. Formulas are available to correct dose based on creatinine clearance, although this does not replace the need for TDM. Monitoring free drug concentrations has been suggested for all highly bound AEDs in patients that are uremic. It is also likely that patients will require supplemental dosing after undergoing hemodialysis [8,40]. With hepatic dysfunction, AED pharmacokinetics is affected by changes in protein status (eg, hypoalbuminemia) and reduced expression of metabolic enzymes (eg, cytochrome P450 isozymes and glucosyltransferases). These consequences may increase the proportion of free drug and may also increase the likelihood of drug–drug interactions [8,40].

Changes in body composition associated with pregnancy and obesity may also require dose optimization, best guided by TDM. In pregnancy, TDM is important to recognize and manage changes in pharmacokinetics that result from the changes in body weight, plasma composition, and physiology. Goals of therapy are to maintain seizure control and minimize excess exposure of drugs to the embryo or fetus, particularly drugs that are recognized teratogens. TDM is recommended throughout pregnancy and at least until AED concentrations return to pre-pregnant concentrations, usually within the first month or two after delivery. Of the newer AEDs, lamotrigine has been studied most extensively in pregnancy [41,42].

Extremes of age represent additional challenges. Most studies of AED use for infants and newborns have been performed with the traditional drugs [43,44]. In general, infants exhibit shorter elimination half-lives than adults, making dose adjustments less predictable than in adults. Newborns may never achieve a steady state due to rapid changes in pharmacokinetics. Therapeutic ranges and toxic thresholds have not been determined in newborns and infants, but TDM is useful for determining an individual patient's pharmacokinetics. Although therapeutic ranges that have been established for adults are referenced for newborns and infants, actual target ranges should be individualized based on clinical response and tolerability. For the elderly, changes in body composition and physiology are associated with age, and they make dose adjustments based on results of TDM important. For example, plasma protein concentrations decrease with age, as renal function also decreases with age. Drug–drug interactions are also amplified by polypharmacy [8].

For obese patients, AEDs may exhibit unpredictable pharmacokinetics, particularly volume of distribution and clearance. Obesity can be an adverse effect of some AEDs as well, such as valproic acid [45]. Calculations to estimate dose based on ideal or lean body weight may be available, but determination of individual pharmacokinetics by TDM will be more accurate [46]. Of additional concern is when a patient undergoes bariatric surgery to treat the obesity. After surgery, it is anticipated that absorption will be reduced, which will affect dose requirements for orally administered drugs due to a truncated area under the curve. TDM is likely to be required to optimize dose of AEDs after surgery [47].

5.2.5 Pharmacogenetics of Antiepileptic Drugs

Pharmacogenetics (characterization of genetic variation in drug response) is a tool that could potentially prevent more than half of all adverse drug events by predicting which patients are good candidates for a particular drug [48]. Pharmacogenetics can predict both pharmacokinetics (eg, drug metabolism) and pharmacodynamics. There are currently no clear guidelines for applications of pharmacogenetics to the newer AEDs. However, pharmacogenetic testing is well recognized for predicting the risk of hypersensitivity reactions such as Stevens–Johnson syndrome and toxic epidermal necrolysis that occurs in some patients treated with carbamazepine, phenytoin, and potentially other AEDs. Thus, the human leukocyte antigen B (*HLA-B*) *15:02* allele is associated with hypersensitivity to these and other drugs. The *HLA-B*15:02* allele is most common in people of Asian descent, with highest occurrence in the Han Chinese, in whom allele frequency has been reported to be as high as 36%. The Clinical Pharmacogenetics Implementation Consortium has recommended that carriers of the *HLA-B*15:02* allele should avoid drugs associated with precipitating this reaction, such as carbamazepine and phenytoin, due to increased risk for hypersensitivity reactions [49,50]. These hypersensitivity reactions are not dose-related, such that TDM is not informative or predictive of risk. Pharmacogenetics testing is required to identify patients at risk.

Regarding pharmacokinetics, TDM can be informative. For phenytoin, the cytochrome P450 (CYP) isozyme 2C9 is critical for inactivation of the drug. As such, people who are genetically poor metabolizers of CYP2C9 substrates are at risk for dose-related toxicity due to accumulation of the active drug. This variation in pharmacokinetics can be managed with TDM, but it is recommended that people who are carriers of poor metabolizer alleles avoid phenytoin unless already on the medication [49]. Many of the newer AEDs presented here are substrates of CYP3A4. Most CYP3A4 substrates are also substrates of CYP3A5, which is not expressed in approximately 84% of Caucasians. Expressors of CYP3A5 may require alternate dosing than non-expressors of CYP3A5, as has been demonstrated for some immunosuppressant drugs [51]. Phenobarbital is an example substrate of CYP2C19 that may require dose adjustment or drug avoidance in patients who are poor metabolizers of CYP2C19 substrates [52]. Studies investigating the role of pharmacogenetic variation in the *ABCB1* gene (also known as *MDR1*), which codes for P-glycoprotein transporters, have exhibited controversy about relevance to AEDs, particularly in drug-resistant patients [53,54]. Contemporary literature should be consulted to better understand these and other potential pharmacogenetic associations with AEDs. Regardless of pharmacogenetic findings, TDM may help to personalize dosing for a patient based on the pharmacokinetic phenotype for that patient.

5.3 NEWER GENERATION OF ANTIEPILEPTIC DRUGS

Since the mid-1990s, 17 new AEDs have gained approval in Europe and/or the United States (for chemical structures, see Fig. 5.2): clobazam, ezogabine (retigabine), eslicarbazepine acetate, felbamate, gabapentin, lacosamide, lamotrigine, levetiracetam, oxcarbazepine, perampanel, pregabalin, rufinamide, stiripentol, tiagabine, topiramate, vigabatrin, and zonisamide [55–57]. Stiripentol is not yet approved in the United States, but it is designated as an "orphan drug" that can be legally prescribed on a compassionate use basis [58].

With the previously presented background, each of the 17 newer AEDs are discussed. With regard to analytical methods, selected numbers of representative references are cited. Table 5.1 summarizes generic and United States trade names, primary seizure indications, and saliva:serum concentration ratios for the newer AEDs. Table 5.2 summarizes therapeutic range, examples of commercially available immunoassay methods, and one commercially available chromatographic method.

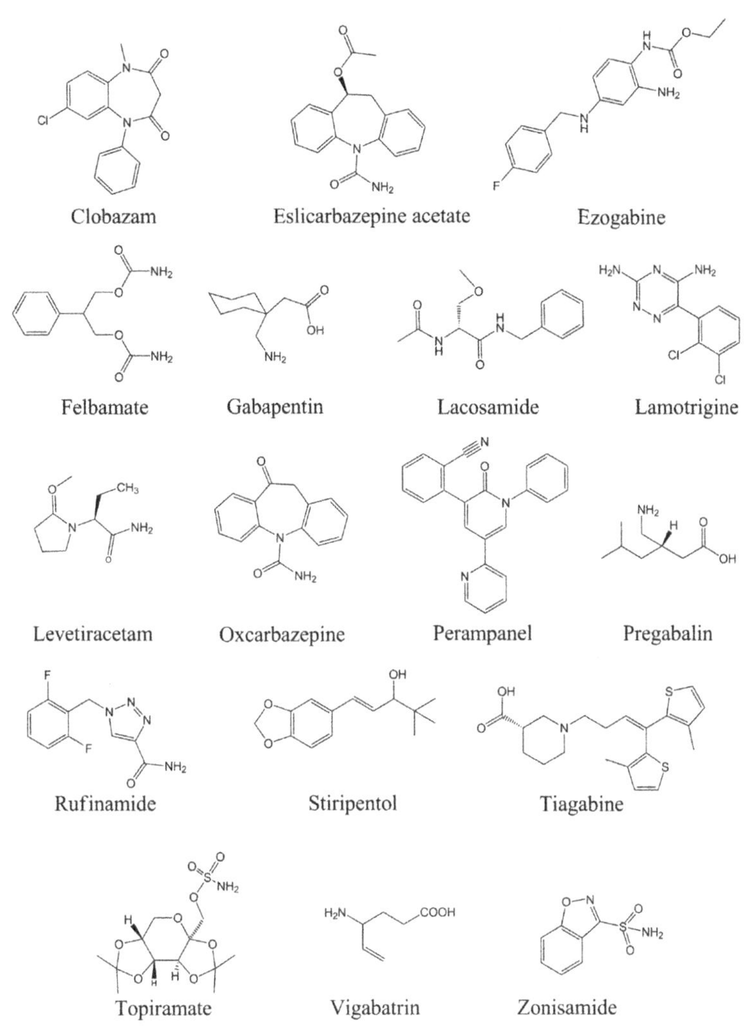

FIGURE 5.2
Chemical structures of the newer-generation antiepileptic drugs.

5.3.1 Clobazam

Clobazam, a 1,5-benzodiazepine drug, was only approved for use as an AED in the United States in 2011 but had been used worldwide starting in the 1970s for anxiolytic and antiepileptic properties [59]. The primary indication is for adjuvant treatment of seizures associated with Lennox–Gastaut syndrome (a rare syndrome associated with epilepsy often refractory to standard AED therapy), but clobazam is also used as adjunctive therapy of partial and generalized seizures, as well as the acute management of status epilepticus.

Clobazam shows rapid absorption following oral administration with a bioavailability exceeding 95%. Clobazam has predictable pharmacokinetics, with peak serum/plasma concentration in 1–3 h and an elimination half-life of 10–30 h.

Activity of clobazam is attributed to both the parent drug and the metabolite *N*-desmethylclobazam, which is generated primarily through a reaction mediated by CYP3A4. The metabolite has a half-life of 36–46 h and is subsequently inactivated by CYP2C19, leading to a recommendation that individuals with genetic polymorphisms of CYP2C19 associated with reduced metabolizing ability ("poor metabolizers") be treated with low doses and followed carefully by TDM [60]. Immunoassays for benzodiazepines (such as used in urine or serum drug of abuse screening assays) [39,61] may detect clobazam, but quantitation is not accurate. TDM of clobazam must therefore be accomplished with specific chromatographic methods [62,63].

5.3.2 Eslicarbazepine Acetate

Eslicarbazepine acetate is a prodrug that is rapidly metabolized by liver esterases to form eslicarbazepine ((*S*)-licarbazepine, (*S*)-10-hydroxycarbazepine), a compound with antiepileptic properties [64,65]. Eslicarbazepine is also one of the active metabolites of oxcarbazepine (itself one of the newer AEDs). Eslicarbazepine acetate gained approval as an AED in Europe in 2009 and in the United States in 2013 [66]. TDM of eslicarbazepine acetate focuses on the primary active metabolite (eslicarbazepine) and not the parent drug. Additional minor (and active) metabolites of eslicarbazepine acetate are (*R*)-licarbazepine and oxcarbazepine. Eslicarbazepine has low binding to serum proteins and an elimination half-life of 20–24 h during chronic administration [67]. Eslicarbazepine exhibits minimal drug–drug interactions [68]. Mild to moderate liver failure has little impact on eslicarbazepine pharmacokinetics [69]. Renal clearance is the main elimination route for eslicarbazepine and its minor metabolites, and hemodialysis effectively clears eslicarbazepine and its active metabolites [70]. The generally predictable pharmacokinetics of eslicarbazepine lead to a minimal role of TDM except when this drug is used in patients with renal impairment. Enantioselective HPLC and LC-MS/MS methods have been reported for measurement of eslicarbazepine in serum/plasma [71,72].

5.3.3 Ezogabine (Retigabine)

Ezogabine (retigabine) is an AED approved in the United States in 2010 and in Europe in 2011 for treatment of partial seizures [66,73]. Ezogabine has an oral bioavailability of approximately 60% due to first-pass metabolism by the liver [74]. The serum half-life is 6–10 h, and the binding of ezogabine to serum proteins is estimated at 80%. Currently, there are no published data on the distribution of ezogabine into saliva [17,66]. A therapeutic range for ezogabine has not yet been established.

Approximately 20–30% of the ezogabine administered dose is excreted unchanged by the kidneys [75]. The remainder is metabolized to an *N*-acetyl metabolite that has weak pharmacologic action. Ezogabine has minimal drug–drug interactions [75,76]. Both renal and hepatic impairment decrease ezogabine clearance rate. The clearance of ezogabine decreases in the elderly, mainly attributable to age-related decreases in renal function [74].

There is minimal information on the role of TDM in management of patients receiving ezogabine [66]. An LC-MS/MS method for determining serum/plasma concentrations of ezogabine has been reported [77].

5.3.4 Felbamate

Felbamate is an AED with limited clinical use due to its association with rare but severe adverse effects. Felbamate was approved in 1993 in the United States for the treatment of partial seizures in adults and for Lennox–Gastaut syndrome [78]. By 1994, rare cases of aplastic anemia (failure of bone marrow to produce blood cells; approximately 1 in 4000) and severe liver failure (approximately 1 in 30,000) were associated with felbamate therapy, each with mortality exceeding 30%. Felbamate now has revised labeling and very restricted use. There is currently no method to predict which individuals are at high risk for felbamate adverse effects.

Felbamate has excellent bioavailability and is metabolized by the liver to multiple inactive metabolites, with significant interindividual variation in metabolism [79]. The clearance of felbamate is 20–65% higher in children than in adults [80]. Valproic acid inhibits the metabolism while inducers of liver metabolism (eg, carbamazepine, phenytoin, phenobarbital, and rifampin) increase the metabolism of felbamate [81,82]. Typical doses of felbamate used in epilepsy management result in serum/plasma concentrations of 30–60 mg/L [83].

The interindividual variability in metabolism, developmental changes in felbamate clearance, and potential for drug–drug interactions favor TDM for felbamate. Unfortunately, monitoring of felbamate serum/plasma concentrations does not predict the rare adverse effects. Multiple analytical methodologies have been reported for the measurement of felbamate in serum/plasma, including GC [84] and HPLC [85]. Regular monitoring of blood counts and liver function is advised during felbamate therapy, with immediate discontinuation of the drug advised if abnormalities are detected.

5.3.5 Gabapentin

Gabapentin was approved in 1994 in the United States as adjuvant therapy for partial seizures but is now used more often for other uses, such as management of chronic pain syndromes [86]. Two extended-release formulations, gabapentin enacarbil (Horizant) and a once-daily formulation (Gralise, 1800 mg), were approved in 2011 and are used for treating chronic pain, neuralgia, and restless legs syndrome. Following oral administration, gabapentin is rapidly absorbed by the L-amino acid transport system [87]. Concentrations of gabapentin in saliva are 10% or less than those in plasma, limiting the potential of saliva as a specimen for TDM [17,88].

Gabapentin shows negligible binding to serum proteins and undergoes little or no metabolism [87]. The main clearance is by the kidneys. Patients with renal failure show a prolonged elimination half-life. Hemodialysis effectively clears gabapentin from the circulation [13]. An approximate therapeutic range of 2–20 mg/L has been proposed for gabapentin for seizure control [89].

TDM has limited clinical utility in gabapentin therapy except in renal compromise and to assess adherence with therapy. Multiple analytical methodologies have been reported for the measurement of gabapentin in plasma/serum, including homogeneous immunoassay [90], HPLC [91], LC-MS/MS [92], and gas chromatography/mass spectrometry (GC/MS) [93].

5.3.6 Lacosamide

Lacosamide is a functionalized amino acid whose mechanism of action is thought to involve enhancement of slow inactivation of voltage-gated sodium channels [94]. Lacosamide was approved in the United States and Europe in 2008 [95]. Lacosamide has excellent oral bioavailability and minimal serum protein binding [66,96]. Approximately 40% of the drug is excreted unchanged by the kidney; the remainder is cleared by liver metabolism, mostly by CYP2C19 to an inactive metabolite. Drug–drug interactions involving lacosamide appear to be minimal [97,98]. A therapeutic range of 5–15 mg/L has been proposed [66]. HPLC and LC-MS/MS procedures for measuring lacosamide concentrations in plasma or serum have been reported [99]. The generally predictable pharmacokinetics of lacosamide limit the routine need for TDM other than to establish individualized therapeutic ranges or to manage therapy in patients with liver and/or kidney failure [100].

5.3.7 Lamotrigine

Lamotrigine was approved in the United States in 1994 as adjunct therapy for partial seizures and has subsequently gained additional approval for management of bipolar disorder [7,8,56]. Lamotrigine has a good safety record in pregnancy (contrasting with the teratogenic effects of first-generation AEDs such as carbamazepine, phenobarbital, phenytoin, and valproic acid), currently making lamotrigine a first-line option for management of epilepsy in pregnancy [14,101]. TDM plays a significant role in lamotrigine therapy largely due to the complex pharmacokinetics of the drug [102].

Lamotrigine has excellent oral bioavailability and approximately 50% binding to serum proteins. Steady-state salivary lamotrigine concentrations correlate well with serum/plasma concentrations (with an average saliva:serum ratio of approximately 0.5), making saliva a viable specimen for lamotrigine TDM [103].

Lamotrigine is extensively metabolized, mainly by glucuronidation to a series of pharmacologically inactive metabolites. Lamotrigine metabolism also exhibits the phenomenon of "autoinduction" (increase in its own metabolism during the course of therapy) [104], similar to the first-generation AED carbamazepine, with an approximately 20% reduction in steady-state serum/plasma concentrations if dosage is not escalated. Autoinduction usually fades by 3 or 4 weeks of therapy.

Lamotrigine exhibits many clinically significant drug–drug interactions. Classic liver enzyme inducers (eg, carbamazepine and phenytoin) increase the metabolism of lamotrigine, reducing serum half-life from 15–35 h (as monotherapy) to 8–20 h [102]. Estradiol-containing oral contraceptives also significantly reduce lamotrigine serum/plasma levels [105]. In contrast, valproic acid (a CYP enzyme inhibitor) reduces metabolism of lamotrigine, increasing serum half-life to 60 h or more [102]. Severe renal failure also increases elimination half-life of lamotrigine. Hemodialysis effectively clears lamotrigine and metabolites from the circulation and may be a therapeutic option in managing lamotrigine overdose [106]. The clearance of lamotrigine is higher in children compared to adults and markedly higher (up to 300% or more) in pregnant versus nonpregnant women [80,107].

A therapeutic range of 3–14 mg/L for lamotrigine has been proposed for management of seizures [108]. Variability of clinical response makes establishment of an individualized therapeutic range beneficial in lamotrigine TDM. Lamotrigine toxicity is more common when serum/plasma levels exceed 15 mg/L [108].

Overall, multiple factors make TDM clinically useful in lamotrigine therapy. Clearance of the drug is affected by other medications, extremes of age, pregnancy, and renal insufficiency. Monitoring of lamotrigine levels can be especially useful when any of these factors change. TDM can also help avoid toxicity associated with supratherapeutic serum/plasma concentrations [108,109]. There are many analytical methodologies for lamotrigine [110], including homogeneous immunoassay [111,112], HPLC [113,114], and LC-MS/MS [115].

5.3.8 Levetiracetam

Levetiracetam is an AED structurally unrelated to other marketed AEDs. Levetiracetam has high oral bioavailability, although co-ingestion with food significantly slows rate of absorption [116]. Salivary levetiracetam concentrations correlate well with those in serum/plasma [117].

Levetiracetam has predictable pharmacokinetics, with nearly 100% of the drug cleared by renal excretion, either as the parent drug or as the metabolite LO57 that is formed by hydrolysis in the blood [118,119]. Levetiracetam has low serum protein binding and should be effectively cleared by hemodialysis [13]. Due to lack of liver metabolism, drug–drug interactions with levetiracetam are uncommon [98]. The clearance of levetiracetam varies with age and during pregnancy, in large part due to variation in renal clearance. Pregnant women show an approximately 60% decrease in serum/plasma concentrations if dose is not adjusted [14]. The serum half-life in neonates is 16–18 h compared to 6–8 h in adults [8]. An important preanalytical factor in performing TDM for levetiracetam is to process samples quickly (ideally separating plasma/serum from cells within hours) to avoid in vitro hydrolysis of the drug [119]. A therapeutic range of 12–46 mg/L has been proposed for levetiracetam in epilepsy [120].

There are multiple analytical methodologies for levetiracetam [110], including homogeneous immunoassay [121–123], GC [124,125], HPLC [126], and LC-MS/MS [127]. Levetiracetam TDM can be beneficial in adjusting dosage for renal insufficiency and to maintain effective levels across changes in age and during pregnancy [128].

5.3.9 Oxcarbazepine

Oxcarbazepine produces much of its effects through an active metabolite, 10-hydroxycarbazepine [129]. Oxcarbazepine is structurally related to carbamazepine but has lower propensity to induce expression of hepatic drug-metabolizing enzymes and thereby cause drug–drug interactions [130]. Following administration, oxcarbazepine is rapidly metabolized to 10-hydroxycarbazepine, at an approximate ratio of 4:1 for the *S*-enantiomer (also termed eslicarbazepine; discussed previously) and the *R*-enantiomer, which is equipotent and equiactive [129]. Additional minor metabolites likely contribute only minimally to therapeutic effect.

For the purposes of TDM, oxcarbazepine is typically treated like a prodrug, with monitoring focusing on the metabolite [8]. Saliva is not an optimal specimen for TDM. The half-life of 10-hydroxycarbazepine in saliva is short, and there is a complicated, dose-dependent relation between salivary and serum/plasma concentration of 10-hydroxycarbazepine [131,132]. The metabolite

10-hydroxycarbazepine is cleared by a combination of renal excretion and liver metabolism. 10-Hydroxycarbazepine and oxcarbazepine have approximately equal potency as anticonvulsants, but 10-hydroxycarbazepine accumulates to higher serum/plasma concentrations during chronic therapy [133]. 10-Hydroxycarbazepine is likely cleared effectively by hemodialysis, although formal studies have not been reported [13].

10-Hydroxycarbazepine clearance is slower in the elderly and in patients with renal insufficiency [80,134]. Drug clearance is increased in young children and during pregnancy and by concomitant use of liver enzyme-inducing drugs [135]. Young children require higher oxcarbazepine doses per body weight than adults [80,107]. 10-Hydroxycarbazepine serum/plasma concentration has a wide therapeutic range of 3–35 mg/L [136]. Toxic side effects are more common when serum/plasma concentrations exceed 35 mg/L [137]. Multiple analytical methodologies have been reported for the measurement of 10-hydroxycarbazepine in plasma/serum, including GC [138], HPLC [139,140], and LC-MS/MS [141].

TDM for oxcarbazepine is generally most helpful when there are changes in renal function, pregnancy, and/or concomitant use of liver enzyme-inducing drugs. Monitoring of 10-hydroxycarbazepine levels can be helpful in achieving and maintaining effective serum/plasma concentrations while avoiding toxicity.

5.3.10 Perampanel

Perampanel is a novel AED approved in 2012 in Europe and the United States for adjunctive therapy of partial seizures [142,143]. Perampanel shows rapid and near complete absorption following oral administration [142]. Currently, there are no data on the distribution of perampanel into saliva [17]. The serum half-life is approximately 60–90 h at steady state in adult men [142]. Perampanel is strongly bound (>95%) to serum proteins. Perampanel is primarily metabolized by CYP3A4 and CYP3A5 oxidation, followed by glucuronidation. Clearance rate is reduced in patients with renal and hepatic insufficiency. Minimal data are available on the role of TDM in perampanel therapy. A therapeutic range has not yet been established.

5.3.11 Pregabalin

Pregabalin is pharmacologically related to gabapentin and similarly used more for management of chronic pain than for treatment of seizure disorders [144]. In the United States, pregabalin has approval for treatment of fibromyalgia [56]. Pregabalin has predictable pharmacokinetics with high bioavailability, no reported drug–drug interactions, and minimal binding to serum proteins [145]. Pregabalin is not metabolized to any appreciable degree and is excreted unchanged in the urine with a clearance that approximates glomerular filtration rate [146]. Reduced dosage is needed for patients with renal insufficiency [147].

TDM is not commonly needed for pregabalin, although TDM may be useful in patients with renal insufficiency. In the treatment of chronic pain, verifying compliance with pregabalin therapy may be useful in ensuring that patients are following therapy before switching to other drugs (eg, opioids such as fentanyl or oxycodone) with higher abuse and toxicity potential. An approximate therapeutic range for pregabalin of 2.8–8.3 mg/L for the management of seizures has been proposed [8]. Pregabalin has a short serum half-life (~6 h), so proper timing of dosing relative to sample draw time is important for TDM [148]. Determination of plasma/serum concentrations of pregabalin can be done by HPLC following sample derivatization [149] or by LC-MS/MS [150].

5.3.12 Rufinamide

Rufinamide is a novel AED approved in Europe in 2007 and in the United States in 2008 for treatment of Lennox–Gastaut syndrome [151]. Rufinamide has high bioavailability (85% or higher), and absorption is increased when the drug is taken with food [152]. Patients are advised to take rufinamide in the same temporal relationship to meals to limit fluctuations in peak levels. Rufinamide is extensively metabolized by carboxyesterases to inactive metabolites that are mainly excreted renally. Rufinamide metabolism is increased by concomitant therapy with liver enzyme inducers such as carbamazepine, phenytoin, phenobarbital, rifampin, and St. John's wort [152]. Concentrations of rufinamide are not affected by lamotrigine, topiramate, or benzodiazepines [152].

TDM for rufinamide is currently not well-defined [151–153]. A wide range of serum/plasma concentrations (3–30 mg/L) have been associated with successful treatment of epilepsy; however, individual patients appear to have a narrower range of concentrations related to good clinical response [154]. Consequently, the use of an individual therapeutic concentration strategy appears to be a reasonable approach [12]. Steady-state concentrations are achieved within approximately 2 days due to the relatively short elimination half-life (6–10 h). Rufinamide serum/plasma concentrations can be determined by HPLC with ultraviolet/visible or MS detection [152,155].

5.3.13 Stiripentol

Stiripentol was originally approved in Europe in 2001 as an orphan drug for the treatment of infantile severe myotonic epilepsy (Dravet spectrum disorders) [156]. Stiripentol is not approved in the United States but is classified as an orphan drug that may be prescribed on a compassionate use basis [58]. Stiripentol has complicated pharmacokinetics similar to those of the classic AED phenytoin. Stiripentol is rapidly absorbed following oral administration but shows extensive first-pass metabolism by the liver. Stiripentol is strongly bound (>99%) to plasma proteins. Measuring the free drug fraction of stiripentol (as is commonly done with phenytoin) might be useful in TDM, although methods to measure free stiripentol concentrations have not yet been reported.

Stiripentol has complex biotransformation involving at least five different metabolic pathways. Significant drug–drug interactions can occur with other AEDs, especially drugs that alter activity of CYP enzymes. Carbamazepine, phenobarbital, and phenytoin decrease stiripentol serum/plasma concentrations by induction of CYP enzymes [157]. Like phenytoin, stiripentol has nonlinear pharmacokinetics, with decreased clearance at higher dosages [157]. A provisional therapeutic range of 4–22 mg/L in serum/plasma for stiripentol has been proposed based on therapy of absence seizures in children [158]. An HPLC method has been reported for the analysis of stiripentol concentrations in serum/plasma [159].

5.3.14 Tiagabine

Tiagabine is approved in Europe and the United States for adjunctive therapy of partial seizures [56]. The use of tiagabine has been limited by side effects, which include seizures and rarely life-threatening, nonconvulsive status epilepticus [160,161]. Tiagabine has excellent bioavailability and a high degree (>95%) of binding to serum proteins [162]. Valproic acid can displace tiagabine from serum protein binding sites, thereby increasing the free fraction of tiagabine [163]. Tiagabine is extensively metabolized by the liver, with higher clearance in children than in adults [164]. Inducers of liver metabolism (eg, carbamazepine, phenytoin, phenobarbital, and rifampin) increase the metabolism of tiagabine, reducing the serum half-life from 5–9 h to 2–4 h [162,165]. Decrease in renal function has little effect on tiagabine clearance [166], whereas severe liver failure can increase half-life to 12–16 h or more [167].

TDM can be useful in tiagabine therapy due to variable metabolism and potential for drug–drug interactions. The tight binding of tiagabine to serum protein binding also suggests that measurement of free drug concentrations may be clinically useful, although analytical methods to measure free tiagabine have not yet been reported [37]. A therapeutic range of 0.02–0.2 mg/L (20–200 ng/mL) has been recommended based on a multicenter study [168]. GC/MS [169], HPLC [170], and LC-MS/MS [171] methods for measuring tiagabine serum/plasma concentrations have been reported.

5.3.15 Topiramate

Topiramate was first approved in the United States in 1996 for treatment of epilepsy [56]. Topiramate has gained additional approvals for treatment of migraine headaches and, in combination with phentermine, for weight loss. Topiramate has high bioavailability and low binding to serum proteins [172]. Salivary topiramate levels correlate well with plasma/serum concentrations, making saliva a viable specimen for TDM [173]. Approximately 50% of the absorbed dose is metabolized by the liver. Inhibitors and inducers of liver drug metabolism can significantly alter the serum half-life of topiramate in adults from 20–30 h to 12 h [174]. Children typically clear topiramate faster than adults [80].

The proposed therapeutic range for topiramate management of epilepsy is 5–20 mg/L [175]. TDM of topiramate can be especially helpful due to the variable liver metabolism, especially when topiramate is used concomitantly with medications that alter liver enzyme metabolism. The availability of homogeneous immunoassays allows for rapid determination of topiramate serum/plasma concentrations [176,177]. Other analytical methodologies include GC [178], GC/MS [179], HPLC [180], and LC-MS/MS [181].

5.3.16 Vigabatrin

Vigabatrin is one of the newer AEDs for which TDM has a limited role. The main drawback with performing TDM is that the drug irreversibly inhibits its molecular target (GABA transaminase), resulting in a poor correlation between serum/plasma concentrations and therapeutic effect [182]. Vigabatrin has good bioavailability and minimal binding to serum proteins. The drug is primarily excreted unchanged in the urine [182]. Hemodialysis effectively clears vigabatrin [36].

A wide range of serum/plasma concentrations (0.8–36 mg/L) are associated with successful treatment of epilepsy [57,182]. TDM may help to assess compliance with therapy or to evaluate possible drug toxicity. Analytical methodologies for measuring vigabatrin include capillary electrophoresis [183], GC/MS [93], and HPLC [184].

5.3.17 Zonisamide

Zonisamide is approved in Australia, Europe, Japan, and the United States for treatment of partial seizures [56,185]. The drug is also used off-label for other conditions, such as bipolar disorder and migraine headaches. Zonisamide is rapidly absorbed after oral administration and only 50% bound to serum proteins. Saliva is a promising specimen for zonisamide given good correlation between saliva and serum/plasma concentrations

[17]. The major pharmacokinetic variability with zonisamide is in the metabolism. Zonisamide is extensively metabolized by acetylation, oxidation (including by CYP3A4), and other pathways [186]. The clearance of zonisamide is significantly altered by drugs that alter CYP enzyme activity. The serum half-life of zonisamide is approximately 60 h as monotherapy; the half-life may be decreased to 25–35 h during concomitant therapy with CYP enzyme inducers such as phenobarbital and phenytoin [187]. Conversely, CYP enzyme inhibitors (eg, cimetidine and valproic acid) may significantly prolong serum half-life. Children clear zonisamide faster than adults and thus require higher doses by weight [80]. Hemodialysis effectively clears zonisamide from the circulation [188].

A serum/plasma therapeutic range of 10–40 mg/L has been proposed for use of zonisamide in epilepsy [189]. The interindividual variability in metabolism makes TDM useful for zonisamide therapy. Toxicity is uncommon at serum/plasma concentrations less than 30 mg/L. Multiple analytical methodologies are available for measurement of zonisamide in plasma/serum, including homogenous immunoassay [190], HPLC [191], LC/MS [192], and LC-MS/MS [193].

5.4 CONCLUSIONS

The newer generation of AEDs has expanded the therapeutic options for managing seizure disorders as well as other neurologic and psychiatric conditions. Overall, the newer AEDs have less adverse effects and wider therapeutic margin compared to the first-generation AEDs such as carbamazepine, phenobarbital, and phenytoin. The newer-generation AEDs with the strongest justifications for TDM are lamotrigine, levetiracetam, oxcarbazepine, stiripentol, and zonisamide. Perampanel, stiripentol, and tiagabine are strongly protein bound and may be candidates for monitoring of the free drug fractions, although analytical methods for this purpose have not yet been reported. For other AEDs, TDM may be clinically useful to assess medication compliance or to adjust dosing in organ failure. Future research is needed to better delineate therapeutic ranges (including for non-epilepsy uses) and to document benefit of TDM in clinical practice.

References

[1] LaRoche SM, Helmers SL. The new antiepileptic drugs: scientific review. J Am Med Assoc 2004;291:605–14.

[2] Golden AS, Haut SR, Moshe SL. Nonepileptic uses of antiepileptic drugs in children and adolescents. Pediatr Neurol 2006;34:421–32.

[3] Johannessen Landmark C. Antiepileptic drugs in non-epilepsy disorders: relations between mechanisms of action and clinical efficacy. CNS Drugs 2008;22:27–47.

[4] Spina E, Perugi G. Antiepileptic drugs: indications other than epilepsy. Epileptic Disord 2004;6:57–75.

[5] Berg AT, Millichap JJ. The 2010 revised classification of seizures and epilepsy. Continuum (Minneap Minn) 2013;19:571–97.

[6] Epilepsy CoCaTotILA. Proposal for revised clinical and electroencephalographic classification of epileptic seizures. Epilepsia 1981;22:489–501.

[7] Neels HM, Sierens AC, Naelerts K, Scharpé SL, Hatfield GM, Lambert WE. Therapeutic drug monitoring of old and newer anti-epileptic drugs. Clin Chem Lab Med 2004;42:1228–55.

[8] Patsalos PN, Berry DJ, Bourgeois BFD, Cloyd JC, Glauser TA, Johannessen SI, et al. Antiepileptic drugs – best practice guidelines for therapeutic drug monitoring: a position paper by the subcommission on therapeutic drug monitoring, ILAE Commission on Therapeutic Strategies. Epilepsia 2008;49:1239–76.

[9] Fröscher W, Eichelbaum M, Gugler R, Hildebrand G, Penin H. A prospective randomized trial on the effect of monitoring plasma anticonvulsant levels in epilepsy. J Neurol 1981;224:193–201.

[10] Januzzi G, Cian P, Fattore C, Gatti G, Bartoli A, Monaco F, et al. A multicenter randomized controlled trial on the clinical impact of therapeutic drug monitoring in patients with newly diagnosed epilepsy. Epilepsia 2000;41:222–30.

[11] Rosenbloom D, Upton AR. Drug treatment of epilepsy: a review. Can Med Assoc J 1983;128:261–70.

[12] Perucca E. Is there a role for therapeutic drug monitoring of new anticonvulsants? Clin Pharmacokinet 2000;38:191–204.

[13] Asconape JJ. Use of antiepileptic drugs in hepatic and renal disease. Handb Clin Neurol 2014;119:417–32.

[14] Tomson T, Landmark CJ, Battino D. Antiepileptic drug treatment in pregnancy: changes in drug disposition and their clinical implications. Epilepsia 2013;54:405–14.

[15] Shah NM, Hawwa AF, Millership JS, Collier PS, Ho P, Tan ML, et al. Adherence to antiepileptic medicines in children: a multiple-methods assessment involving dried blood spot sampling. Epilepsia 2013;54:1020–7.

[16] Nilsson L, Bergman U, Diwan V, Farahmand BY, Persson PG, Tomson T. Antiepileptic drug therapy and its management in sudden unexpected death in epilepsy: a case-control study. Epilepsia 2001;42:667–73.

[17] Patsalos PN, Berry DJ. Therapeutic drug monitoring of antiepileptic drugs by use of saliva. Ther Drug Monit 2013;35:4–29.

[18] Jones MD, Ryan M, Miles MV, Tang PH, Fakhoury TA, Degrauw TJ, et al. Stability of salivary concentrations of the newer antiepileptic drugs in the postal system. Ther Drug Monit 2005;27:576–9.

[19] Ruiz ME, Conforti P, Fagiolino P, Volonte MG. The use of saliva as a biological fluid in relative bioavailability studies: comparison and correlation with plasma results. Biopharm Drug Dispos 2010;31:476–85.

[20] Shah NM, Hawwa AF, Millership JS, Collier PS, McElnay JC. A simple bioanalytical method for the quantification of antiepileptic drugs in dried blood spots. J Chromatogr B Analyt Technol Biomed Life Sci 2013;923–924:65–73.

[21] Kong ST, Lim SH, Chan E, Ho PC. Estimation and comparison of carbamazepine population pharmacokinetics using dried blood spot and plasma concentrations from people with epilepsy: the clinical implication. J Clin Pharmacol 2014;54:225–33.

[22] Wilhelm AJ, den Burger JC, Swart EL. Therapeutic drug monitoring by dried blood spot: progress to date and future directions. Clin Pharmacokinet 2014;53:961–73.

[23] Kolocouri F, Dotsikas Y, Loukas YL. Dried plasma spots as an alternative sample collection technique for the quantitative LC-MS/MS determination of gabapentin. Anal Bioanal Chem 2010;398:1339–47.

[24] Luo W, Kong ST, Yang S, Chan BC, Ho PC. A simple assay for determination of levetiracetam in rat dried blood spots by LC-MS/MS. Bioanalysis 2013;5:1843–51.

[25] la Marca G, Malvagia S, Filippi L, Fiorini P, Innocenti M, Luceri F, et al. Rapid assay of topiramate in dried blood spots by a new liquid chromatography-tandem mass spectrometric method. J Pharm Biomed Anal 2008;48:1392–6.

[26] la Marca G, Malvagia S, Filippi L, Innocenti M, Rosati A, Falchi M, et al. Rapid assay of rufinamide in dried blood spots by a new liquid chromatography-tandem mass spectrometric method. J Pharm Biomed Anal 2011;54:192–7.

[27] den Burger JC, Wilhelm AJ, Chahbouni AC, Vos RM, Sinjewel A, Swart EL. Haematocrit corrected analysis of creatinine in dried blood spots through potassium measurement. Anal Bioanal Chem 2014.

[28] Rambeck B, Jurgens UH, May TW, Pannek HW, Behne F, Ebner A, et al. Comparison of brain extracellular fluid, brain tissue, cerebrospinal fluid, and serum concentrations of antiepileptic drugs measured intraoperatively in patients with intractable epilepsy. Epilepsia 2006;47:681–94.

[29] Shannon RJ, Carpenter KL, Guilfoyle MR, Helmy A, Hutchinson PJ. Cerebral microdialysis in clinical studies of drugs: pharmacokinetic applications. J Pharmacokinet Pharmacodyn 2013;40:343–58.

[30] Krasowski MD, Siam MG, Iyer M, Ekins S. Molecular similarity methods for predicting cross-reactivity with therapeutic drug monitoring immunoassays. Ther Drug Monit 2009; 31:337–44.

[31] Dasgupta A. Impact of interferences including metabolite crossreactivity on therapeutic drug monitoring results. Ther Drug Monit 2012;34:496–506.

[32] Saint-Marcoux F, Sauvage FL, Marquet P. Current role of LC-MS in therapeutic drug monitoring. Anal Bioanal Chem 2007;388:1327–49.

[33] Rodriguez J, Castaneda G, Munoz L. Direct determination of pregabalin in human urine by nonaqueous CE-TOF-MS. Electrophoresis 2013;34:1429–36.

[34] Korman E, Langman LJ, Jannetto PJ. High-throughput method for the quantification of lacosamide in serum using ultra-fast SPE-MS/MS. Ther Drug Monit 2015;37(1):126–31.

[35] Fortuna A, Alves G, Falcao A. Chiral chromatographic resolution of antiepileptic drugs and their metabolites: a challenge from the optimization to the application. Biomed Chromatogr 2014;28:27–58.

[36] Jacqz-Aigrain E, Guillonneau M, Rey E, Macher MA, Montes C, Chiron C, et al. Pharmacokinetics of the S(+) and R(−) enantiomers of vigabatrin during chronic dosing in a patient with renal failure. Br J Clin Pharmacol 1997;44:183–5.

[37] Dasgupta A. Usefulness of monitoring free (unbound) concentrations of therapeutic drugs in patient management. Clin Chim Acta 2007;377:1–13.

[38] Chan K, Beran RG. Value of therapeutic drug level monitoring and unbound (free) levels. Seizure 2008;17:572–5.

[39] Krasowski MD, McMillin GA. Advances in anti-epileptic drug testing. Clin Chim Acta 2014;436:224–36.

[40] Lacerda G, Krummel T, Sabourdy C, Ryvlin P, Hirsch E. Optimizing therapy of seizures in patients with renal or hepatic dysfunction. Neurology 2006;67:S28–33.

[41] Pirie DA, Al Wattar BH, Pirie AM, Houston V, Siddiqua A, Doug M, et al. Effects of monitoring strategies on seizures in pregnant women on lamotrigine: a meta-analysis. Eur J Obstet Gynecol Reprod Biol 2014;172:26–31.

[42] Reisinger TL, Newman M, Loring DW, Pennell PB, Meador KJ. Antiepileptic drug clearance and seizure frequency during pregnancy in women with epilepsy. Epilepsy Behav 2013; 29:13–18.

[43] Pellock J. Antiepileptic drugs trials: neonates and infants. Epilepsy Res 2006;68:42–5.

[44] Walson PD. Role of therapeutic drug monitoring (TDM) in pediatric anti-convulsant drug dosing. Brain Dev 1994;16:23–6.

[45] Fang J, Chen S, Tong N, Chen L, An D, Mu J, et al. Metabolic syndrome among Chinese obese patients with epilepsy on sodium valproate. Seizure 2012;21:578–82.

[46] Martin JH, Saleem M, Looke D. Therapeutic drug monitoring to adjust dosing in morbid obesity – a new use for an old methodology. Br J Clin Pharmacol 2012;73:685–90.

[47] Padwal R, Brocks D, Sharma AM. A systematic review of drug absorption following bariatric surgery and its theoretical implications. Obes Rev 2010;11:41–50.

[48] Arnaout R, Buck TP, Roulette P, Sukhatme VP. Predicting the cost and pace of pharmacogenomic advances: an evidence-based study. Clin Chem 2013;59:649–57.

[49] Caudle KE, Rettie AE, Whirl-Carrillo M, Smith LH, Mintzer S, Lee MT, et al. Clinical pharmacogenetics implementation consortium guidelines for CYP2C9 and HLA-B genotypes and phenytoin dosing. Clin Pharmacol Ther 2014;96:542–8.

[50] Leckband SG, Kelsoe JR, Dunnenberger HM, George Jr. AL, Tran E, Berger R, et al. Clinical pharmacogenetics implementation consortium guidelines for HLA-B genotype and carbamazepine dosing. Clin Pharmacol Ther 2013;94:324–8.

[51] MacPhee IA, Holt DW. A pharmacogenetic strategy for immunosuppression based on the CYP3A5 genotype. Transplantation 2008;85:163–5.

[52] Lopez-Garcia MA, Feria-Romero IA, Fernando-Serrano H, Escalante-Santiago D, Grijalva I, Orozco-Suarez S. Genetic polymorphisms associated with antiepileptic metabolism. Front Biosci (Elite Ed) 2014;6:377–86.

[53] Saygi S, Alehan F, Atac FB, Erol I, Verdi H, Erdem R. Multidrug resistance 1 (MDR1) 3435C/T genotyping in childhood drug-resistant epilepsy. Brain Dev 2014;36:137–42.

[54] Shaheen U, Prasad DK, Sharma V, Suryaprabha T, Ahuja YR, Jyothy A, et al. Significance of MDR1 gene polymorphism C3435T in predicting drug response in epilepsy. Epilepsy Res 2014;108:251–6.

[55] Bialer M, Johannessen SI, Levy RH, Perucca E, Tomson T, White HS. Progress report on new antiepileptic drugs: a summary of the Eleventh Eilat Conference (EILAT XI). Epilepsy Res 2013;103:2–30.

[56] LaRoche SM, Helmers SL. The new antiepileptic drugs: clinical applications. J Am Med Assoc 2004;291:615–20.

[57] Patsalos PN. New antiepileptic drugs. Ann Clin Biochem 1999;36:10–19.

[58] Nabbout R, Chiron C. Stiripentol: an example of antiepileptic drug development in childhood epilepsies. Eur J Paediatr Neurol 2012;16(Suppl. 1):S13–17.

[59] Giarratano M, Standley K, Benbadis SR. Clobazam for treatment of epilepsy. Expert Opin Pharmacother 2012;13:227–33.

[60] Yamamoto Y, Takahashi Y, Imai K, Miyakawa K, Nishimura S, Kasai R, et al. Influence of CYP2C19 polymorphism and concomitant antiepileptic drugs on serum clobazam and N-desmethyl clobazam concentrations in patients with epilepsy. Ther Drug Monit 2013;35:305–12.

[61] Krasowski MD, Siam MG, Iyer M, Pizon AF, Giannoutsos S, Ekins S. Chemoinformatic methods for predicting interference in drug of abuse/toxicology immunoassays. Clin Chem 2009;55:1203–13.

[62] Bolner A, Tagliaro F, Lomeo A. Optimised determination of clobazam in human plasma with extraction and high-performance liquid chromatography analysis. J Chromatogr B Biomed Sci Appl 2001;750:177–80.

[63] de Leon J, Spina E, Diaz FJ. Clobazam therapeutic drug monitoring: a comprehensive review of the literature with proposals to improve future studies. Ther Drug Monit 2013;35:30–47.

[64] Ambrosio AF, Silva AP, Malva JO, Soares-da-Silva P, Carvalho AP, Carvalho CM. Inhibition of glutamate release by BIA 2-093 and BIA 2-024, two novel derivatives of carbamazepine, due to blockade of sodium but not calcium channels. Biochem Pharmacol 2001;61: 1271–5.

[65] Maia J, Vaz-da-Silva M, Almeida L, Falcao A, Silveira P, Guimaraes S, et al. Effect of food on the pharmacokinetic profile of eslicarbazepine acetate (BIA 2-093). Drugs R D 2005;6: 201–6.

[66] Patsalos PN, Berry DJ. Pharmacotherapy of the third-generation AEDs: lacosamide, retigabine and eslicarbazepine acetate. Expert Opin Pharmacother 2012;13:699–715.

[67] Almeida L, Falcao A, Maia J, Mazur D, Gellert M, Soares-da-Silva P. Single-dose and steady-state pharmacokinetics of eslicarbazepine acetate (BIA 2-093) in healthy elderly and young subjects. J Clin Pharmacol 2005;45:1062–6.

[68] Almeida L, Nunes T, Sicard E, Rocha JF, Falcao A, Brunet JS, et al. Pharmacokinetic interaction study between eslicarbazepine acetate and lamotrigine in healthy subjects. Acta Neurol Scand 2010;121:257–64.

[69] Almeida L, Potgieter JH, Maia J, Potgieter MA, Mota F, Soares-da-Silva P. Pharmacokinetics of eslicarbazepine acetate in patients with moderate hepatic impairment. Eur J Clin Pharmacol 2008;64:267–73.

[70] Maia J, Almeida L, Falcao A, Soares E, Mota F, Potgieter MA, et al. Effect of renal impairment on the pharmacokinetics of eslicarbazepine acetate. Int J Clin Pharmacol Ther 2008;46:119–30.

[71] Alves G, Fortuna A, Sousa J, Direito R, Almeida A, Rocha M, et al. Enantioselective assay for therapeutic drug monitoring of eslicarbazepine acetate: no interference with carbamazepine and its metabolites. Ther Drug Monit 2010;32:512–16.

[72] Loureiro AI, Fernandes-Lopes C, Wright LC, Soares-da-Silva P. Development and validation of an enantioselective liquid-chromatography/tandem mass spectrometry method for the separation and quantification of eslicarbazepine acetate, eslicarbazepine, R-licarbazepine and oxcarbazepine in human plasma. J Chromatogr B Analyt Technol Biomed Life Sci 2011;879:2611–18.

[73] Porter RJ, Nohria V, Rundfeldt C. Retigabine. Neurotherapeutics 2007;4:149–54.

[74] Hermann R, Ferron GM, Erb K, Knebel N, Ruus P, Paul J, et al. Effects of age and sex on the disposition of retigabine. Clin Pharmacol Ther 2003;73:61–70.

[75] Hempel R, Schupke H, McNeilly PJ, Heinecke K, Kronbach C, Grunwald C, et al. Metabolism of retigabine (D-23129), a novel anticonvulsant. Drug Metab Dispos 1999;27: 613–22.

[76] Ferron GM, Patat A, Parks V, Rolan P, Troy SM. Lack of pharmacokinetic interaction between retigabine and phenobarbitone at steady-state in healthy subjects. Br J Clin Pharmacol 2003;56:39–45.

[77] Knebel NG, Grieb S, Leisenheimer S, Locher M. Determination of retigabine and its acetyl metabolite in biological matrices by on-line solid-phase extraction (column switching) liquid chromatography with tandem mass spectrometry. J Chromatogr B Biomed Sci Appl 2000;748:97–111.

[78] Pellock JM, Faught E, Leppik IE, Shinnar S, Zupanc ML. Felbamate: consensus of current clinical experience. Epilepsy Res 2006;71:89–101.

[79] Pellock JM. Felbamate in epilepsy therapy: evaluating the risks. Drug Safety 1999;21: 225–39.

[80] Perucca E. Clinical pharmacokinetics of new-generation antiepileptic drugs at the extremes of age. Clin Pharmacokinet 2006;45:351–64.

[81] Sachdeo R, Narang-Sachdeo SK, Shumaker RC, Perhach JL, Lyness WH, Rosenberg A. Tolerability and pharmacokinetics of monotherapy felbamate doses of 1200–6000 mg/day in subjects with epilepsy. Epilepsia 1997;38:887–92.

[82] Ward DL, Wagner ML, Perhach JL, Kramer L, Graves N, Leppik I, et al. Felbamate steady-state pharmacokinetics during co-administration of valproate. Epilepsia 1991;32:8.

[83] Sachdeo RC, Kramer LD, Rosenberg A, Sachdeo S. Felbamate monotherapy: controlled trial in patients with partial onset seizures. Ann Neurol 1992;32:386–92.

[84] Poquette MA. Isothermal gas chromatographic method for the rapid determination of felbamate concentration in human serum. Ther Drug Monit 1995;17:168–73.

[85] Annesley TM, Clayton LT. Determination of felbamate in human serum by high-performance liquid chromatography. Ther Drug Monit 1994;16:419–24.

[86] Ettinger AB, Argoff CE. Use of antiepileptic drugs for nonepileptic conditions: psychiatric disorders and chronic pain. Neurotherapeutics 2007;4:75–83.

[87] Vollmer KO, von Hodenberg A, Kölle EU. Pharmacokinetics and metabolism of gabapentin in rat, dog and man. Arzneimittelforschung 1988;36:830–9.

[88] Berry DJ, Beran RG, Plunkeft MJ, Clarke LA, Hung WT. The absorption of gabapentin following high dose escalation. Seizure 2003;12:28–36.

[89] Lindberger M, Luhr O, Johannessen SI, Larsson S, Tomson T. Serum concentrations and effects of gabapentin and vigabatrin: observations from a dose titration study. Ther Drug Monit 2003;25:457–62.

[90] Juenke JM, Wienhoff KA, Anderson BL, McMillin GA, Johnson-Davis KL. Performance characteristics of the ARK diagnostics gabapentin immunoassay. Ther Drug Monit 2011;33: 398–401.

[91] Juenke JM, Brown PI, McMillin GA, Urry FM. Procedure for the monitoring of gabapentin with 2,4,6-trinitrobenzene sulfonic acid derivatization followed by HPLC with ultraviolet detection. Clin Chem 2003;49:1198–201.

[92] Ifa DR, Falci M, Moraes ME, Bezerra FA, Moraes MO, de Nucci G. Gabapentin quantification in human plasma by high-performance liquid chromatography coupled to electrospray tandem mass spectrometry. Application to bioequivalence study. J Mass Spectrom 2001;36:188–94.

[93] Borrey DC, Godderis KO, Engelrelst VI, Bernard DR, Langlois MR. Quantitative determination of vigabatrin and gabapentin in human serum by gas chromatography-mass spectrometry. Clin Chim Acta 2005;354:147–51.

[94] Curia G, Biagini G, Perucca E, Avoli M. Lacosamide: a new approach to target voltage-gated sodium currents in epileptic disorders. CNS Drugs 2009;23:555–68.

[95] Chung S, Sperling MR, Biton V, Krauss G, Hebert D, Rudd GD, et al. Lacosamide as adjunctive therapy for partial-onset seizures: a randomized controlled trial. Epilepsia 2010.

[96] Ben-Menachem E, Biton V, Jatuzis D, Abou-Khalil B, Doty P, Rudd GD. Efficacy and safety of oral lacosamide as adjunctive therapy in adults with partial-onset seizures. Epilepsia 2007;48:1308–17.

[97] Cawello W, Nickel B, Eggert-Formella A. No pharmacokinetic interaction between lacosamide and carbamazepine in healthy volunteers. J Clin Pharmacol 2010;50:459–71.

[98] Johannessen Landmark C, Patsalos PN. Drug interactions involving the new second- and third-generation antiepileptic drugs. Expert Rev Neurother 2010;10:119–40.

[99] Greenaway C, Ratnaraj N, Sander JW, Patsalos PN. A high-performance liquid chromatography assay to monitor the new antiepileptic drug lacosamide in patients with epilepsy. Ther Drug Monit 2010.

[100] Halford JJ, Lapointe M. Clinical perspectives on lacosamide. Epilepsy Curr 2009;9:1–9.

[101] Sabers A, Tomson T. Managing antiepileptic drugs during pregnancy and lactation. Curr Opin Neurol 2009;22:157–61.

[102] Biton V. Pharmacokinetics, toxicology and safety of lamotrigine in epilepsy. Expert Opin Drug Metab Toxicol 2006;2:1009–18.

[103] Malone SA, Eadie MJ, Addison RS, Wright AW, Dickinson RG. Monitoring salivary lamotrigine concentrations. J Clin Neurosci 2006;13:902–7.

[104] Hussein Z, Posner J. Population pharmacokinetics of lamotrigine monotherapy in patients with epilepsy: retrospective analysis of routine monitoring data. Br J Clin Pharmacol 1997;43:457–64.

[105] Sabers A, Buchholt JM, Uldall P, Hansen EL. Lamotrigine plasma levels reduced by oral contraceptives. Epilepsy Res 2001;47:151–4.

[106] Fillastre JP, Taburet AM, Fialaire A, Etienne I, Bidault R, Singlas E. Pharmacokinetics of lamotrigine in patients with renal impairment: influence of haemodialysis. Drugs Exp Clin Res 1993;19:25–32.

[107] Battino D, Estienne M, Avanzini G. Clinical pharmacokinetics of antiepileptic drugs in pediatric patients. Part II. Phenytoin, carbamazepine, sulthiame, lamotrigine, vigabatrin, oxcarbazepine and felbamate. Clin Pharmacokinet 1995;29:341–69.

[108] Morris RG, Black AB, Harris AL, Batty AB, Sallustio BC. Lamotrigine and therapeutic drug monitoring: retrospective survey following the introduction of a routine service. Br J Clin Pharmacol 1998;46:547–51.

[109] Pennell PB, Peng L, Newport DJ, Ritchie JC, Koganti A, Holley DK, et al. Lamotrigine in pregnancy: clearance, therapeutic drug monitoring, and seizure frequency. Neurology 2008;70:2130–6.

[110] Krasowski MD. Therapeutic drug monitoring of the newer anti-epilepsy medications. Pharmaceuticals (Basel) 2010;3:1909–35.

[111] Juenke JM, Miller KA, Ford MA, McMillin GA, Johnson-Davis KL. A comparison of two FDA approved lamotrigine immunoassays with ultra-high performance liquid chromatography tandem mass spectrometry. Clin Chim Acta 2011;412:1879–82.

[112] Westley IS, Morris RG. Seradyn quantitative microsphere system lamotrigine immunoassay on a Hitachi 911 analyzer compared with HPLC-UV. Ther Drug Monit 2008;30:634–7.

[113] Forssblad E, Eriksson AS, Beck O. Liquid chromatographic determination of plasma lamotrigine in pediatric samples. J Pharm Biomed Anal 1996;14:755–8.

[114] Greiner-Sosanko E, Lower DR, Virji MA, Krasowski MD. Simultaneous determination of lamotrigine, zonisamide, and carbamazepine in human plasma by high-performance liquid chromatography. Biomed Chromatogr 2007;21:225–8.

[115] Lee W, Kim JH, Kim HS, Kwon OH, Lee BI, Heo K. Determination of lamotrigine in human serum by high-performance liquid chromatography-tandem mass spectrometry. Neurol Sci 2010;31:717–20.

[116] Fay MA, Sheth RD, Gidal BE. Oral absorption kinetics of levetiracetam: the effect of mixing with food or enteral nutrition formulas. Clin Ther 2005;27:594–8.

[117] Grim SA, Ryan M, Miles MV, Tang PH, Strawsburg RH, de Grauw TJ, et al. Correlation of levetiracetam concentrations between serum and plasma. Ther Drug Monit 2003;25:61–6.

[118] Patsalos PN. Clinical pharmacokinetics of levetiracetam. Clin Pharmacokinet 2004;43: 707–24.

[119] Patsalos PN, Ghattaura S, Ratnaraj N, Sander JW. In situ metabolism of levetiracetam in blood of patients with epilepsy. Epilepsia 2006;47:1818–21.

[120] Leppik IE, Rarick JO, Walczak TS, Tran TA, White JR, Gumnit RJ. Effective levetiracetam doses and serum concentrations: age effects. Epilepsia 2002;43:240.

[121] Bianchi V, Arfini C, Vidali M. Therapeutic drug monitoring of levetiracetam: comparison of a novel immunoassay with an HPLC method. Ther Drug Monit 2014;36(5):681–5.

[122] Juenke JM, McGraw JP, McMillin GA, Johnson-Davis KL. Performance characteristics and patient comparison of the ARK Diagnostics levetiracetam immunoassay with an ultra-high performance liquid chromatography with tandem mass spectrometry detection method. Clin Chim Acta 2012;413:529–31.

[123] Reineks EZ, Lawson SE, Lembright KE, Wang S. Performance characteristics of a new levetiracetam immunoassay and method comparison with a high-performance liquid chromatography method. Ther Drug Monit 2011;33:124–7.

[124] Greiner-Sosanko E, Giannoutsos S, Lower DR, Virji MA, Krasowski MD. Drug monitoring: simultaneous analysis of lamotrigine, oxcarbazepine, 10-hydroxycarbazepine, and zonisamide by HPLC-UV and a rapid GC method using a nitrogen-phosphorus detector for levetiracetam. J Chromatogr Sci 2007;45:616–22.

[125] Vermeij TA, Edelbroek PM. High-performance liquid chromatographic and megabore gas-liquid chromatographic determination of levetiracetam (ucb L059) in human serum after solid-phase extraction. J Chromatogr B Biomed Appl 1994;662:134–9.

[126] Pucci V, Bugamelli F, Mandrioli R, Ferranti A, Kenndler E, Raggi MA. High-performance liquid chromatographic determination of Levetiracetam in human plasma: comparison of different sample clean-up procedures. Biomed Chromatogr 2004;18:37–44.

[127] Guo T, Oswald LM, Mendu DR, Soldin SJ. Determination of levetiracetam in human plasma/serum/saliva by liquid chromatography-electrospray tandem mass spectrometry. Clin Chim Acta 2007;375:115–18.

[128] Radtke RA. Pharmacokinetics of levetiracetam. Epilepsia 2001;42:24–7.

[129] May TW, Korn-Merker E, Rambeck B. Clinical pharmacokinetics of oxcarbazepine. Clin Pharmacokinet 2003;42:1023–42.

[130] Larkin JG, McKee PJ, Forrest G, Beastall GH, Park BK, Lowrie JI, et al. Lack of enzyme induction with oxcarbazepine (600 mg daily) in healthy subjects. Br J Clin Pharmacol 1991;31:65–71.

[131] Cardot JM, Degen P, Flesch G, Menge P, Dieterle W. Comparison of plasma and saliva concentrations of the active monohydroxy metabolite of oxcarbazepine in patients at steady state. Biopharm Drug Dispos 1995;16:603–14.

[132] Miles MV, Tang PH, Ryan MA, Grim SA, Fakhoury TA, Strawsburg RH, et al. Feasibility and limitations of oxcarbazepine monitoring using salivary monohydroxycarbamazepine (MHD). Ther Drug Monit 2004;26:300–4.

[133] Lloyd P, Flesch G, Dieterle W. Clinical pharmacology and pharmacokinetics of oxcarbazepine. Epilepsia 1994;(Suppl. 3):10–13.

[134] Rouan MC, Lecaillon JB, Godbillon J, Menard F, Darragon T, Meyer P, et al. The effect of renal impairment on the pharmacokinetics of oxcarbazepine and its metabolites. Eur J Clin Pharmacol 1994;47:161–7.

[135] Mazzucchelli I, Onat FY, Ozkara C, Atakli D, Specchio LM, Neve AL, et al. Changes in the disposition of oxcarbazepine and its metabolites during pregnancy and the puerperium. Epilepsia 2006;47:504–9.

[136] Friis ML, Kristensen O, Boas J, Dalby M, Deth SH, Gram L, et al. Therapeutic experiences with 947 epileptic out-patients in oxcarbazepine treatment. Acta Neurol Scand 1993;87:224–7.

[137] Striano S, Striano P, Di Nocera P, Italiano D, Fasiello C, Ruosi P, et al. Relationship between serum mono-hydroxy-carbazepine concentrations and adverse effects in patients with epilepsy on high-dose oxcabazepine therapy. Epilepsy Res 2006;69:170–6.

[138] von Unruh GE, Paar WD. Gas chromatographic assay for oxcarbazepine and its main metabolites in plasma. J Chromatogr 1985;345:67–76.

[139] Juenke JM, Brown PI, Urry FM, McMillin GA. Drug monitoring and toxicology: a procedure for the monitoring of oxcarbazepine metabolite by HPLC-UV. J Chromatogr Sci 2006;44:45–8.

[140] Vermeij TA, Edelbroek PM. Robust isocratic high performance liquid chromatographic method for simultaneous determination of seven antiepileptic drugs including lamotrigine, oxcarbazepine and zonisamide in serum after solid-phase extraction. J Chromatogr B Analyt Technol Biomed Life Sci 2007;857:40–6.

[141] Breton H, Cociglio M, Bressolle F, Peyriere H, Blayac JP, Hillaire-Buys D. Liquid chromatography-electrospray mass spectrometry determination of carbamazepine, oxcarbazepine and eight of their metabolites in human plasma. J Chromatogr B Analyt Technol Biomed Life Sci 2005;828:80–90.

[142] Faulkner MA. Perampanel: a new agent for adjunctive treatment of partial seizures. Am J Health Syst Pharm 2014;71:191–8.

[143] Plosker GL. Perampanel: as adjunctive therapy in patients with partial-onset seizures. CNS Drugs 2012;26:1085–96.

[144] Selak I. Pregabalin (Pfizer). Curr Opin Invest Drugs 2001;2:828–34.

[145] Busch JA, Strand JC, Posvar EL, Bockbrader HN, Radulovic LL. Pregabalin (CI-1008) single-dose pharmacokinetics and safety/tolerance in healthy subjects after oral administration of pregabalin solution or capsule doses. Epilepsia 1998;39:58.

[146] Corrigan BW, Poole WF, Posvar EL, Strand JC, Alvey CW, Radulovic LL. Metabolic disposition of pregabalin in healthy volunteers. Clin Pharmacol Ther 2001;69:P18.

[147] Randinitis EJ, Posvar EL, Alvey CW, Sedman AJ, Cook JA, Bockbrader HN. Pharmacokinetics of pregabalin in subjects with various degrees of renal functions. J Clin Pharmacol 2003;43:277–83.

[148] Bockbrader HN, Hunt T, Strand J, Posvar EL, Sedman A. Pregabalin pharmacokinetics and safety in health volunteers: results from two phase I studies. Neurology 2000;11:412.

[149] Berry D, Millington C. Analysis of pregabalin at therapeutic concentrations in human plasma/serum by reversed-phase HPLC. Ther Drug Monit 2005;27:451–6.

[150] Nirogi R, Kandikere V, Mudigonda K, Komarneni P, Aleti R. Liquid chromatography atmospheric pressure chemical ionization tandem mass spectrometry method for the quantification of pregabalin in human plasma. J Chromatogr B Analyt Technol Biomed Life Sci 2009;877:3899–906.

[151] Wheless JW, Vazquez B. Rufinamide: a novel broad-spectrum antiepileptic drug. Epilepsy Curr 2010;10:1–6.

[152] Perucca E, Cloyd J, Critchley D, Fuseau E. Rufinamide: clinical pharmacokinetics and concentration-response relationships in patients with epilepsy. Epilepsia 2008;49: 1123–41.

[153] Luszczki JJ. Third-generation antiepileptic drugs: mechanisms of action, pharmacokinetics and interactions. Pharmacol Rep 2009;61:197–216.

[154] May TW, Boor R, Rambeck B, Jurgens U, Korn-Merker E, Brandt C. Serum concentrations of rufinamide in children and adults with epilepsy: the influence of dose, age, and comedication. Ther Drug Monit 2011;33:214–21.

[155] Contin M, Mohamed S, Candela C, Albani F, Riva R, Baruzzi A. Simultaneous HPLC-UV analysis of rufinamide, zonisamide, lamotrigine, oxcarbazepine monohydroxy derivative and felbamate in deproteinized plasma of patients with epilepsy. J Chromatogr B Analyt Technol Biomed Life Sci 2010;878:461–5.

[156] Chiron C. Stiripentol. Neurotherapeutics 2007;4:123–5.

[157] May TW, Boor R, Mayer T, Jurgens U, Rambeck B, Holert N, et al. Concentrations of stiripentol in children and adults with epilepsy: the influence of dose, age, and comedication. Ther Drug Monit 2012;34:390–7.

[158] Farwell JR, Anderson GD, Kerr BM, Tor JA, Levy RH. Stiripentol in atypical absence seizures in children: an open trial. Epilepsia 1993;34:305–11.

[159] Arends RH, Zhang K, Levy RH, Baillie TA, Shen DD. Stereoselective pharmacokinetics of stiripentol: an explanation for the development of tolerance to anticonvulsant effect. Epilepsy Res 1994;18:91–6.

[160] Balslev T, Uldall P, Buchholt J. Provocation of non-convulsive status epilepticus by tiagabine in three adolescent patients. Eur J Paediatr Neurol 2000;4:169–70.

[161] Schapel G, Chadwick D. Tiagabine and non-convulsive status epilepticus. Seizure 1996;5: 153–6.

[162] Gustavson LE, Mengel HB. Pharmacokinetics of tiagabine, a γ-aminobutyric acid-uptake inhibitor, in healthy subjects after single and multiple doses. Epilepsia 1995;36:605–11.

[163] Patsalos PN, Elyas AA, Ratnaraj N, Iley J. Concentration-dependent displacement of tiagabine by valproic acid. Epilepsia 2002;43:143.

[164] Gustavson LE, Boellner SW, Granneman GR, Qian JX, Guenther HJ, el-Shourbagy T, et al. A single-dose study to define tiagabine pharmacokinetics in pediatric patients with complex partial seizures. Neurology 1997;48:1032–7.

[165] So EL, Wolff D, Graves NM, Leppik IE, Cascino GD, Pixton GC, et al. Pharmacokinetics of tiagabine as add-on therapy in patients taking enzyme-inducing antiepilepsy drugs. Epilepsy Res 1995;22:221–6.

[166] Cato III A, Gustavson LE, Qian J, El-Shourbagy T, Kelly EA. Effect of renal impairment on the pharmacokinetics and tolerability of tiagabine. Epilepsia 1998;39:43–7.

[167] Lau AH, Gustavson LE, Sperelakis R, Lam NP, El-Shourbagy T, Qian JX, et al. Pharmacokinetics and safety of tiagabine in subjects with various degrees of hepatic function. Epilepsia 1997;38:445–51.

[168] Uthman BM, Rowan AJ, Ahmann PA, Leppik IE, Schachter SC, Sommerville KW, et al. Tiagabine for complex partial seizures: a randomized, add-on, dose-response trial. Arch Neurol 1998;55:56–62.

[169] Chollet DF, Castella E, Goumaz L, Anderegg G. Gas chromatography-mass spectrometry assay method for the therapeutic drug monitoring of the antiepileptic drug tiagabine. J Pharm Biomed Anal 1999;21:641–6.

[170] Wang X, Ratnaraj N, Patsalos PN. The pharmacokinetic inter-relationship of tiagabine in blood, cerebrospinal fluid and brain extracellular fluid (frontal cortex and hippocampus). Seizure 2004;13:574–81.

[171] Shibata M, Hashi S, Nakanishi H, Masuda S, Katsura T, Yano I. Detection of 22 antiepileptic drugs by ultra-performance liquid chromatography coupled with tandem mass spectrometry applicable to routine therapeutic drug monitoring. Biomed Chromatogr 2012;26:1519–28.

[172] Easterling DE, Zakszewski T, Moyer MD, Margul BL, Marriott TB, Nayak RK. Plasma pharmacokinetics of topiramate, a new anticonvulsants in humans. Epilepsia 1988;29:662.

[173] Miles MV, Tang PH, Glauser TA, Ryan MA, Grim SA, Strawsburg RH, et al. Topiramate concentration in saliva: an alternative to serum monitoring. Pediatr Neurol 2003;29:143–7.

[174] Britzi MP, Perucca E, Soback S, Levy RH, Fattore C, Crema F, et al. Pharmacokinetic and metabolic investigation of topiramate disposition in healthy subjects in the absence and in the presence of enzyme induction by carbamazepine. Epilepsia 2005;46:378–84.

[175] Johannessen SI, Battino D, Berry DJ, Bialer M, Kramer G, Tomson T, et al. Therapeutic drug monitoring of the newer antiepileptic drugs. Ther Drug Monit 2003;25:347–63.

[176] Berry DJ, Patsalos PN. Comparison of topiramate concentrations in plasma and serum by fluorescence polarization immunoassay. Ther Drug Monit 2000;22:460–4.

[177] Snozek CL, Rollins LA, Peterson PW, Langman LJ. Comparison of a new serum topiramate immunoassay to fluorescence polarization immunoassay. Ther Drug Monit 2010;32:107–11.

[178] Riffitts JM, Gisclon LG, Stubbs RJ, Palmer ME. A capillary gas chromatographic assay with nitrogen phosphorus detection for the quantification of topiramate in human plasma, urine and whole blood. J Pharm Biomed Anal 1999;19:363–71.

[179] Mozayani A, Carter J, Nix R. Distribution of topiramate in a medical examiner's case. J Anal Toxicol 1999;23:556–8.

[180] Bahrami G, Mirzaeei S, Kiani A. Sensitive analytical method for Topiramate in human serum by HPLC with pre-column fluorescent derivatization and its application in human pharmacokinetic studies. J Chromatogr B Analyt Technol Biomed Life Sci 2004;813:175–80.

[181] Christensen J, Hojskov CS, Poulsen JH. Liquid chromatography tandem mass spectrometry assay for topiramate analysis in plasma and cerebrospinal fluid: validation and comparison with fluorescence-polarization immunoassay. Ther Drug Monit 2002;24:658–64.

[182] Rey E, Pons G, Olive G. Vigabatrin. Clinical pharmacokinetics. Clin Pharmacokinet 1992;23:267–78.

[183] Chang SY, Lin WC. Determination of vigabatrin by capillary electrophoresis with laser-induced fluorescence detection. J Chromatogr B Analyt Technol Biomed Life Sci 2003;794:17–22.

[184] Erturk S, Aktas ES, Atmaca S. Determination of vigabatrin in human plasma and urine by high-performance liquid chromatography with fluorescence detection. J Chromatogr B Biomed Sci Appl 2001;760:207–12.

[185] Mimaki T. Clinical pharmacology and therapeutic drug monitoring of zonisamide. Ther Drug Monit 1998;20:593–7.

[186] Buchanan R, Bockbrader HN, Chang T, Sedman AJ. Single- and multiple-dose pharmacokinetics of zonisamide. Epilepsia 1996;37:172.

[187] Perucca E, Bialer M. The clinical pharmacokinetics of the newer antiepileptic drugs. Focus on topiramate, zonisamide and tiagabine. Clin Pharmacokinet 1996;31:29–46.

[188] Ijiri Y, Inoue T, Fukuda F, Suzuki K, Kobayashi T, Shibahara N, et al. Dialyzability of the antiepileptic drug zonisamide in patients undergoing hemodialysis. Epilepsia 2004;45: 924–7.

[189] Berent S, Sackellares JC, Giordani B, Wagner JG, Donofrio PD, Abou-Khalil B. Zonisamide (CI-912) and cognition: results from preliminary study. Epilepsia 1987;28:61–7.

[190] 510(k) Substantial Equivalence Determination Decision Summary. Zonisamide. Homogeneous enzyme immunoassay. Available from: <http://www.ark-tdm.com/pdfs/K091884_FDA_Summary_Zonisamide.pdf>.

[191] Juenke J, Brown PI, Urry FM, McMillin GA. Drug monitoring and toxicology: a procedure for the monitoring of levetiracetam and zonisamide by HPLC-UV. J Anal Toxicol 2006;30: 27–30.

[192] Subramanian M, Birnbaum AK, Remmel RP. High-speed simultaneous determination of nine antiepileptic drugs using liquid chromatography-mass spectrometry. Ther Drug Monit 2008;30:347–56.

[193] Kim KB, Seo KA, Kim SE, Bae SK, Kim DH, Shin JG. Simple and accurate quantitative analysis of ten antiepileptic drugs in human plasma by liquid chromatography/tandem mass spectrometry. J Pharm Biomed Anal 2011;56:771–7.

Therapeutic Drug Monitoring of Antiretrovirals

Mark A. Marzinke[1,2]
[1]Division of Clinical Pharmacology, Department of Medicine, Johns Hopkins University School of Medicine, Baltimore, MD, United States
[2]Division of Clinical Chemistry, Department of Pathology, Johns Hopkins University School of Medicine, Baltimore, MD, United States

CONTENTS

6.1 Introduction 135
6.2 Rationale for Therapeutic Drug Monitoring of Antiretroviral Agents . 137
6.3 Antiretroviral Drug Classes 137
6.3.1 Nucleoside/Nucleotide Reverse Transcriptase Inhibitors 137
6.3.2 Non-Nucleoside Reverse Transcriptase Inhibitors 140
6.3.3 Integrase Strand Transfer Inhibitors 141
6.3.4 Protease Inhibitors 143
6.3.5 Other Antiretroviral Classes 147
6.4 Analytical Methodologies 148
6.5 Clinical Trials 149
6.6 Challenges and Limitations of Therapeutic Drug Monitoring for Antiretroviral Agents . 152
6.6.1 Recommendations 152
6.7 Conclusions 153
References 154

6.1 INTRODUCTION

Recent estimates indicate that 34 million people are currently living with HIV/AIDS worldwide, with approximately 2.5 million new infections occurring annually [1]. The virus is transmitted through the exchange of virus-containing fluids, including blood, breast milk, semen, and genital secretions [2–4]. Routes of viral infection include sexual contact; injection drug use; from mother to child during pregnancy, childbirth, or breast-feeding; and exposure of infected body fluids to exposed membranes or tissue [4,5]. Antiretroviral therapy (ART) is the primary modality for the treatment and management of the disease and can substantially reduce HIV-related morbidity and mortality [6–8]. Recently, two randomized controlled trials have demonstrated the clinical benefits of early initiation of ART in HIV-infected individuals, in which treatment was initiated in asymptomatic patients with $CD4^+$ T cell counts greater than 500 cells/mm^3 [9,10]. Based on the results from these studies, ART is strongly recommended for all HIV-infected individuals, regardless of pretreatment $CD4^+$ T cell count. Furthermore, ART has shown efficacy not only in disease management but also in viral prevention as pre-exposure prophylaxis in high-risk populations [11–14].

There are currently more than 25 antiretroviral (ARV) agents approved for HIV treatment by the U.S. Food and Drug Administration (FDA) in both single- and multi-drug formulations [15]. Combinatorial ART regimens are typically required for the sustained suppression of viral replication and clinical benefit [16]. Currently, more than 100 regimens exist for the treatment of HIV [17]. ARVs elicit their therapeutic effects through the targeted inhibition of various stages of the viral infection cycle. Thus, drug classes are stratified

W. Clarke & A. Dasgupta (Eds): Clinical Challenges in Therapeutic Drug Monitoring. DOI: http://dx.doi.org/10.1016/B978-0-12-802025-8.00006-4

as CCR5 antagonists, viral fusion inhibitors, nucleoside/nucleotide reverse transcriptase inhibitors (NRTIs/NtRTIs), non-nucleoside reverse transcriptase inhibitors (NNRTIs), integrase strand transfer inhibitors (INSTIs), and protease inhibitors (PIs). Many combinatorial ART regimens incorporate drugs from more than one ARV class, and the U.S. Department of Health and Human Services (DHHS) has indicated recommended and alternative regimens for disease management [18]. A list of approved ARVs within each class is provided in Table 6.1. To achieve therapeutic efficacy, ARVs must be maintained at sufficient inhibitory concentrations to prevent continued viral replication. Typically, drug concentrations should be higher than the 50% inhibitory concentration (IC_{50}) for HIV throughout dosing, as determined via in vitro pharmacokinetic–pharmacodynamic studies [19,20].

Although ART is the primary mechanism for disease management, treatment failure may still occur. Lack of therapeutic efficacy is assessed pharmacodynamically via the manifestation of increasing HIV viral loads and decreasing $CD4^+$ T cell counts. Causes of ART failure may be multifactorial and attributable to inadequate regimen adherence, high baseline viral loads, diminished drug potency, the emergence of drug-resistant virus, pharmacokinetic variability, as well as potential ARV interactions with comedications and genetic disposition. Notably, several studies have demonstrated that ART adherence is a critical driver of therapeutic efficacy and response and, in conjunction with $CD4^+$ T cell counts, is a predictor of disease progression and mortality [21–23]. Although response heterogeneity may be due to interindividual behavioral or pharmacologic variation, therapeutic drug monitoring (TDM) is not routinely performed for these drugs. However, studies have demonstrated vast interindividual variability in various ARV classes, including both PIs and NNRTIs [24,25]. This chapter discusses ARV drug classes and provides an overview of clinical trials highlighting the potential utility of TDM for ARVs, as well as current recommendations for the practice of TDM in the management of HIV/AIDS.

Table 6.1 Antiretrovirals Approved by the U.S. Food and Drug Administration (FDA) for the Management of HIV/AIDS

ARV Class	FDA-Approved ARVs
CCR5 antagonist	Maraviroc
Fusion inhibitor	Enfuvirtide
NRTI	Abacavir, didanosine, emtricitabine, lamivudine, stavudine, tenofovir[a], zalcitabine, zidovudine
NNRTI	Delavirdine, efavirenz, etravirine, nevirapine, rilpivirine
INSTI	Dolutegravir, elvitegravir, raltegravir
PI	Atazanavir, darunavir, fosamprenavir, indinavir, lopinavir, nelfinavir, ritonavir, saquinavir, tipranavir

NRTI, nucleotide reverse transcriptase inhibitor; NNRTI, non-nucleoside reverse transcriptase inhibitor; INSTI, integrase strand transfer inhibitor; PI, protease inhibitor.
[a]Tenofovir is administered as the prodrug tenofovir disoproxil fumarate (TDF).

6.2 RATIONALE FOR THERAPEUTIC DRUG MONITORING OF ANTIRETROVIRAL AGENTS

TDM is appropriate and indicated for drugs with specific clinical characteristics; consequently, not all drugs necessitate TDM for clinical management. In addition to the availability of highly specific analytical assays for the robust quantification of drugs of interest, candidate drugs must also exhibit a dose–response or concentration–response relationship. Furthermore, TDM is indicated for drugs that have narrow windows between therapeutic and toxic concentrations, and blood drug concentrations should exhibit stable intraindividual variability but wide interindividual variability [20, 26]. If such criteria are met, TDM may be indicated for a drug or drug class.

Many of the aforementioned attributes are relevant to drugs used in the management of HIV/AIDS. ARVs demonstrate high interindividual variability in many pharmacokinetic parameters, including highest (C_{max}) and lowest, or trough (C_{min}), drug concentrations, the concentration–time curve (area under the curve), and the formation and kinetics of relevant intracellular metabolites [20,27–29]. Both in vitro and in vivo studies have demonstrated concentration–response relationships for several ARVs, including many traditional PIs and NNRTIs, and target therapeutic ranges are indicated in published literature as well as in the European AIDS Clinical Society's guidelines for disease management [30,31]. For more comprehensively evaluated ARVs, studies have demonstrated narrow therapeutic windows, and drug-associated toxicities may range from nausea and vomiting to pancreatitis, renal or hepatic toxicity, and neurologic effects, depending on the specific ARV agent [32–40]. Furthermore, the inability to maintain drug concentrations above the mean inhibitory concentration can lead to the generation and propagation of drug-resistant viral isolates, which may result in treatment failure. These characteristics suggest the potential utility of TDM for ARVs in an infected population. However, in order to understand the role of TDM in the management of HIV, the drug classes involved in treatment regimens must be reviewed.

6.3 ANTIRETROVIRAL DRUG CLASSES

There are several classes of ARV drugs. These are addressed in this section.

6.3.1 Nucleoside/Nucleotide Reverse Transcriptase Inhibitors

NRTIs and NtRTIs are analogs of thymidine, cytidine, guanosine, and adenosine, and they elicit their mechanism of action through the competitive inhibition of the viral reverse transcriptase enzyme. Biochemically, these compounds are activated intracellularly via phosphorylation of their deoxyribose moiety, and phosphorylated metabolites prevent viral reverse transcription [28]. Metabolic bioconversion may be impacted by a host of factors, including kinase activity and genetic polymorphisms, making it challenging to interpret systemic ARV concentrations alone with regard to therapeutic efficacy [41,42]. NRTI/NtRTI structures are illustrated in Fig. 6.1. These drugs are commonly part of combinatorial ART regimens, forming the backbone of many treatment regimens [18,43]. Members of this drug class include zidovudine (ZDV), which was the first FDA-approved drug for HIV treatment in the United States, abacavir (ABC), and the adenosine analog tenofovir (TFV) [44]. Nonadherence to ART regimens containing NRTIs/NtRTIs may lead to drug-specific or global NRTI/NtRTI drug resistance.

NRTIs are associated with several class-associated toxicities, thereby affecting a number of cellular processes and organ systems. NRTIs have been shown to induce mitochondrial dysfunction through the inhibition of

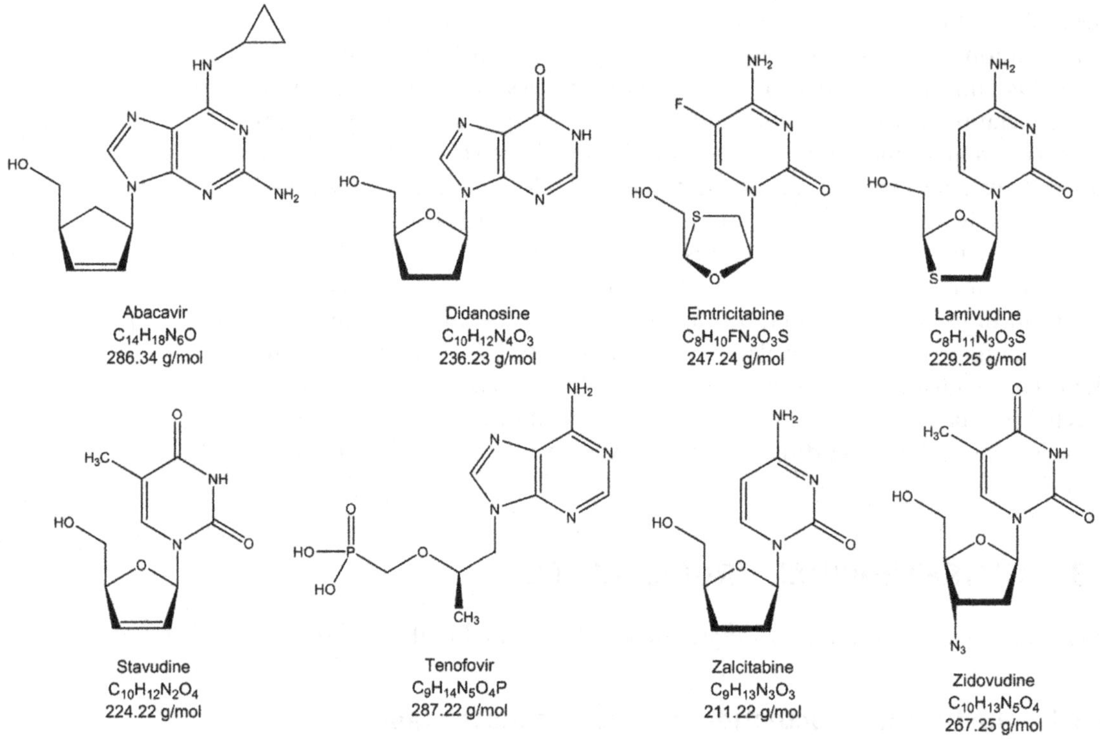

Abacavir
$C_{14}H_{18}N_6O$
286.34 g/mol

Didanosine
$C_{10}H_{12}N_4O_3$
236.23 g/mol

Emtricitabine
$C_8H_{10}FN_3O_3S$
247.24 g/mol

Lamivudine
$C_8H_{11}N_3O_3S$
229.25 g/mol

Stavudine
$C_{10}H_{12}N_2O_4$
224.22 g/mol

Tenofovir
$C_9H_{14}N_5O_4P$
287.22 g/mol

Zalcitabine
$C_9H_{13}N_3O_3$
211.22 g/mol

Zidovudine
$C_{10}H_{13}N_5O_4$
267.25 g/mol

FIGURE 6.1
Structures, chemical formulas, and molecular weights of approved NRTIs/NtRTIs.

mitochondrial DNA polymerase gamma (mtDNA γ) [45–47]. The inhibition of mtDNA γ can lead to decreased bioavailability of ATP and is associated with hepatic steatosis and lactic acidosis [46,48,49]. In vitro studies have demonstrated that the NRTIs zalcitabine (which is no longer available for treatment), didanosine, and stavudine have the highest affinity for mtDNA γ [50]. Furthermore, regimens containing stavudine and didanosine have also been associated with increased lactic acidosis during pregnancy, and they may have implications in utero and postpartum; thus, caution must be used in treating pregnant women with regimens containing these drugs [48,49,51]. Pancreatitis, cardiomyopathy, peripheral neuropathy, and lipodystrophy have also been associated with specific NRTIs due to mitochondrial enzyme inhibition [46,52]. An overview of primary drug-associated toxicities for NRTIs is provided in Table 6.2.

Although the NRTI tenofovir is part of a number of recommended first-line ART regimens as the prodrug tenofovir disoproxil fumarate, and has shown efficacy as a preexposure prophylactic agent in high-risk populations, it is also associated with renal toxicities [59]. TFV is structurally similar to acyclic nucleotide analogues associated with renal dysfunction, and in vitro and animal studies demonstrated cytotoxic effects on renal cells and overall tubular toxicity at higher drug concentrations [54,58]. Findings of proximal tubulopathy, impaired glomerular filtration, decreased bone density, and nephrotoxicity have all been reported in individuals using TFV [55–57,59,60]. However, in 2010, Cooper and colleagues conducted a meta-analysis of renal toxicities associated with TFV use in HIV-infected patients to characterize the extent of renal dysfunction. Although the analysis demonstrated a loss of renal function with tenofovir treatment, there was only a modest impact in

Table 6.2 NRTI Primary Drug-Associated Toxicities

NRTI	Primary Drug-Associated Toxicities	References
Abacavir	Hypersensitivity	[53]
Didanosine	Pancreatitis, peripheral neuropathy, strong association with mitochondrial toxicity	[40,46]
Emtricitabine	Headache, gastrointestinal anomalies, rash	[40]
Lamivudine	Headache, nausea, malaise	[40]
Stavudine	Lipoatrophy, pancreatitis, peripheral neuropathy, strong association with mitochondrial toxicity	[46,52]
Tenofovir	Fanconi's syndrome, decreased bone density, renal insufficiency, gastrointestinal anomalies	[54–60]
Zalcitabine[a]	Thrombocytopenia, anemia, pancreatitis, cardiomyopathy, peripheral neuropathy, lactic acidosis, strong association with mitochondrial toxicity	[46,61]
Zidovudine	Myelosuppression, lipodystrophy, moderate association with mitochondrial toxicity	[62]

[a]*Zalcitabine has been discontinued and is not part of any therapeutic regimen.*

terms of clinical care and management. They concluded that restriction of TFV-containing regimens in infected individuals with no history of renal disease was unnecessary [59].

6.3.2 Non-Nucleoside Reverse Transcriptase Inhibitors

NNRTIs noncompetitively bind to the reverse transcriptase enzyme, altering its conformation to prevent DNA binding [63]. Approved NNRTIs include delavirdine, efavirenz, etravirine, nevirapine, and rilpivirine (Fig. 6.2). NNRTIs demonstrate many attributes making them viable candidates for TDM. Pharmacokinetically, absorption and distribution of NNRTIs are highly dependent on drug transporter activity, particularly P-glycoprotein [64]. NNRTIs interact with a host of other antiretrovirals, including CCR5 antagonists and protease inhibitors, as well as other drug classes including antidepressants, analgesics, antifungals, anticoagulants, and antiarrhythmics [17,18]. Blood plasma concentrations of efavirenz and nevirapine have correlated with drug-associated toxicities including hepatoxicity, rash, and central nervous system-related toxicity [65–67]. An overview of NNRTI-associated toxicities and metabolic modulators is provided in Table 6.3.

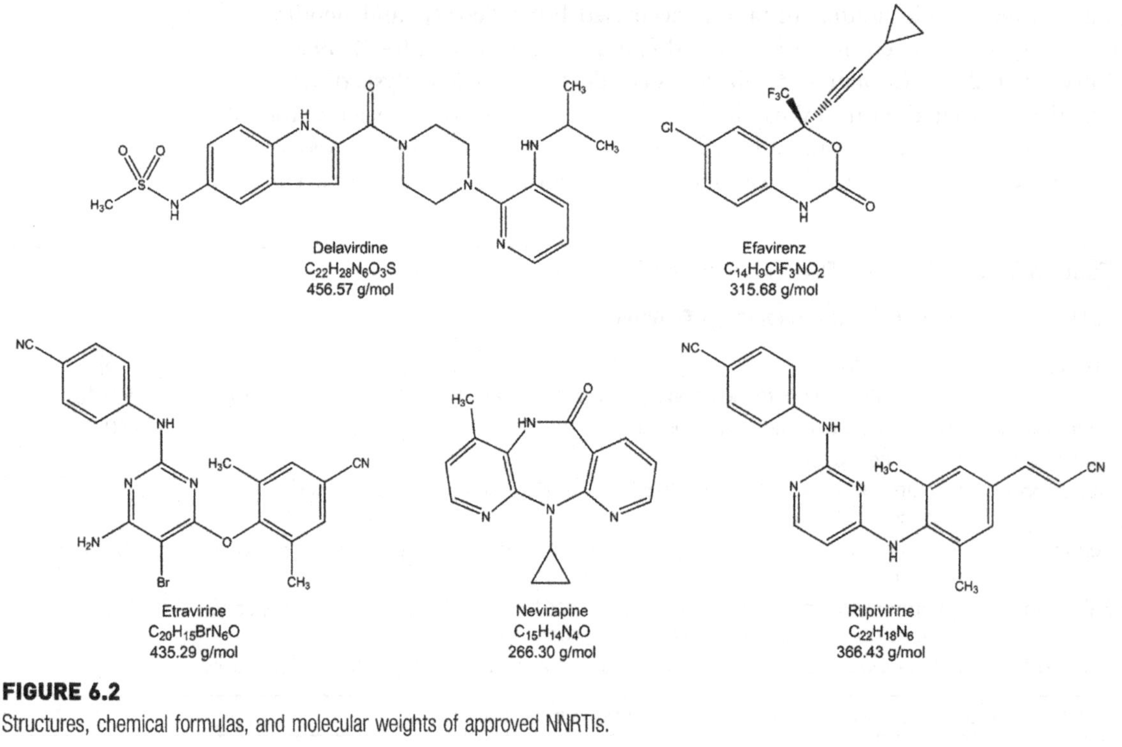

Delavirdine
$C_{22}H_{28}N_6O_3S$
456.57 g/mol

Efavirenz
$C_{14}H_9ClF_3NO_2$
315.68 g/mol

Etravirine
$C_{20}H_{15}BrN_6O$
435.29 g/mol

Nevirapine
$C_{15}H_{14}N_4O$
266.30 g/mol

Rilpivirine
$C_{22}H_{18}N_6$
366.43 g/mol

FIGURE 6.2
Structures, chemical formulas, and molecular weights of approved NNRTIs.

Table 6.3 NNRTI Primary Drug-Associated Toxicities, Mechanism of Metabolism, and Proteins Induced or Inhibited by Each NNRTI

NNRTI	Primary Drug-Associated Toxicities	Metabolism	Protein Activity Induced	Protein Activity Inhibited	References
Delavirdine	Rash, neutropenia	CYP3A4	—	CYP3A4	[68,69]
Efavirenz	Rash, central nervous system toxicities	CYP2B6, CYP3A4	CYP3A4	CYP2C9, CYP2C19	[64–67,69–71]
Etravirine	Rash, dyslipidemia, hepatitis	CYP3A4, CYP2C9, CYP2C19	CYP3A4	CYP2C9, CYP2C19, P-glycoprotein	[69,72,73]
Nevirapine	Rash, hepatitis	CYP3A4, CYP2B6	CYP3A4, CYP2B6	CYP3A4	[66,69,70]
Rilpivirine	Rash, hepatitis, depressive disorders	CYP3A4	—	—	[74]

NNRTIs induce activity of several isoenzymes of the cytochrome P450 (CYP450) family of drug-metabolizing enzymes, including CYP2B6, 2C9, 2C19, and 3A4, which may influence not only ARV drug concentrations but also blood concentrations of comedications [69,73]. Clinical studies have demonstrated exposure–response relationships for several NNRTIs, including nevirapine and, most notably, efavirenz. A study conducted by the AIDS Clinical Trials Group (ACTG; A5097) demonstrated that the presence of a genetic polymorphism in *CYP2B6* (516 G>T mutation) was found more commonly in African Americans and associated with higher efavirenz concentrations, which may correlate to higher susceptibility of central nervous system toxicities [75]. Another ACTG study (ACTG 384) demonstrated the contribution of multilocus genetic interactions involving both *CYP2B6* G516T and the transporter gene *ABCB1* (*ABCB1* 2677 G>T mutation) in higher efavirenz concentrations in a Caucasian cohort, whereas only the *CYP2B6* mutation contributed to increased drug levels in African Americans [76]. Recently, it was suggested that *CYP2B6 *6* (containing the haplotype G516T as well as the 785 A>G mutation), *CYP2B6 *18* (938 T>C), as well as gender and weight were predictors of efavirenz pharmacokinetics and treatment response in an HIV/tuberculosis (TB) co-infected African cohort [77].

6.3.3 Integrase Strand Transfer Inhibitors

As their name implies, INSTIs are viral integrase inhibitors, preventing incorporation of viral genetic information into the host cell DNA, thereby abrogating viral replication [78]. Raltegravir was the first INSTI approved by the FDA in 2007 following the reporting of results from two double-blind,

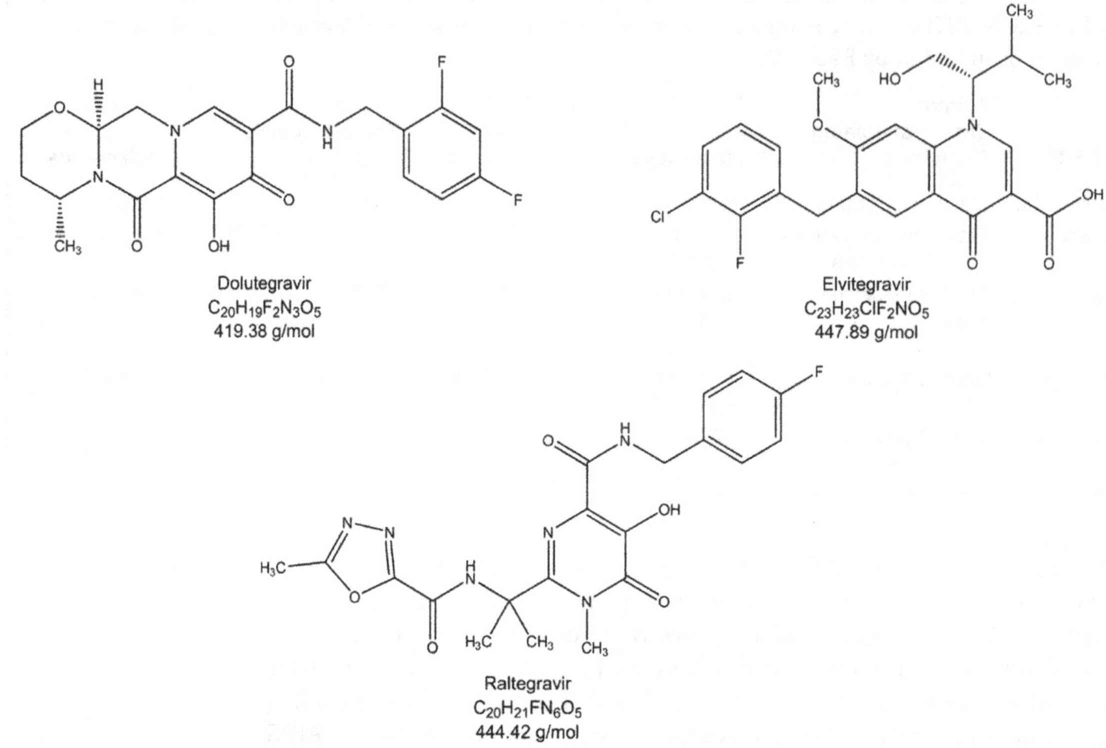

Dolutegravir
$C_{20}H_{19}F_2N_3O_5$
419.38 g/mol

Elvitegravir
$C_{23}H_{23}ClF_2NO_5$
447.89 g/mol

Raltegravir
$C_{20}H_{21}FN_6O_5$
444.42 g/mol

FIGURE 6.3
Structures, chemical formulas, and molecular weights of approved INSTIs.

placebo-controlled trials that demonstrated the therapeutic efficacy of raltegravir in ART-experienced HIV-infected individuals in conjunction with background therapies [79,80]. Recently, the use of raltegravir was expanded to children, adolescents, and ART-naive infected individuals [81,82].

In addition to raltegravir, two other INSTIs (dolutegravir and elvitegravir) have been approved by the FDA, both of which are components of multidrug ART regimens [18]. Chemical structures of these drugs are shown in Fig. 6.3. Notably, recommended regimens were updated in 2015; consequently, INSTIs are currently part of first-line therapies for treatment of ART-naive infected individuals [18]. Recommended regimens are described in Table 6.4. The transition from NNRTI- and PI-centered regimens to INSTI-centered therapies was due to the virologic efficacy, superior safety and tolerability profiles, and low number of drug–drug interactions, including no known CYP3A4–drug interactions, of INSTIs. Furthermore, INSTI-containing therapies showed improved efficacy with regard to viral suppression compared to ritonavir-boosted darunavir therapy [83,84].

Table 6.4 U.S. DHHS Recommended and Alternative Regimens for Treatment of an HIV-Infected ART-Naive Patient

Recommended ART Regimens for Treatment Naive Patients			
Regimen	**Drug Class Components**	**Indication**	**U.S. DHHS Recommendation**
Dolutegravir/abacavir/lamivudine	INSTI, NRTI	For patients who are HLA-B*5701 negative	Strong
Dolutegravir/tenofovir disoproxil fumarate/emtricitabine	INSTI, NRTI		Strong
Elvitegravir (cobicistat-boosted)/tenofovir disoproxil fumarate/emtricitabine	INSTI, NRTI	For patients with CrCl $\geq$70 mL/min	Strong
Raltegravir/tenofovir disoproxil fumarate/emtricitabine	INSTI, NRTI		Strong
Darunavir (ritonavir-boosted)/tenofovir disoproxil fumarate/emtricitabine	PI, NRTI		Strong
Efavirenz/tenofovir disoproxil fumarate/emtricitabine	NRTI, NNRTI		Moderate
Rilpivirine/tenofovir disoproxil fumarate/emtricitabine	NRTI, NNRTI	For patients with HIV RNA $<$100,000 copies/mL and CD4 cell counts $>$200 cells/mm^3	Moderate
Atazanavir (cobicistat-boosted)/tenofovir disoproxil fumarate/emtricitabine	PI, NRTI	For patients with CrCl $\geq$70 mL/min	Moderate
Atazanavir (ritonavir-boosted)/tenofovir disoproxil fumarate/emtricitabine	PI, NRTI		Moderate
Darunavir (ritonavir- or cobicistat-boosted)/abacavir/lamivudine	PI, NRTI	For patients who are HLA-B*5701 negative	Moderate
Darunavir (cobicistat-boosted)/tenofovir disoproxil fumarate/emtricitabine	PI, NRTI	For patients with CrCl $\geq$70 mL/min	Moderate

DHHS, Department of Health and Human Services.
Source: Panel on Antiretroviral Guidelines for Adults and Adolescents. Guidelines for the use of antiretroviral agents in HIV-1-infected adults and adolescents. Department of Health and Human Services. Available at http://www.aidsinfo.nih.gov/ContentFiles/AdultandAdolescentGL.pdf. Updated April 8th, 2015; accessed August 6th, 2015.

Although part of a number of recommended regimens, limited pharmacokinetic (PK) data are available for these drugs. Therefore, therapeutic thresholds have not been established to direct or target treatments. Furthermore, approved INSTIs are not associated with many adverse effects, although patients may experience diarrhea, nausea, insomnia, and headaches [85].

6.3.4 Protease Inhibitors

PIs block viral replication and propagation through inhibition of viral HIV-1 protease activity and the production of viral particles. Following approval of the PI saquinavir by the FDA in the mid-1990s, PIs have been a major component within a number of combinatorial therapies for HIV management.

Currently, there are nine PIs approved for use by the FDA, primarily as part of combinatorial therapies [18]. Chemical structures for approved PIs are shown in Fig. 6.4. Similar to NNRTIs, PIs exhibit many characteristics that could benefit from a TDM approach. As highly lipophilic molecules, PIs are transported in systemic circulation primarily bound to α_1 -acid glycoprotein (AAG) and elicit their antiviral activity in the free, or unbound, form. Thus, variability in AAG concentrations can lead to wide interindividual variability in PI activity and, consequently, therapeutic efficacy [86].

Metabolically, PIs act as both substrates and modulators (inducers or inhibitors) of several CYP450 isoenzymes, including CYP1A2, 2C9, 2C19, 2D6, and 3A4, as well as drug transporters [87]. All PIs inhibit CYP3A4 activity to some degree; this has been exploited therapeutically through the incorporation of lower doses of potent CYP3A4 PI inhibitors such as ritonavir and atazanavir, as well as the non-PI cobicistat, into ART regimens. The inclusion of nontherapeutic doses of the aforementioned drugs leads to inhibition of the metabolic conversion of other CYP3A4-mediated ARVs, leading to higher bioavailability as well as increased and sustained drug concentrations. Ritonavir-boosted regimens have been effective in increasing concentrations of a number of ARVs, including fusion inhibitors, NNRTIs, INSTIs, and other PIs [88–91]. The presence of boosting PIs can also increase concentrations of concomitant medications also metabolized via the CYP3A4 enzyme, including rifamycins, amiodarone, and immunosuppressants such as everolimus and tacrolimus, which may increase pharmacologic variability of these medications as well as the potential for adverse drug reactions [87].

In addition to the potential drug–drug interactions associated with PIs, drugs within this class also exhibit concentration-dependent toxicities. Although common side effects include diarrhea, nausea, vomiting, abdominal pain, fever, headache, and fatigue, several PIs have also demonstrated an impact on metabolic processes, including hyperlipidemia, lipodystrophy, and glucose intolerance. ART regimens containing lopinavir, ritonavir, tipranavir, and indinavir are associated with increased cholesterol and triglyceride concentrations, the latter of which may play a role in the development of pancreatitis [87,92–96]. Furthermore, several PIs have also been implicated in modulating metabolic pathways that regulate glycemic control, including glucose transporter type-4 inhibition, leading to increased glucose concentrations [87,97–99]. An overview of PI-associated toxicities and metabolic modulators is provided in Table 6.5.

In ART-experienced patients, PI resistance due to accumulation of drug or class-specific mutations may necessitate the need for dose adjustments or regimen selection. However, drug resistance may be overcome at higher drug concentrations. A more nuanced way of determining a potential dose

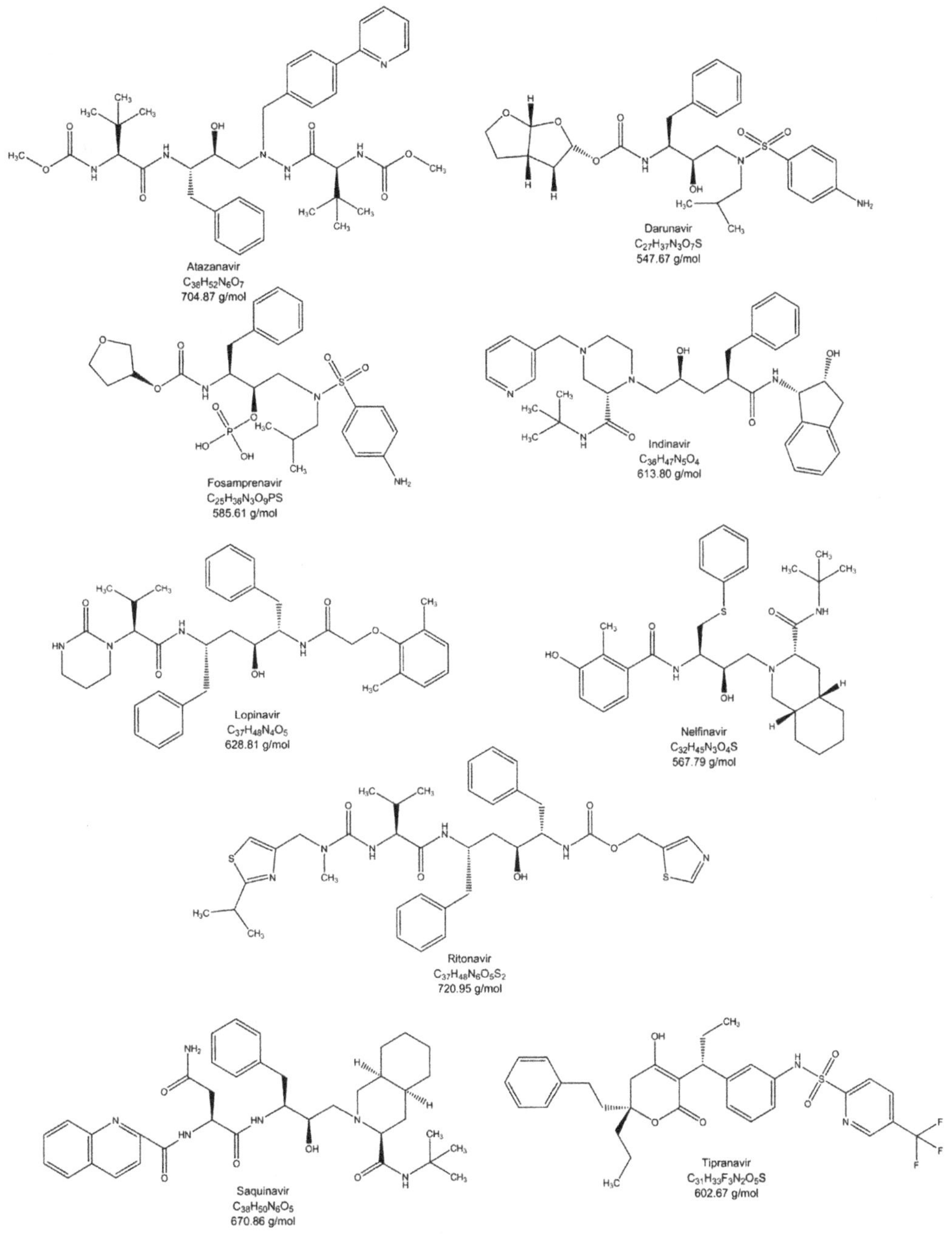

Atazanavir
$C_{38}H_{52}N_6O_7$
704.87 g/mol

Darunavir
$C_{27}H_{37}N_3O_7S$
547.67 g/mol

Fosamprenavir
$C_{25}H_{36}N_3O_9PS$
585.61 g/mol

Indinavir
$C_{36}H_{47}N_5O_4$
613.80 g/mol

Lopinavir
$C_{37}H_{48}N_4O_5$
628.81 g/mol

Nelfinavir
$C_{32}H_{45}N_3O_4S$
567.79 g/mol

Ritonavir
$C_{37}H_{48}N_6O_5S_2$
720.95 g/mol

Saquinavir
$C_{38}H_{50}N_6O_5$
670.86 g/mol

Tipranavir
$C_{31}H_{33}F_3N_2O_5S$
602.67 g/mol

FIGURE 6.4

Structures, chemical formulas, and molecular weights of approved PIs.

Table 6.5 PI Primary Drug-Associated Toxicities, Mechanism of Metabolism, and Proteins Induced or Inhibited by Each PI

PI	Primary Drug-Associated Toxicities	Metabolism	Protein Activity Induced	Protein Activity Inhibited	References
Atazanavir	Abdominal pain, nausea, diarrhea, increased total bilirubin	CYP3A4	–	CYP3A4	[92,94,96,100]
Darunavir	Abdominal pain, nausea, diarrhea, rash	CYP3A4	–	CYP3A4, P-glycoprotein	[40,101,102]
Fosamprenavir	Abdominal pain, nausea, diarrhea, rash, risk for myocardial infarction	CYP3A4	P-glycoprotein	CYP3A4	[103]
Indinavir	Abdominal pain, nausea, diarrhea, increased total bilirubin, dyslipidemia, nephrolithiasis	CYP3A4	–	CYP3A4	[104–106]
Lopinavir	Abdominal pain, nausea, diarrhea, risk for myocardial infarction, dyslipidemia	CYP3A4	UGT, CYP1A2	CYP3A4, CYP2D6	[87,92,93,95,107]
Nelfinavir	Abdominal pain, nausea, diarrhea	CYP2C19, CYP2D6	UGT, CYP1A2, CYP2C9, CYP3A4, P-glycoprotein	CYP3A4	[87,108]
Ritonavir	Abdominal pain, nausea, diarrhea, dyslipidemia, risk for mycocardial infarction	CYP3A4, CYP2D6	CYP1A2	CYP3A4, CYP2D6	[87,92,93,95,107]
Saquinavir	Abdominal pain, nausea, diarrhea, dyslipidemia	CYP3A4	–	CYP3A4	[40,87]
Tipranavir	Abdominal pain, nausea, diarrhea, rash, dyslipidemia, risk for intracranial hemorrhage	CYP3A4	CYP2C19, P-glycoprotein	CYP3A4, CYP2D6	[109,110]

adjustment is via the calculation and application of the inhibitory quotient (IQ), which is a ratio of the trough ARV concentration divided by the 50% inhibitory concentration (C_{min}/IC_{50}). This quotient could be further derived to incorporate the viral genotype (virtual inhibitory quotient, which is based on a genotype-derived virtual phenotype and IC_{50}) as well as normalization to reference virtual inhibitory quotients (normalized virtual inhibitory quotient (NIQ)) and the genotypic inhibitory quotient (C_{min}/number of PI-related mutations) [86,111,112]. In the Can Resistance Enhance Selection of Therapy? (CREST) study, the use of both a baseline viral load and the NIQ was shown to be predictive of virologic response over a 48-week period in a treatment-experienced population [113]. However, with the advent of ARV regimens focused around other drug classes, including NRTIs and INSTIs, the use and application of the IQ or its derivatives is not commonly employed if TDM of ARVs is indicated.

6.3.5 Other Antiretroviral Classes

Maraviroc and enfuvirtide act in preventing viral entry into a host cell. Maraviroc is a CCR5 antagonist that abrogates further interaction with the gp120–CD4 receptor complex [114], and it is approved in combination with other ARVs as a therapy for both treatment-experienced and -naive patients [115]. The drug, which is well tolerated, is metabolized by CYP3A4 and the polymorphic CYP3A5 enzymes; genetic polymorphisms may lead to increased interindividual variability in drug concentrations [116]. Enfuvirtide is a biometric peptide that binds to the viral protein gp41 and inhibits viral membrane fusion to the host cell [117,118]. It is approved in conjunction with other ARVs for disease management in ART-experienced patients. Limited PK data are available for these drugs, and therapeutic thresholds have not been suggested for TDM. Drug structures are shown in Fig. 6.5.

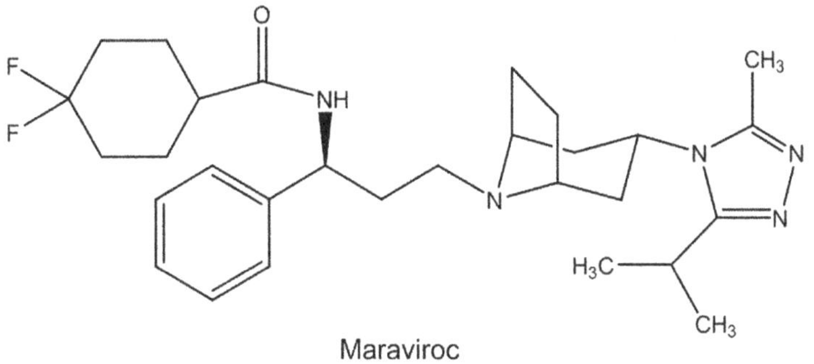

Maraviroc
$C_{29}H_{41}F_2N_5O$
513.68 g/mol

Ac-Tyr-Thr-Ser-Leu-Ile-His-Ser-Leu-Ile-Glu-Glu-Ser-Gln-Asn-Gln-Gln-Glu-Lys
Asn
NH_2-Phe-Trp-Asn-Trp-Leu-Ser-Ala-Trp-Lys-Asp-Leu-Glu-Leu-Leu-Glu-Gln-Glu

Enfuvirtide
$C_{204}H_{301}N_{51}O_{64}$
4492.1 g/mol

FIGURE 6.5
Structure (maraviroc) and amino acid sequence (enfuvirtide), chemical formulas, and molecular weights of maraviroc and enfuvirtide.

6.4 ANALYTICAL METHODOLOGIES

Currently, there is no FDA-approved immunoassay for TDM of antiretroviral drugs. Although targeted immunoassays are available for select PIs and NNRTIs, such testing is not approved for clinical diagnostic use, and such assays are available for research use only. The primary methods used to support pharmacokinetic studies for antiretrovirals include liquid chromatographic assays with ultraviolet, photodiode array, or mass spectrometric detection [119]. Today, most assays employ liquid chromatographic–tandem mass spectrometric (LC-MS/MS) strategies for the selective or multiplexed detection of various ARV drugs using a variety of sample preparation techniques, including protein precipitation and solid phase extraction approaches [119–123]. More recently, a multiplexed liquid chromatographic–high-resolution mass spectrometric (HRMS) assay was developed and validated for the qualitative detection of 15 ARVs, including PIs, NRTIs, and NNRTIs, in serum [124]. This HRMS approach has been applied to assess both qualitative identification and off-study drug use in a number of clinical trials [125–127].

To accommodate studies performed in developing areas and to assist in sample collection, storage, and transport, dried blood spots have been assessed and shown promise as an alternative collection strategy [128]. Assays in seminal plasma, cervicovaginal fluid, rectal fluid, breast milk, saliva, and hair have all been developed and utilized [129–135]. These alternative sampling and testing strategies can provide more information on localized drug PK (seminal plasma, breast milk, and luminal fluids) and evaluate noninvasive techniques for adherence assessment (hair, oral fluid, and dried blood spots). However, of note, understanding the pharmacokinetic relationships among biological compartments is critical in the interpretation of PK results and ARV concentrations in any of the aforementioned specimen sources. Analytical detection of an ARV agent does not necessarily denote adherence to an ART regimen.

With the development and validation of bioanalytical tools to support drug quantification, availability of reagents for the generation of calibrators and quality controls are also needed. Whereas some reagents and internal standards are available through commercial vendors, another resource for compounds is the National Institutes of Health (NIH) AIDS Reagent Program, which is directed by Division of AIDS (DAIDS), National Institute of Allergy and Infectious Diseases (NIAID), NIH, and operated by Fisher Bioservices. In order to participate in the program, registration is required, and participation must be agreed upon by the registrant's institution. In addition, with the number of bioanalytical methods found in the literature and lack of available commercial assays, issues surrounding harmonization and standardization also exist. However, one way that clinical pharmacology laboratories performing analyses for clinical trial may assess the performance of their results is through participation in the Clinical Pharmacology Quality Assurance program, which is overseen by the UB Pharmacotherapy Research Center at the University of Buffalo and is sponsored by DAIDS, NIAID, NIH. Notably, the program offers semiannual target-based proficiency challenges for many ARVs [136–138]. Support of this quality assurance initiative speaks to the importance of oversight of bioanalytical methods used in clinical trial PK analyses, and it can provide laboratories with a way of assessing assay performance against both peer and target values.

6.5 CLINICAL TRIALS

Clinical trials have been the primary resource for evaluating the role of TDM in the management of HIV infection. During approximately the past 15 years, a number of prospective and retrospective studies have assessed the impact of correlating ARV drug concentrations with viral suppression. Although all of the studies have focused on the relationship between TDM-guided therapies and therapeutic outcomes, they all have limitations.

In order to mitigate the aforementioned NNRTI- and PI-related drug toxicities or improve subtherapeutic concentrations, TDM was applied to a small cohort of patients ($N = 137$) during a 3-year period in an outpatient clinic setting [139]. Approximately one-third (36%) of patients had drug concentrations associated with toxicity, whereas approximately another one-third (37%) were associated with nonprotective drug concentrations. Of note, patients who were at subtherapeutic concentrations also showed phenotypic failure, as indicated by nonsuppression of HIV RNA levels. As a result, modifications in dosing regimens led to improvement of virologic response as well as drug-related adverse reactions in 80% of cases.

In the AIDS Therapy Evaluation in The Netherlands (ATHENA) prospective randomized controlled trial, ART-naive patients were randomized to a TDM-directed or standard of care arm and treated with ART regimens containing the PIs indinavir or nelfinavir [140]. Outcomes of the study included indinavir- or nelfinavir-related toxicities and renal dysregulation; therapeutically, virologic suppression was assessed. At 48 weeks of follow-up, those randomized to the TDM-directed arm showed a higher proportion of virologic suppression (HIV RNA levels <500 copies/mL; 78.2% vs. 55.1% in the standard of care arm), as well as decreased frequencies of indinavir-related nephrolithiasis. Although the study demonstrated the utility of TDM for individuals being treated using ART regimens containing indinavir or nelfinavir, it must be noted that the trial was conducted in an ART-naive population and extension of these findings to ART-experienced patients or those with viral resistance may not yield the same benefits.

In 2002, Fletcher and colleagues published results from a prospective, randomized, open-label clinical trial to assess the clinical feasibility of concentration-controlled zidovudine, lamivudine, and indinavir therapies for HIV treatment [141]. The primary endpoint for the study was HIV RNA levels <50 copies/mL after 52 weeks on study. Notably, similar to the ATHENA study, the study cohort included ART-naive individuals. Intensive PK sampling was conducted post-drug administration at weeks 2 and 28 for indinavir, lamivudine, and zidovudine quantification. PK parameters were determined, and dose adjustments were made based on these findings. Using this approach, dose adjustments were made between 31% and 81% of individuals randomized to the concentration-controlled arm, and 88–100% of study arm participants were within therapeutic target ranges at the end of the study. Although a small sampling size (N = 33), at 52 weeks of follow–up, 94% of individuals randomized to the concentration-controlled study had HIV RNA levels <50 copies/mL compared to 53% in the standard of care arm. There were no differences in adverse events between the study arms.

Although the ATHENA and concentration-controlled studies demonstrated both feasibility and a rationale for TDM-directed therapies in ART-naive populations for specific antiretrovirals, four other clinical studies published at approximately the same time did not support the use of TDM to direct therapies for all patients requiring or already prescribing to an ART regimen.

The PharmAdapt study was a randomized prospective clinical trial conducted in an ART-experienced, HIV-infected population in which previous therapies failed [142]. Study participants were randomized to undergo treatment modifications based on either genotype resistance alone (control arm) or in conjunction with PI trough concentrations (TDM arm). During the course of this 12-week study, plasma trough PI concentrations were determined at week 4, with dose adjustments occurring at week 8 in the TDM arm. Approximately two-thirds of regimens in both arms included ritonavir-boosting regimens. Although physician- and protocol-driven PI modifications were made in 7% and 23.5% of the control and TDM arm, respectively, at week 12 there was no statistically significant difference in virologic outcomes (viral load suppression) between the two arms. Follow-up studies at week 32 corroborated the initial week 12 findings. Unlike the studies performed in ART-naive patients, these data demonstrate that TDM-directed dosing modifications in ART-experienced patients did not improve phenotypic outcomes.

The findings of the PharmAdapt study were recapitulated in a number of other trials, including the GENOPHAR and Resistance and Dosage Adapted Regimens (RADAR) trials, as well as a randomized controlled trial including both ART-naive and -experienced participants [143]. With a similar study design and patient population as those in the PharmAdapt study, the

GENOPHAR trial demonstrated no added benefit of pharmacologically driven dose adjustments in reducing viral loads during the course of the study [144]. Equivalent percentages of participants from both study arms demonstrated viral loads less than 200 copies/mL (52% in the genotype control arm vs. 60% in the genotype pharmacological-driven arm) at 24 weeks; these findings were not statistically significant. Similar findings were found in a number of other trials as well [145,146].

The Pharmacologic Optimizations of PIs and NNRTIs (POPIN) clinical study, which was conducted in the United Kingdom, corroborated aforementioned characteristics associated with ARVs (ie, high interindividual variability and lower intraindividual variability). However, the study did not show a correlation between subtherapeutic drug concentrations and virologic failure (HIV RNA loads <50 copies/mL) at 24 weeks [147]. Furthermore, there were no statistical differences in the frequency of drug-related toxicities between the control and the study arm. However, fluctuating intraindividual drug concentrations over time reflected intermittent or poor adherence to study regimens.

Although a number of randomized controlled trials did not show improved phenotypic outcomes using a concentration-controlled approach to ARV dosing, there were limitations associated with the studies. Shortcomings include nonadherence throughout the study (as indicated in the POPIN trial), the short duration of the study post-pharmacological adjustments (GENOPHAR and PharmAdapt trials), the inclusion of patients with treatment-resistant viral strains, and different primary endpoints with regard to HIV RNA levels [29]. In addition, in many of the studies, the potential lack of statistical power may have contributed to the conclusion that there was no statistical benefit to pharmacologically driven ART regimen management. However, the conclusions of these studies have provided evidence for a recommendation to not perform TDM of ARVs in routine disease management [148].

Despite doubts regarding the clinical utility of TDM for ARVs, a clinical study was performed that examined the impact of TDM for ARVs in improving hospital visits and associated health care costs, in addition to clinical outcomes [23]. The 2-year retrospective analysis included HIV-infected patients stratified as to having received drug concentrations at some point during their treatment. Consistent with prospective randomized controlled trials, there was no statistical difference between the TDM^+ and TDM^- groups with regard to the frequency of patients with HIV RNA levels less than 50 copies/mL (85.3% vs 81.4%); however, there was a marked decrease in HIV-related hospitalization stays in the TDM^+ group (mean duration, 7.21 vs 29.5 days). Although hospitalization-associated costs were lower in the TDM^+ arm, overall costs for management of this group were higher primarily due to incremental use of ambulatory specialists and adherence assessment [23]. This highlights other parameters to be evaluated when assessing the applicability or feasibility of concentration-controlled therapies for ARVs.

6.6 CHALLENGES AND LIMITATIONS OF THERAPEUTIC DRUG MONITORING FOR ANTIRETROVIRAL AGENTS

There are a number of challenges in the implementation and interpretation of TDM for ARVs. First, therapeutic concentration ranges have not been established for all ARVs [20]. Furthermore, combinatorial therapies include drugs from various ARV classes that differ in their pharmacokinetic profiles. Thus, monitoring a single drug within the regimen may not comprehensively inform therapeutic efficacy, toxicity, or adherence, and numerous studies have demonstrated drug–drug interactions related to ARVs [119]. In addition, ART regimens are administered as fixed-dose formulations irrespective of pharmacologic or biological factors that influence ARV pharmacokinetics. Whereas TDM may be useful in dose adjustment for other drug classes, in the case of ARVs, the potential utility may lie in the selection of another combinatorial ART regimen, understanding causality of treatment failure, or as an objective assessment of adherence.

6.6.1 Recommendations

The DHHS Panel on Antiretroviral Guidelines for Adults and Adolescents is the most comprehensive resource for current recommendations with regard to baseline assessment of an infected individual, regimen selection in both ART-naive and -experienced patients, management of adverse effects, treatment failure and potential drug–drug interactions, as well as maintenance and monitoring of treatment. Notably, the guidelines do not recommend TDM for ARVs in the routine management of HIV-infected patients [18]. Updated in April 2015, the panel's working group took this stance based on an overall lack of prospective studies demonstrating that generalized and routine use of TDM for ARVs improves clinical outcomes. In addition, there were concerns regarding uncertain therapeutic thresholds for ARVs, intra- and interpatient variability in drug concentrations achieved, and lack of commercial assays available for testing [20,31,148]. However, although commercial assays are not available for the quantification of ARVs, LC with ultraviolet or LC-MS/MS methods (as previously described) are available for drug quantification, similar to many other therapeutic agents for which TDM is indicated. In addition, although therapeutic ranges may not be established for all ARVs, thresholds have been determined for many of the PIs and NNRTIs used in ART regimens; a review by Pretorius and colleagues in 2011 includes population-based average trough concentrations for both traditional and newer ARVs [31].

Although TDM is not indicated in routine patient management, the DHHS does suggest the potential utility of ARV drug measurements to assist in patient management for the following conditions or scenarios: suspicion of significant drug–drug or drug–food interactions; disease states that may alter gastrointestinal, hepatic, or renal function that could impact drug pharmacokinetics; pregnant women who are not virally suppressed and at risk of further virologic failure; treatment-experienced patients who may have developed drug-resistant virus; implementation of alternative HIV ART regimens of unknown therapeutic efficacy; drug-associated toxicities; and treatment failure in adherent patients [18]. Each of these clinical conditions or scenarios may result in significant alterations in drug absorption, distribution, metabolism, or clearance, and a provider should take into account such comorbidities in the management of an ART-naive or -experienced HIV-infected patient.

Considerations should also be taken into account in the management of those co-infected with hepatitis B or hepatitis C, as well as tuberculosis, because drug–drug interactions may exist, decreasing the efficacy of ARVs or concomitant medications [18]. For example, simeprevir is a hepatitis C virus protease inhibitor that inhibits CYP3A4, P-glycoprotein activity, and the drug transporter OATP1B1/3 [149]. Coadministration of this drug with PIs, as well as elvitegravir/cobicistat/tenofovir disoproxil fumarate/emtricitabine, etravirine, and efavirenz, is not recommended [18]. Rifamycins are used in tuberculosis management, as are inducers of CYP3A4 activity; therefore, appropriate ART regimens must be taken into account. The DHHS has provided recommendations for treatment options in individuals co-infected with HIV and TB [18].

6.7 CONCLUSIONS

Although TDM is not indicated for the routine management of all patients, there are clearly circumstances in which drug concentrations could assist in the management of a patient. One area in which drug levels may become important in the longitudinal management of an ART-naive or -experienced patient is to assess nonadherence. Noncompliance is a major cause of treatment failure, and drug levels may help to differentiate nonadherence from drug-resistant virus in non-virologically suppressed individuals. Nonadherence may be due to behavioral, psychosocial, and generational barriers, and treatment failure can lead to the rise and potential transmission of drug-resistant virus [150–152]. Barriers also exist in the use of ART as prevention. Determination of drug concentrations will also be needed in less well-described populations, including obese populations and individuals who undergo roux-en-Y gastric bypass surgery, or other inventions that may potentially result in drug absorption and distribution. Furthermore, as focus also expands to prevention strategies, drug measurements may become more important not only in managing uninfected individuals but also in protecting individuals who are at risk of acquiring the virus. Regardless, as HIV infection has shifted from being an acute disease with poor prognosis to a chronic disease with an improved quality of life, having robust bioanalytical tools to properly manage these patients is imperative.

References

[1] Global report: UNAIDS report on the global AIDS epidemic 2012, <http://www.unaids.org/en/resources/publications/2012>; 2012 [accessed 08.06.15].

[2] Zagury D, Bernard J, Leibowitch J, Safai B, Groopman JE, Feldman M, et al. HTLV-III in cells cultured from semen of two patients with AIDS. Science 1984;226(4673):449–51.

[3] Vogt MW, Witt DJ, Craven DE, Byington R, Crawford DF, Schooley RT, et al. Isolation of HTLV-III/LAV from cervical secretions of women at risk for AIDS. Lancet 1986;1 (8480):525–7.

[4] Friedland GH, Klein RS. Transmission of the human immunodeficiency virus. N Engl J Med 1987;317:1125–35.

[5] Chermann JC. Sexual and mother-to-child transmission of the human immunodeficiency virus type 1: a review. Am J Reprod Immunol 1998;40(3):183–6.

[6] Mocroft A, Vella S, Benfield TL, Chiesi A, Miller V, Gargalianos P, et al. Changing patterns of mortality across Europe in patients infected with HIV-1. EuroSIDA Study Group. Lancet 1998;45(10):1093–9.

[7] Palella FJ, Delaney KM, Moorman AC, Loveless MO, Fuhrer J, Satten GA, et al. Declining morbidity and mortality among patients with advanced human immunodeficiency virus infection. HIV Outpatient Study Investigators. N Engl J Med 1998;338(13):853–60.

[8] Vittinghoff E, Scheer S, O'Malley P, Colfax G, Holmberg SD, Buchbinder SP. Combination antiretroviral therapy and recent declines in AIDS incidence and mortality. J Infect Dis 1999;179(3):717–20.

[9] INSIGHT START Study Group, Lundgren JD, Babiker AG, Gordin F, Emery S, Grund B, et al. Initiation of antiretroviral therapy in early asymptomatic HIV infection. N Engl J Med 2015;373(9):795–807.

[10] TEMPRANO ANRS 12136 Study Group, Danel C, Moh R, Gabillard D, Badje A, Le Carrou J, et al. A trial of early antiretrovirals and isoniazid preventative therapy in Africa. N Engl J Med 2015;373(9):808–22.

[11] Grant RM, Lama JR, Anderson PL, McMahan V, Liu AY, Vargas L, et al. Preexposure chemoprophylaxis for HIV prevention in men who have sex with men. N Engl J Med 2010;363 (27):2587–99.

[12] Cohen MS, Chen YQ, McCauley M, Gamble T, Hosseinipour MC, Kumarasamy N, et al. Prevention of HIV-1 infection with early antiretroviral therapy. N Engl J Med 2011;365 (6):493–505.

[13] Centers for Disease Control and Prevention (CDC). Interim guidance: preexposure prophylaxis for the prevention of HIV infection in men who have sex with men. MMWR Morb Mortal Wkly Rep 2011;60(3):65–8.

[14] Baeten JM, Donnell D, Ndase P, Mugo NR, Campbell JD, Wangisi J, et al. Antiretroviral prophylaxis for HIV prevention in heterosexual men and women. N Engl J Med 2012;367(5):399–410.

[15] Food, Drug Administration, FDA-approved ARV drugs, <http://www.fda.gov/ForPatients/Illness/HIVAIDS/Treatment/ucm118915.htm>; 2014 [accessed 06.08.15].

[16] Volberding PA, Deeks SG. Antiretroviral therapy and management of HIV infections. Lancet 2010;376:49–62.

[17] Capetti A, Astuti N, Cossu MV, Rizzardini G, Carenzi L. The role of therapeutic drug monitoring and pharmacogenetic testing in the management of HIV infection: a review. J Aids Clin Res 2015;6:5.

[18] Panel on Antiretroviral Guidelines for Adults and Adolescents. Guidelines for the use of antiretroviral agents in HIV-1-infected adults and adolescents. Department of Health and Human Services. Updated April 8th, 2015. Available at: <http://www.aidsinfo.nih.gov/ContentFiles/AdultandAdolescentGL.pdf>; [accessed 06.08.15].

[19] Acosta EP, Gerber JG, Adult Pharmacology Committee of the AIDS Clinical Trials Group. Position paper on therapeutic drug monitoring of antiretroviral agents. AIDS Res Hum Retroviruses 2002;18(12):825–34.

[20] Acosta EP, Gerber JG, Kuritzkes DR. Antiretroviral pharmacokinetics, resistance testing, and therapeutic drug monitoring. In: Libman H, Makadon HJ, editors. HIV. 3rd ed. American College of Physicians; 2007. p. 115–36.

[21] Knobel H, Carmona A, Grau S, Pedro-Botet J, Diez A. Adherence and effectiveness of highly active antiretroviral therapy. Arch Intern Med 1998;128(17):1953.

[22] Weiss L, French T, Finkelstein R, Waters M, Mukherjee R, Agins B. HIV-related knowledge and adherence to HAART. AIDS Care 2003;15(5):673–9.

[23] Perrone V, Cattaneo D, Radice S, Sangiorgi D, Federici AB, Gismondo MR. Impact of therapeutic drug monitoring of antiretroviral drugs in routine clinical management of patients infected with human immunodeficiency virus and related health care costs: a real-life study in a large cohort of patients. ClinicoEcon Outcomes Res 2014;6:341–8.

[24] Guiard-Schmid JB, Poirier JM, Meynard JL, Bonnard P, Gbadoe AH, Amiel C, et al. High variability of plasma drug concentrations in dual protease inhibitor regimens. Antimicrob Agents Chemother 2003;47(3):986–90.

[25] Molto J, Blanco A, Miranda C, Miranda J, Puig J, Valle M, et al. Variability in non-nucleoside reverse transcriptase and protease inhibitors concentrations among HIV-infected adults in routine clinical practice. Br J Clin Pharmacol 2007;63(6):715–21.

[26] Rayner C, Dooley M, Nation R, et al. Antivirals for HIV. In: Burton M, Schentag J, Shaw L, Evans W, editors. Applied pharmacokinetics and pharmacodynamics: principles of therapeutic drug monitoring. 4th ed. Baltimore (MD): Lippincott Williams and Wilkins; 2006. p. 355–409.

[27] Fabbiani M, Di Giambenedetto S, Bracciale L, Bacarelli A, Ragazzoni E, Cauda R, et al. Pharmacokinetic variability of antiretroviral drugs and correlation with virological outcome: 2 years of experience in routine clinical practice. J Antimicrob Chemother 2009;64(1):109–17.

[28] Fletcher CV, Kawle SP, Kakuda TN, Anderson PL, Weller D, Bushman LR, et al. Zidovudine triphosphate and lamivudine triphosphate concentration-response relationships in HIV-infected persons. AIDS 2000;14(14):2137–44.

[29] Higgins N, Tseng A, Sheehan NL, la Porte CJ. Antiretroviral therapeutic drug monitoring in Canada: current status and recommendations for clinical practice. Can J Hosp Pharm 2009;63(6):715–21.

[30] Clumeck N, Pozniak A, Raffi F. European AIDS Clinical Society (EACS) guidelines for the clinical management and treatment of HIV-infected adults. HIV Med 2008;9(2):65–71.

[31] Pretorius E, Klinker H, Rosenkranz B. The role of therapeutic drug monitoring in the management of patients with human immunodeficiency virus infection. Ther Drug Monit 2011;33(3):265–74.

[32] Back DJ, Khoo SH, Gibbons SE, Barry MG, Merry C. Therapeutic drug monitoring of antiretrovirals in human immunodeficiency virus infection. Ther Drug Monit 2000;22(1):122–6.

[33] Marzolini C, Telenti A, Decosterd LA, Greub G, Biollaz J, Buclin T. Efavirenz plasma levels can predict treatment failure and central nervous side effects in HIV-1-infected patients. AIDS 2001;15(1):71–5.

[34] Glesby MJ. Perspective: overview of mitochondrial toxicity of nucleoside reverse transcriptase inhibitors. Topics in HIV Med 2002;10(1):42–6.

[35] Van Heeswijk RP. Critical issues in therapeutic drug monitoring of antiretroviral drugs. Ther Drug Monit 2002;24(3):323–31.

[36] Rakhmanina NY, van den Anker JN, Soldin SJ. Therapeutic drug monitoring of antiretroviral therapy. AIDS Patient Care STDS 2004;18(1):7–14.

[37] Santini-Oliveira M, Grinsztein B. Adverse drug reactions associated with antiretroviral therapy during pregnancy. Expert Opin Drug Saf 2014;13(12):1623–52.

[38] Yombi JC, Pozniak A, Boffito M, Jones R, Khoo S, Levy J, et al. Antiretrovirals and the kidney in current clinical practice: renal pharmacokinetics, alterations of renal function and renal toxicity. AIDS 2014;28(5):621–32.

[39] Abers MS, Shandera WX, Kass JS. Neurological and psychiatric adverse effects of antiretroviral drugs. CNS Drugs 2014;28(2):131–45.

[40] Margolis AM, Heverling H, Pham PA, Stolbach A. A review of the toxicity of HIV medications. J Med Toxicol 2014;10(1):26–39.

[41] Becher F, Landman R, Mboup S, Kane CN, Canestri A, Liegeois F, et al. Monitoring of didanosine and stavudine intracellular trisphosphorylated anabolite concentrations in HIV-infected patients. AIDS 2004;18(2):181–7.

[42] Liu X, Ma Q, Zhang F. Therapeutic drug monitoring in highly active antiretroviral therapy. Expert Opin Drug Saf 2010;9(5):743–58.

[43] Thompson MA, Aberg JA, Cahn P, Montaner JS, Rizzardini G, Telenti A, et al. Antiretroviral treatment of adult HIV infection: 2010 recommendations of the International AIDS Society-USA panel. J Am Med Assoc 2010;304(3):321–33.

[44] Wright K. AIDS therapy. First tentative signs of therapeutic promise. Nature 1986;323 (6086):283.

[45] Brinkman K, ter Hofstede HJ, Burger DM, Smeitink JA, Koopmans PP. Adverse effects of reverse transcriptase inhibitors: mitochondrial toxicity as common pathway. AIDS 1998;12 (14):1735–44.

[46] Lewis W, Day BJ, Copeland WC. Mitochondrial toxicity of NRTI antiviral drugs: an integrated cellular perspective. Nat Rev Drug Discov 2003;2(10):812–22.

[47] Fleischer R, Boxwell D, Sherman KE. Nucleoside analogues and mitochondrial toxicity. Clin Infect Dis 2004;38(8):e79–80.

[48] Currier JS. Sex differences in antiretroviral therapy toxicity: lactic acidosis, stavudine, and women. Clin Infect Dis 2007;45(2):261–2.

[49] Bolhaar MG, Karstaedt AS. A high incidence of lactic acidosis and symptomatic hyperlactatemia in women receiving highly active antiretroviral therapy in Soweto, South Africa. Clin Infect Dis 2007;45(2):254–60.

[50] Birkus G, Hitchcock MJ, Cihlar T. Assessment of mitochondrial toxicity in human cells treated with tenofovir: comparison with other nucleoside reverse transcriptase inhibitors. Antimicrob Agents Chemother 2002;46(3):716–23.

[51] Mandelbrot L, Kermarrec N, Marcollet A, Lafanechere A, Longuet P, Chosidow D. Case report: nucleoside analogue-induced lactic acidosis in the third trimester of pregnancy. AIDS 2003;17(2):272–3.

[52] Hurst M, Noble S. Stavudine: an update of its use in the treatment of HIV infection. Drugs 1999;58(5):919–49.

[53] Mallal S, Nolan D, Witt C, Masel G, Martin AM, Moore C, et al. Association between presence of HLA-B*5701, HLA-DR7, and HLA-DQ3 and hypersensitivity to HIV-1 reverse-transcriptase inhibitor abacavir. Lancet 2002;359(9308):727–32.

[54] Daugas E, Rougier JP, Hill G. HAART-related nephropathies in HIV-infected patients. Kidney Int 2005;67:393–403.

[55] Mathew G, Knaus SJ. Acquired Fanconi's syndrome associated with tenofovir therapy. J Gen Intern Med 2006;21(11):C3–5.

[56] Sax PE, Gallant JE, Klotman PE. Renal safety of tenofovir disoproxil fumarate. AIDS Read 2007;17:90–2 9-104, C3.

[57] Fux CA, Simcock M, Wolbers M, Bucher HC, Hirschel B, Opravil M, et al. Tenofovir use is associated with a reduction in calculated glomerular filtration rates in the Swiss HIV Cohort Study. Antivir Ther 2007;12(8):1165–73.

[58] Kohler JJ, Hosseini SH, Hoying-Brandt A, Green E, Johnson DM, Russ R. Tenofovir renal toxicity targets mitochondria of renal proximal tubules. Lab Invest 2009;89:513–19.

[59] Cooper RD, Wiebe N, Smith N, Keiser P, Naicker S, Tonelli M. Systematic review and meta-analysis: renal safety of tenofovir disoproxil fumarate in HIV-infected patients. Clin Infect Dis 2010;51(5):496–505.

[60] Scherzer R, Estrella M, Li Y, Choi AI, Deeks SG, Grunfeld C, et al. Association of tenofovir exposure with kidney disease risk in HIV infection. AIDS 2012;26(7):867–75.

[61] D:A:D Study Group, Sabin CA, Worm SW, Weber R, et al. Use of nucleoside reverse transcriptase inhibitors and risk of myocardial infarction in HIV-infected patients enrolled in the D:A:D study: a multi-cohort collaboration. Lancet 2008;371(9622):1417.

[62] Dudley MN. Clinical pharmacokinetics of nucleoside antiretroviral agents. J Infect Dis 1995;171:S99–112.

[63] De Clerq E. The role of non-nucleoside reverse transcriptase inhibitors (NNRTIs) in the therapy of HIV-1 infection. Antiviral Res 1998;38(3):153–79.

[64] Marzolini C, Paus E, Buclin T, Kim RB. Polymorphisms in human MDR1 (P-glycoprotein): recent advances and clinical relevance. Clin Pharmacol Ther 2004;75:13–33.

[65] Adkins JC, Noble S. Efavirenz. Drugs 1998;56(6):1055–64.

[66] Gonzalez de Requena D, Nunez M, Jiminez-Nacher I, Soriano V. Liver toxicity caused by nevirapine. AIDS 2002;16(2):290–1.

[67] Puzantian T. Central nervous system adverse effects with efavirenz: case report and review. Pharmacotherapy 2002;22(7):930–3.

[68] Scott LJ, Perry CM. Delavirdine: a review of its use in HIV infection. Drugs 2000;60 (6):1411–44.

[69] Von Moltke LL, Greenblatt DJ, Granda BW, Giancarlo GM, Duan SX, Daily JP. Inhibition of human cytochrome P450 isoforms by nonnucleoside reverse transcriptase inhibitors. J Clin Pharmacol 2001;41(1):85–91.

[70] Sheran M. The nonnucleoside reverse transcriptase inhibitors efavirenz and nevirapine in the treatment of HIV. HIV Clin Trials 2005;6(3):158–68.

[71] Blanch J, Corbella B, Garcia F, Parellada E, Gatell JM. Manic syndrome associated with efavirenz overdose. Clin Infect Dis 2001;33(2):270–1.

[72] Deeks ED, Keating GM. Etravirine. Drugs 2008;68(16):2357–72.

[73] Kakuda TN, van Solingen-Ristea RM, Onkelinx J, Stevens T, Aharchi F, de Smedt G, et al. The effect of single- and multiple-dose etravirine on a drug cocktail of representative cytochrome P450 probes and digoxin in healthy subjects. J Clin Pharmacol 2014;54 (4):422–31.

[74] Crauwels H, van Jeeswijk RP, Stevens M, Buelens A, Vangeggel S, Boven K, et al. Clinical perspective on drug-drug interactions with the non-nucleoside reverse transcriptase inhibitor rilpivirine. AIDS Rev 2013;15(2):87–101.

[75] Haas DW, Ribaudo HJ, Kim RB, Tierney C, Wilkinson GR, Gulick RM, et al. Pharmacogenetics of efavirenz and central nervous system side effects: an Adult AIDS Clinical Trials Group study. AIDS 2004;18(18):2391–400.

[76] Motsinger AA, Ritchie MD, Shafer RW, Robbins GK, Morse GD, Labbe L, et al. Multilocus genetic interactions and response to efavirenz-containing regimens: an adult AIDS clinical trials group study. Pharmacogenet Genomics 2006;16(11):837–45.

[77] Dhoro M, Zvada S, Ngara B, Nhachi C, Kadzirange G, Chonzi P, et al. CYP2B6*6, CYP2B6*18, Body weight and sex are predictors of efavirenz pharmacokinetics and treatment response: population pharmacokinetic modeling in an HIV/AIDS and TB cohort in Zimbabwe. BMC Pharmacol Toxicol 2015;16:4.

[78] Savarino A. A historical sketch of the discovery and development of HIV-1 integrase inhibitors. Expert Opin Investig Drugs 2006;15(12):1507–22.

[79] Steigbigel RT, Cooper DA, Kumar PN, Eron JE, Schechter M, Markowitz M, et al. Raltegravir with optimized background therapy for resistant HIV-1 infection. N Engl J Med 2008;359 (4):339–54.

[80] Steigbigel RT, Cooper DA, Teppler H, Eron JJ, Gatell JM, Kumar PN, et al. Long-term efficacy and safety of Raltegravir combined with optimized background therapy in treatment-experienced patients with drug-resistant HIV infection: week 96 results of the BENCHMRK 1 and 2 Phase III trials. Clin Infect Dis 2010;50(4):605–12.

[81] DeJesus E, Rockstroh JK, Lennox JL, Saag MS, Lazzarin A, Zhao J, et al. Efficacy of raltegravir versus efavirenz when combined with tenofovir/emtricitabine in treatment-naïve HIV-1-infected patients: week-192 overall and subgroup analyses from STARTMRK. HIV Clin Trials 2010;13(4):228–32.

[82] Raffi F, Jaeger H, Quiros-Roldan E, Albrecht H, Belonosova E, Gatell JM, et al. Once-daily dolutegravir versus twice-daily raltegravir in antiretroviral-naive adults with HIV-1 infection (SPRING-2 study): 96 week results from a randomised, double-blind, non-inferiority trial. Lancet Infect Dis 2013;13(11):927–35.

[83] Lennox JL, Landovitz RJ, Ribaudo HJ, Ofotokun I, Na LH, Godfrey C, et al. Efficacy and tolerability of 3 nonnucleoside reverse transcriptase inhibitor-sparing antiretroviral regimens for treatment-naive volunteers infected with HIV-1: a randomized, controlled equivalence trial. Ann Intern Med 2014;161(7):461–71.

[84] Clotet B, Feinberg J, van Lunzen J, Khoung-Josses MA, Antinori A, Dumitru I, et al. Once-daily dolutegravir versus darunavir plus ritonavir in antiretroviral-naive adults with HIV-1

infection (FLAMINGO): 48 week results from the randomised open-label phase 3b study. Lancet 2014;383(9936):2222–31.

[85] Wohl DA, Cohen C, Gallant JE, Mills A, Sax PE, Dejesus E, et al. A randomized, double-blind comparison of single-tablet regimen elvitegravir/cobicistat/emtricitabine/tenofovir DF versus single-tablet regimen efavirenz/emtricitabine/tenofovir DF for initial treatment of HIV-1 infection: analysis of week 144 results. J Acquir Immune Defic Syndr 2014;65(3): e118–20.

[86] Morse GD, Catanzaro LM, Acosta EP. Clinical pharmacodynamics of HIV-1 protease inhibitors: use of inhibitory quotients to optimise pharmacotherapy. Lancet Infect Dis 2006;6 (4):215–25.

[87] Hughes PJ, Cretton-Scott E, Teague A, Wensel TM. Protease inhibitors for patients with HIV-1 infection: a comparative overview. P T. 2011;36(6):332–45.

[88] Llibre JM. First-line boosted protease inhibitor-based regimens in treatment-naïve HIV-1-infected patients-making a good thing better. AIDS Rev 2009;11(4):215–22.

[89] Bierman WF, van Agtmael MA, Nijhuis M, Danner SA, Boucher CA. HIV monotherapy with ritonavir-boosted protease inhibitors: a systematic review. AIDS 2009;23(3):279–91.

[90] Hornberger J, Simpson K, Shewade A, Dietz B, Baran R, Podsadecki T. Broadening the perspective when assessing evidence on boosted protease inhibitor-based regimens for initial antiretroviral therapy. Adv Ther 2010;27(11):763–73.

[91] Baril J, Conway B, Giguere P, Ferko N, Hollmann S, Angel JB. A meta-analysis of the efficacy and safety of unbooted atazanavir compared with ritonavir-boosted protease inhibitor maintenance therapy in HIV-infected adults with established virological suppression after induction. HIV Med 2014;15(5):301–10.

[92] Haerter G, Manfras BJ, Mueller M, Kern P, Trein A. Regression of lipodystrophy in HIV-infected patients under therapy with the new protease inhibitor atazanavir. AIDS 2004;18:952–5.

[93] Boffito M, Acosta E, Burger D, Fletcher CV, Flexner C, Garaffo R, et al. Therapeutic drug monitoring and drug-drug interactions involving antiretroviral drugs. Antivir Ther 2005;10 (4):469–77.

[94] Guffanti M, Caumo A, Galli L, Bigoloni A, Galli A, Dagba G, et al. Switching to unboosted atazanavir improves glucose tolerance in highly pretreated HIV-1 infected subjects. Eur J Endocrinol 2007;156:503–9.

[95] Stanley TL, Joy T, Hadigan CM, Liebau JG, Makimura H, Chen CY, et al. Effects of switching from lopinavir/ritonavir to atazanavir/ritonavir on muscle glucose uptake and visceral fate in HIV-infected patients. AIDS 2009;23:1349–57.

[96] Murphy RL, Berzins B, Zala C, Fichtenbaum C, Dube MP, Guaraldi G, et al. Change to atazanavir/ritonavir treatment removes lipids but not endothelial function in patients on stable antiretroviral therapy. AIDS 2010;24:885–90.

[97] Calza L, Manfredi R, Colangeli V, Pocaterra D, Rosseti N, Pavoni M, et al. Efficacy and safety of atazanavir-ritonavir plus abacavir-lamivudine or tenofovir-emtricitabine in patients with hyperlipidaemia switched from a stable protease inhibitor-based regimen including one thymidine analogue. AIDS Patient Care STDS 2009;23(9):691–7.

[98] Naggie S, Hicks C. Protease inhibitor-based antiretroviral therapy in treatment-naive HIV-1-infected patients: the evidence behind the options. J Antimicrob Chemother 2010;65 (6):1094–9.

[99] Tsiodras S, Mantzoros C, Hammer S, Samore M. Effects of protease inhibitors on hyperglycemia, hyperlipidemia, and lipodystrophy: a 5-year cohort study. Arch Intern Med 2000;45 (10):1093–9.

[100] Croom KF, Dhillon S, Keam SJ. Atazanavir: a review of its use in the management of HIV-1 infection. Drugs 2009;69(8):1107–40.

[101] Clotet B, Bellos N, Molina JM, POWER 1 and 2 study groups, et al. Efficacy and safety of darunavir-ritonavir at week 48 in treatment-experienced patients with HIV-1 infection in POWER 1 and 2: a pooled subgroup analysis of data from two randomized trials. Lancet 2007;369(9568):1169–78.

[102] Stolbach A, Paziana K, Heverling H, Pham P. A review of the toxicity of HIV medications II: interactions with drugs and complementary and alternative medicine products. J Med Toxicol 2015;11(3):326–41.

[103] Arvieux C, Tribut O. Aprenavir or fosamprenavir plus ritonavir in HIV infection: pharmacology, efficacy and tolerability profile. Drugs 2005;65(5):633–59.

[104] Kopp JB, Miller KD, Mican JA, Feuerstein IM, Vaughan E, Baker C, et al. Crystalluria and urinary tract abnormalities associated with indinavir. Ann Intern Med 1997;127(2):119–25.

[105] Plosker GL, Noble S. Indinavir: a review of its use in the management of HIV infection. Drugs 1999;58(6):1165–203.

[106] Huynh J, Hever A, Tom T, Sim JJ. Indinavir-induced nephrolithiasis three and one-half years after cessation of indinavir therapy. Int Urol Nephrol 2011;43(2):571–3.

[107] Murphy RL, da Silva BA, Hicks CB, Eron JJ, Gulick RM, Thompson MA, et al. Seven-year efficacy of a lopinavir/ritonavir based regimen in antiretroviral-naïve HIV-1 infected patients. HIV Clin Trials 2008;9(1):1–10.

[108] Bardsley-Elliot A, Plosker GL. Nelfinavir: an update on its use in HIV infection. Drugs 2000;59(3):581–620.

[109] Orman JS, Perry CM. Tipranavir: a review of its use in the management of HIV infection. Drugs 2008;68(10):1435–63.

[110] Justice AC, Zingmond DS, Gordon KS, Fultz SL, Goulet JL, King JT, et al. Drug toxicity, HIV progression, or comorbidity of aging: does tipranavir use increase the rise of intracranial hemorrhage? Clin Infect Dis 2008;47(9):1226–30.

[111] Neu HC, Ellner PD. The inhibitory quotient. Bull NY Acad Med 1983;59:430–42.

[112] Ellner PD, Neu HC. The inhibitory quotient. A method for interpreting minimum inhibitory concentration data. J Am Med Assoc 1981;246:1575–8.

[113] Winston A, Hales G, Amin J, van Schaick E, Cooper DA, Emery S, on behalf of CREST investigators. The normalized inhibitory quotient of boosted protease inhibitors is predictive of viral load response in treatment-experienced HIV-1-infected individuals. AIDS 2005;19:1393–9.

[114] Parra J, Portilla J, Pulido F, Sánchez-de la Rosa R, Alonso-Villaverde C, Berenguer J, et al. Clinical utility of maraviroc. Clin Drug Investig 2011;31(8):527–42.

[115] Hardy WD, Gulick RM, Mayer H, Fatkenheuer G, Nelson M, Heera J, et al. Two year safety and virologic efficacy of maraviroc in treatment-experienced patients with CCR5-tropic HIV-1 infection: 96-week combined analysis of MOTIVATE 1 and 2. J Acquir Immune Defic Syndr 2010;55(5):558–64.

[116] Lu Y, Hendrix CW, Bumpus NN. Cytochrome P450 3A5 plays a prominent role in the oxidative metabolism of the anti-human immunodeficiency virus drug maraviroc. Drug Metab Dispos 2012;40(12):2221–30.

[117] Lalezari JP, Eron JJ, Carlson M, Cohen C, DeJesus E, Arduino RC, et al. A phase II clinical study of the long-term safety and antiviral activity of enfuvirtide-based antiretroviral therapy. AIDS 2003;17(5):691–8.

[118] Lazzarin A, Clotet B, Cooper D, Reynes J, Arasteh K, Nelson M, et al. Efficacy of enfuvirtide in patients infected with drug-resistant HIV-1 in Europe and Australia. N Engl J Med 2003;348(22):2186–95.

[119] Rakhmanina N, Laporte CJL. Therapeutic drug monitoring of antiretroviral drugs in the management of human immunodeficiency virus infection. In: Dasgupta A, editor. Therapeutic drug monitoring: newer drugs and biomarkers. 1st ed. Elsevier Inc.; 2012. p. 373–96.

[120] Elens L, Veriter S, Yombi JC, Difazio V, Vanbinst R, Lison D, et al. Validation and clinical application of a high performance liquid chromatography tandem mass spectrometry (LC-MS/MS) method for the quantitative determination of 10 anti-retrovirals in human peripheral blood mononuclear cells. J Chromatogr B Analyt Technol Biomed Life Sci 2009;877(20–21):1805–14.

[121] Watanabe K, Varesio E, Hopfgartner G. Parallel ultra high pressure liquid chromatography-mass spectrometry for the quantification of HIV protease inhibitors using dried spot sample collection format. J Chromatogr B Analyt Technol Biomed Life Sci 2014;965:244–53.

[122] Koehn J, Ho RJ. Novel liquid chromatography-tandem mass spectrometry method for simultaneous detection of anti-HIV drugs Lopinavir, Ritonavir, and Tenofovir in plasma. Antimicrob Agents Chemother 2014;58(5):2675–80.

[123] Else L, Watson V, Tjia J, Hughes A, Siccardi M, Khoo S, et al. Validation of a rapid and sensitive high-performance liquid chromatography-tandem mass spectrometry (HPLC-MS/MS) assay for the simultaneous determination of existing and new antiretroviral compounds. J Chromatogr B Analyt Technol Biomed Life Sci 2010;878(19):1455–65.

[124] Marzinke MA, Breaud A, Parsons TL, Cohen MS, Piwowar-Manning E, Eshleman SH, et al. The development and validation of a method using high-resolution mass spectrometry (HRMS) for the qualitative detection of antiretroviral agents in human blood. Clin Chim Acta 2014;433:157–68.

[125] Marzinke MA, Clarke W, Wang L, Cummings V, Liu TY, Piwowar-Manning E, et al. Nondisclosure of HIV status in a clinical trial setting: antiretroviral drug screening can help distinguish between newly diagnosed and previously diagnosed HIV infection. Clin Infect Dis 2014;58(1):117–20.

[126] Fogel JM, Piwowar-Manning E, Donohue K, Cummings V, Marzinke MA, Clarke W, et al. Determination of HIV status in African adults with discordant HIV rapid tests. J Acquir Immune Defic Syndr 2015;69(4):430–8.

[127] Chen I, Clarke W, Ou SS, Marzinke MA, Breaud A, Emel LM, et al. Antiretroviral drug use in a cohort of HIV-Uninfected women in the United States: HIV prevention trials network 064. PLoS One 2015;10(10):e0140074.

[128] Koal T, Burhenne H, Romling R, Svoboda M, Resch K, Kaever V. Quantification of antiretroviral drugs in dried blood spot samples by means of liquid chromatography/tandem mass spectrometry. Rapid Commun Mass Spectrom. 2005;19(21):2995–3001.

[129] Kromdijk W, Mulder JW, Smit PM, Ter Heine R, Beijnen JH, Huitema AD. Therapeutic drug monitoring of antiretroviral drugs at home using dried blood spots: a proof-of-concept study. Antivir Ther 2013;18(6):821–5.

[130] Olagunju A, Bolaji O, Amara A, Waitt C, Else L, Adejuyigbe E, et al. Breast milk pharmacokinetics of efavirenz and breastfed infants' exposure in genetically defined subgroups of mother-infant pairs: an observational study. Clin Infect Dis 2015;61(3):453–63.

[131] Lowe SH, van Leeuwen E, Droste JA, van der Veen F, Reiss P, Lange JM, et al. Semen quality and drug concentrations in seminal plasma of patients using a didanosine or

didanosine plus tenofovir containing antiretroviral regimen. Ther Drug Monit 2007;29 (5):566–70.

[132] Yamada E, Takagi R, Sudo K, Kato S. Determination of abacavir, tenofovir, darunavir, and raltegravir in human plasma and saliva using liquid chromatography coupled with tandem mass spectrometry. J Pharm Biomed Anal 2015;114:390–7.

[133] Hickey MD, Salmen CR, Tessler RA, Omollo D, Bacchetti P, Magerenge R, et al. Antiretroviral concentrations in small hair samples as a feasible marker of adherence in rural Kenya. J Acquir Immune Defic Syndr 2014;66(3):311–15.

[134] Olds PK, Kiwanuka JP, Nansera D, Huang Y, Bacchetti P, Jin C, et al. Assessment of HIV antiretroviral therapy adherence by measuring drug concentrations in hair among children in rural Uganda. AIDS Care 2015;27(3):327–32.

[135] Parsons TL, Emory JF, Seserko LA, Aung WS, Marzinke MA. Dual quantification of dapivirine and maraviroc in cervicovaginal secretions from ophthalmic tear strips and polyester-based swabs via liquid chromatographic-tandem mass spectrometric (LC-MS/MS) analysis. J Pharm Biomed Anal 2014;98:407–16.

[136] DiFrancesco R, Rosenkranz SL, Taylor CR, Pande PG, Siminski SM, Jenny RW, et al. Clinical pharmacology quality assurance program: models for longitudinal analysis of antiretroviral proficiency testing for international laboratories. Ther Drug Monit 2013;35 (5):631–42.

[137] DiFrancesco R, Tooley K, Rosenkranz SL, Siminski S, Taylor CR, Pande PG, et al. Clinical pharmacology quality assurance for HIV and related infectious diseases research. Clin Pharmacol Ther 2013;93(6):479–82.

[138] DiFrancesco R, Taylor CR, Rosenkranz SL, Tooley KM, Pande PG, Siminski SM, et al. Adding value to antiretroviral proficiency testing. Bioanalysis 2014;6(20):2721–32.

[139] Rendon A, Nunez M, Jiminez-Nacher I, Gonzalez de Requena D, Gonzlez-Lahoz J, Soriano V. Clinical benefit of interventions driven by therapeutic drug monitoring. HIV Med 2005;6(5):360–5.

[140] Burger D, Hugen P, Reiss P, Gyssens I, Schneider M, Kroon F, et al. Therapeutic drug monitoring of nelfinavir and indinavir in treatment-naive HIV-1-infected individuals. AIDS 2003;17(8):1157–65.

[141] Fletcher CV, Anderson PL, Kakuda TN, Schacker TW, Henry K, Gross CR, et al. Concentration-controlled compared with conventional antiretroviral therapy for HIV infection. AIDS 2002;16(4):551–60.

[142] Clevenbergh P, Garraffo R, Durant J, Dellamonica P. PharmAdapt: a randomized prospective study to evaluate the benefit of therapeutic monitoring of protease inhibitors: 12 week results. AIDS 2002;16(17):2311–15.

[143] Best BM, Goicoechea M, Witt MD, Miller L, Daar ES, Diamond C, et al. A randomized controlled trial of therapeutic drug monitoring in treatment-naive and -experienced HIV-1-infected patients. J Acquir Immune Defic Syndr 2007;46(4):433–42.

[144] Bossi P, Peytavin G, Ait-Mohand H, Delaugerre C, Ktorza N, Paris L, et al. GENOPHAR: a randomized study of plasma drug measurements in association with genotypic resistance testing and expert advice to optimize therapy in patients failing antiretroviral therapy. HIV Med 2004;5(5):352–9.

[145] Demeter LM, Jiang H, Mukherjee AL, Morse GD, DiFrancesco R, DiCenzo R, et al. A randomized trial of therapeutic drug monitoring of protease inhibitors in antiretroviral-experienced, HIV-1-infected patients. AIDS 2009;23(3):357–68.

[146] Torti C, Quiros-Roldan E, Regazzi M, De Luca A, Mazzotta F, Antinori A, et al. A randomized controlled trial to evaluate antiretroviral salvage therapy guided by rules-based or phenotype-driven HIV-1 genotypic drug-resistance interpretation with or without

concentration-controlled intervention: the Resistance and Dosage Adapted Regimens (RADAR) study. Clin Infect Dis 2005;40(12):1828–36.

[147] Khoo SH, Lloyd J, Dalton M, Bonington A, Hart E, Gibbons S, et al. Pharmacologic optimization of protease inhibitors and nonnucleoside reverse transcriptase inhibitors (POPIN) – a randomized controlled trial of therapeutic drug monitoring and adherence support. J Acquir Immune Defic Syndr 2006;41(4):461–7.

[148] Kredo T, Van der Walt JS, Siegfried N, Cohen K. Therapeutic drug monitoring of antiretrovirals for people with HIV. Cochrane Database Syst Rev 2009;(3):CD007268.

[149] OLYSIO [package insert]. Label Food and Drug Administration. Available at: <http://www.accessdata.fda.gov/drugsatfda_docs/label/2013/205123s001lbl.pdf>; 2013 [accessed 10.10.15].

[150] Carr RL, Gramling LF. Stigma: a health barrier for women with HIV/AIDS. J Assoc Nurses AIDS Care 2004;15(5):30–9.

[151] Halkitis PN, Shrem MT, Zade DD, Wilton L. The physical, emotional and interpersonal impact of HAART: exploring the realities of HIV seropositive individuals on combination therapy. J Health Psychol 2005;10(3):345–58.

[152] Stirratt MJ, Remjen RH, Smith A, Copeland OQ, Dolezal C, Krieger D, et al. The role of HIV serostatus disclosure in antiretroviral medication adherence. AIDS Behav 2006;10(5):483–93.

CHAPTER 7

Therapeutic Drug Monitoring in Infants and Children

Uttam Garg and Clinton Frazee
Department of Pathology and Laboratory Medicine, Children's Mercy Hospitals and Clinics, University of Missouri School of Medicine, Kansas City, MO, United States

CONTENTS

7.1 Introduction 165

7.2 Basic Concepts of TDM as It Relates to Infants and Children..166

7.2.1 Absorption............166

7.2.2 Distribution...........166

7.2.3 Metabolism...........167

7.2.4 Excretion..............168

7.3 Pharmacokinetic Calculations................168

7.4 Sample Collection, Analysis, and Interpretation 171

7.5 Specific Drug Classes 173

7.5.1 Antibiotics.............173

7.5.2 Anticonvulsants....176

7.5.3 Immunosuppressive Drugs 177

7.5.4 Antidepressants ...178

7.5.5 Cardiovascular Drugs 178

7.5.6 Bronchodilators....179

7.5.7 Antineoplastic Drugs 180

7.6 Conclusions 181

References 181

7.1 INTRODUCTION

A drug may be described as any substance or compound introduced into the body with the intent of producing a desired physiological effect. Depending on the dose, the drug may be ineffective or it may produce a therapeutic or toxic effect. Paracelsus, a Swiss physician who lived during the Renaissance era, documented some of the first known data about dose–response relationships. Paracelsus understood that achieving the desired physiological response was directly dependent on the dose received, and he is perhaps most famously quoted as saying, "All substances are poisons; there is none which is not a poison. The right dose differentiates a poison and a remedy." Unlike during Paracelsus' time, scientific investigations have established therapeutic indexes, or "safe windows," for most drugs that compare how much drug is required to produce a therapeutic effect to the amount of drug that will produce toxicity. If a drug has a very large therapeutic index, it is generally considered safe and does not require monitoring. Drugs that have narrow therapeutic indexes, however, must be monitored more closely to prevent toxicity.

Therapeutic drug monitoring (TDM) is a specialty area of clinical laboratory science primarily concerned with monitoring concentrations of those drugs that have very narrow therapeutic indexes. The ratio between the minimum toxic concentration and the minimum effective concentration is defined as therapeutic index. Through careful monitoring of drug concentrations over a defined period of time following drug administration, subsequent doses of the drug may be adjusted to maintain the desired level of drug in the bloodstream. It is assumed and generally true that drug levels that are in the established therapeutic range will produce the desired therapeutic effect. Drugs produce therapeutic effects through interactions with receptors. Receptors may be cell membrane proteins or other proteins and also

W. Clarke & A. Dasgupta (Eds): Clinical Challenges in Therapeutic Drug Monitoring. DOI: http://dx.doi.org/10.1016/B978-0-12-802025-8.00007-6

structures such as enzymes and transporters. An increase in drug concentration typically results in an increased response, but it does not always parallel the concentration of drug at the receptor site.

7.2 BASIC CONCEPTS OF TDM AS IT RELATES TO INFANTS AND CHILDREN

The "movement" of drug through the body varies from drug to drug, and even within drug classes, and it can be best understood by evaluating each drug's pharmacokinetic properties, particularly drug absorption, distribution, metabolism, and elimination. These pharmacokinetic properties can vary significantly in children compared to adults due to very rapid growth in early life. For example, body weight doubles in the first 5 months of life and triples in 1 year, and body length increases by 50%, body surface area doubles, and caloric intake increases three or four times in the first year of life [1].

7.2.1 Absorption

Intravenous drug administration has nearly 100% bioavailability, and it does not differ significantly between children and adults. However, when the drug is given intramuscularly or orally, bioavailability can vary significantly among neonates, children, and adults. Vasomotor instability, less muscular mass and subcutaneous fat, and a higher proportion of water in neonates can affect intramuscular drug bioavailability. Most of the drugs are administered orally, and absorption from the gastrointestinal tract can vary significantly in children. Drugs are mostly absorbed in their unionized form. Therefore, acidic drugs are preferably absorbed from the stomach, and basic drugs are absorbed from the small intestine. Many other factors, such as intestinal motility, bile acid formation, intestinal drug-metabolizing enzymes, bowel length, and the presence of other drugs and compounds in the intestine can influence drug absorption, and they vary significantly at different ages. In newborns, gastric pH is neutral and may not reach the normal adult stable pH of 1.5–3.0 for several months to 1 or 2 years. Also in newborns, relatively alkaline milk decreases gastric acidity. Lack of acidic pH results in increased bioavailability of acid-labile drugs such as penicillin, ampicillin, and nafcillin [2] and decreased bioavailability of weakly acidic drugs such as barbiturates. Furthermore, reduced bile salt formation in infants decreases bioavailability of lipophilic drugs such as benzodiazepines and tricyclic antidepressants (TCAs). Therefore, a drug's pharmacokinetic profile must be evaluated differently for neonates than for adults and children.

7.2.2 Distribution

Once absorbed from the gastrointestinal tract, the drug is transported to the liver through the portal vein and undergoes first-pass metabolism before entering into general circulation. During first-pass metabolism, the drug is biotransformed. Depending on the degree of biotransformation, very little unchanged drug may reach general circulation. The amount of unchanged drug available in circulation refers to the drug's bioavailability. Because bioavailability and further distribution of the drug is a measure of the extent to which the drug reaches the site of action, factors that affect the drug's bioavailability and distribution should be carefully considered when determining the drug's dosage. Lipophilic nonpolar drugs such as phenytoin and carbamazepine cross the cell membrane more easily than lipophobic drugs such as oxcarbazepine and lamotrigine. Factors that can affect drug distribution, and thus the drug's bioavailability, include the physiochemical properties of the drug (lipophilic vs lipophobic, molecular weight, etc.), blood pH, protein binding, and compartmentalization of body water. These factors continuously change during the first few years of life. Neonates have a lower albumin and protein concentration, and high concentrations of bilirubin can displace drugs from proteins, resulting in a higher fraction of free (active) drugs. Also, lower blood pH and higher free fatty acid concentrations in neonates can affect the free fraction of a drug [3,4]. Furthermore, water makes up 75–85% of body weight in infants compared to 50–55% in adults. This results in a higher volume of distribution of water-soluble drugs in infants compared to older children and adults. As a result, decreased peak concentrations of water-soluble drugs such as aminoglycosides, theophylline, and furosemide can be observed in infants, whereas the lipid-soluble drugs may show higher plasma peak concentrations.

7.2.3 Metabolism

Drug metabolism, sometimes called xenobiotic metabolism, is the process of biotransforming less polar compounds into more polar compounds that can be excreted more easily. Biotransformation typically occurs through a series of enzymatic reactions involving the cytochrome P450 system. The cytochrome P450 system is a family of isozymes that biotransforms drugs via oxidation, reduction, or hydrolysis reactions. The important gene families of these isoenzymes are CYP1, CYP2, and CYP3. Although the majority of these reactions occur in the liver, cytochrome P450 isozymes have been identified in other tissues, including kidney, skin, lungs, and gastrointestinal tract. Biotransformation occurs in two phases that serve to help clear the drug from the body. Phase I metabolism involves the cytochrome P450 system, in which the drug is chemically altered by an enzyme-induced oxidation, reduction, or hydrolysis reaction. The resulting drug metabolite can further undergo enzymatic alterations such as glucuronidation and sulfation. These enzyme reactions typically follow first-order kinetics, in which the amount of biotransformation that occurs is directly proportional to the amount of drug present. If too much drug is present, as can occur during overdose situations, the enzyme(s) can become saturated. When this happens, zero-order kinetics will follow. With zero-order kinetics, biotransformation is constant and independent of the drug concentration. However, some drugs, such as acetaminophen and ethanol, routinely undergo zero kinetics regardless of concentration. During phase II metabolism, the drug and/or its metabolites undergo glucuronidation or sulfation reactions that make them more water soluble for excretion. The expression of the cytochrome P450 isoenzymes varies significantly with age. In neonates, phase I activity is low and increases progressively during the first 6 months of life. For certain drugs, the enzyme activity in the first few years of life exceeds that of adult rates. Due to these differences, drug metabolism can be markedly different in infants and children compared to adults [5,6]. For example, the enzyme CYP2D6 transforms codeine into morphine, the active form of drug. In neonates, the activity of this enzyme is very low, and it takes more than 5 years to develop fully. A study of 3- to 12-year-olds found that 46% were poor metabolizers of codeine. However, CYP3A4 that is involved in metabolism of midazolam, diazepam, and acetaminophen matures rapidly to adult levels in the first 6–12 months of life.

7.2.4 Excretion

The drug and drug metabolites are excreted primarily by the kidneys. However, other excretion pathways exist, including bile, sweat, tears, saliva, breast milk, and feces. As with drug absorption, the ionized and nonionized forms of the drug and its metabolites are important for effective excretion. In general, more ionized or polar compounds will be excreted, but nonionized and protein-bound drug will be reabsorbed. This concept becomes important when managing overdose cases. Alkalinizing the urine can help facilitate excretion of weakly acidic drugs, whereas acidifying the urine can help facilitate excretion of the weakly basic drugs. Excretion becomes more challenging for neonates. Because less tubular excretion occurs and glomerular filtration can be more than 50% lower in neonates than adults, drugs are not as easily cleared. This further complicates drug dosing, but as with all drug administration, pharmacokinetic calculations can be used to calculate a dosing regimen.

7.3 PHARMACOKINETIC CALCULATIONS

The pharmacokinetic calculations that accompany xenobiotics are fairly well understood and are not discussed in-depth here. However, the calculations are important for predicting the drug dosing required for achieving the desired drug concentrations in the target tissues. These calculations typically involve assumptions about the drug distribution and can be expressed as compartmental models.

One of the most basic compartmental models assumes that the drug exhibits 100% bioavailability and is uniformly distributed throughout the body. By assuming that the drug is distributed primarily in the body fluids as occurs with hydrophilic drugs and alcohols, the one-compartmental model helps to simplify the pharmacokinetic calculations. For the majority of drugs, however, the distribution of drug between blood and tissue makes this model unrealistic. Hence, a two-compartmental model can be used that accounts for both the blood and the tissues. The two-compartmental model assumes that the initial drug concentration will decline rapidly following intravenous administration and delivery to the vascular tissues. The concentration of the drug in blood, therefore, is relative to its concentrations in the tissues and can be described by the volume of distribution (V_d):

$$V_d = (\text{Dose/blood drug concentration at "time zero"})$$

Volume of distribution is a ratio that estimates the total amount of drug in the body to its concentration in the plasma at time zero. Because the volume of distribution is an estimation and not a true physiological parameter, the calculated V_d can be greater than the actual body volume. As would be expected, hydrophilic drugs have a lower V_d than the nonpolar drugs that are more distributed in the tissues. Thus, V_d is an important parameter when calculating the loading dose and can help determine treatment in overdose cases.

As previously indicated, drugs are cleared by either first-order or zero-order kinetics. With first-order kinetics, drug concentration will decline in a linear exponential manner (Fig. 7.1). The rate of elimination of the drug is expressed as the elimination rate constant:

$$k = (\ln Ct_1 - \ln Ct_2)/(t_2 - t_1)$$

where Ct_1 and Ct_2 are blood drug concentrations at time 1 and time 2, respectively.

When the elimination constant, k, is known, the drug's half-life can be calculated. The half-life is the time it takes to clear 50% of the drug from the plasma and is expressed by the following equation:

$$k = 0.693/T_{1/2}$$

Once the elimination and half-life are determined, drug clearance (Cl) can be calculated. Drug clearance is the total plasma volume of drug distribution that is being cleared per unit time:

$$Cl = V_d \times k$$

$$Cl = (0.693 \times V_d)/T_{1/2}$$

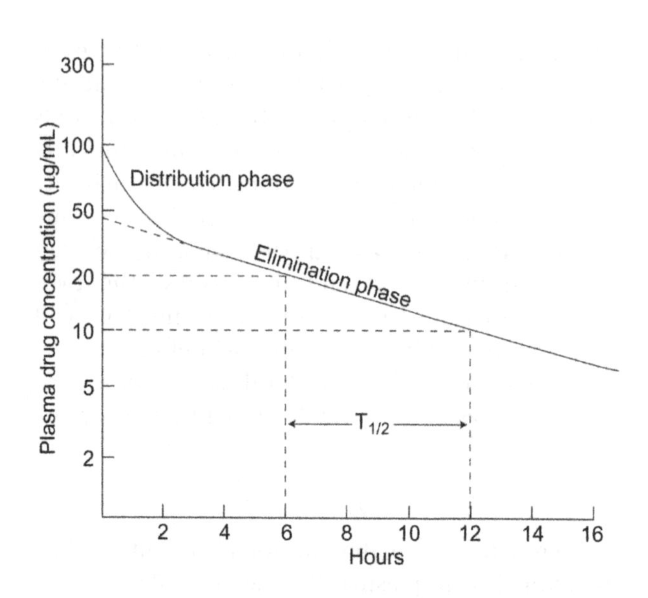

FIGURE 7.1

Drug concentration—time relationship for a two-compartment model system, plotted as log concentration versus time. The initial rapid drop in drug concentration represents the distribution phase, and the linear slow decline represents the elimination phase. *Source: Reprinted with permission from Garg U, Sandritter TL, Leeder JS. Pediatric therapeutic drug monitoring, toxicology, and pharmacogenetics. In: Dietzen DJ, Bennett MJ, Wong ECC, editors. Biochemical and molecular basis of pediatric disease. 4th ed. Washington, DC: AACC Press; 2010. p. 531–60 [6].*

Drug clearance is more generally calculated from the concentration versus time curve (Fig. 7.1).

Cl = Dose(corrected for bioavailability)/area under the curve (AUC)

When the drug half-life is known, drug dosing is typically administered in intervals that are approximately equivalent to the half-life. As the drug continues to be administered intravenously and distributed throughout the body, a steady-state concentration will eventually be reached in which drug administration and drug clearance are in equilibrium. Generally, it takes approximately five half-lives to reach steady state. Once at steady state, the drug concentrations vary between maximum ($C_{ss,\ max}$) and minimum ($C_{ss,\ min}$) concentrations (Fig. 7.2). The steady-state concentrations can be determined by the following equations:

$$C_{ss,max} = (\text{Dose})/[V_d(1 - e^{-kt})]$$

$$C_{ss,min} = (C_{ss,max}) \times e^{-kt}$$

where $(1 - e^{-kt})$ is drug fraction cleared, e^{-kt} is drug fraction left, and t is dosing interval.

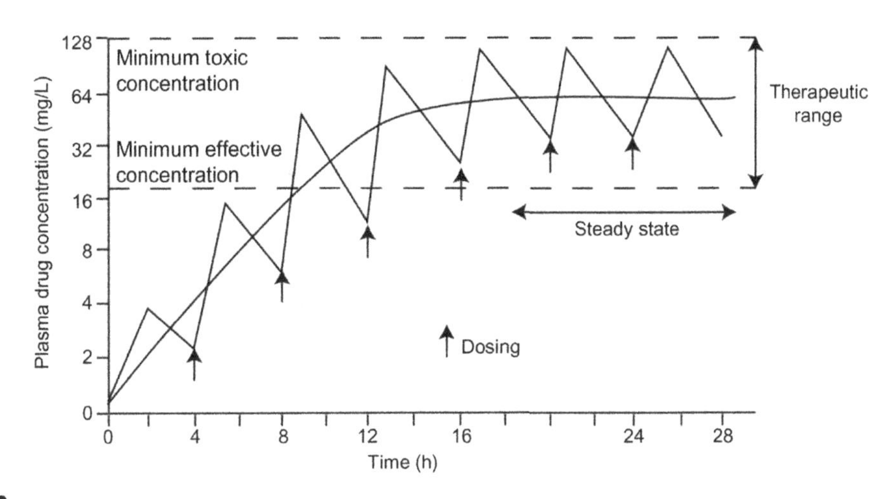

FIGURE 7.2
Concentration versus time profile following multiple doses. Minimum effective and toxic concentrations are shown. To reach steady state, it takes five to seven drug half-lives. *Source: Reprinted with permission from Garg U, Sandritter TL, Leeder JS. Pediatric therapeutic drug monitoring, toxicology, and pharmacogenetics. In: Dietzen DJ, Bennett MJ, Wong ECC, editors. Biochemical and molecular basis of pediatric disease. 4th ed. Washington, DC: AACC Press; 2010. p. 531–60 [6].*

By rearranging the previous steady-state equations and taking into account the drug clearance (Cl), a dosing regimen can be established:

Maintenance dose or rate of infusion $= \mathrm{Cl} \times C_{ss}$

Loading dose $= V_d \times C_{ss}$

Because the parameters discussed previously change rapidly with age, pharmacokinetic calculations are often challenging in infants and children.

7.4 SAMPLE COLLECTION, ANALYSIS, AND INTERPRETATION

Although most of the drugs measured in children and adults following equivalent drug dosing are the same, sample collection, analysis, and interpretation can vary significantly in infants and children. In infants, blood is often collected from the heel. Sample collection from a heel requires special attention, such as selection of proper collection site, warming of heel to increase blood flow, and use of proper lancet [7]. Intravenous sample collection in infants and small children also requires special attention. Small-gauge needles and syringes are generally used for sample collection from infants; vacuum devices are not suitable for blood collection in infants because they may cause venous collapse. Plasma and serum are the preferred specimen matrices, but whole blood and cerebrospinal fluid may also be used for the

TDM. When the sample is collected before the next dose (ie, at the end of the previous dosing interval), the calculated drug concentration is called the trough level. For most drugs, trough levels are ideal for establishing a therapeutic regimen that provides optimal patient benefit. Drug dosing regimens for other drugs, however, may be dependent on clinical outcome. Determining accurate drug levels is critical and, thus, sample collection in the correct collection tube is paramount. Use of gel tubes, for example, is strongly discouraged because many drugs are known to bind to the gel and artificially lower the calculated drug concentration. Efforts should always be employed that reduce error in the TDM process. For example, an intravenous line that has not been flushed following drug administration can lead to significant drug contamination during sample collection. Contamination that goes unchecked can result in significantly elevated drug level determinations and adversely affect the dosing regimen.

Accurately calculating the drug concentration is contingent upon methodologies that are precise, accurate, specific, and reproducible and that span the therapeutic range of the administered drug. Immunoassay methods are frequently favored because they are very fast and require very little sample volume and preparation. Unfortunately, immunoassays often lack specificity and are prone to cross-reactivity from other drugs and compounds that are similar in chemical structure to the measured drug. Immunoassays can be more challenging on samples collected from infants who have a higher rate of hemolysis and increased bilirubin [8]. Also, these samples, particularly the ones collected by heel stick, may have microclots that can plug the sampling probes on automated analyzers.

Due to these reasons or unavailability of immunoassays for certain drugs, chromatographic techniques are frequently used for the analysis of certain drugs. Chromatographic techniques are much more specific, precise, and accurate, and they allow for measuring multiple drugs and their metabolites in a single assay. However, most chromatographic techniques require sample preparation and often sample derivatization to alter the drug's volatility, and they have longer overall analysis times compared to immunoassays. The most commonly used chromatographic techniques are gas chromatography (GC) and high-performance liquid chromatography (HPLC) [9]. GC is coupled with a detector that has been selected to maximize the sensitivity and specificity of the drug or compound being tested. Although flame ionization and nitrogen-phosphorous detectors are still used by some laboratories, mass spectrometers provide the most reliable specificity among chromatographic detectors. HPLC with an ultraviolet detector is also widely used in TDM. When HPLC is coupled to a tandem mass spectrometry system (LC-MS/MS), however, sensitivity and specificity are generally maximized. Very little sample preparation is required for LC-MS/MS analysis, and total testing time tends to be faster. LC-MS/MS systems have become favored for TDM assessment, but they are expensive and require additional skills and expertise compared to the other available methodologies.

Regardless of the methodology used for testing, a clinically interpretable correlation must be established for the drug of interest. The serum drug concentrations must correlate to the intended pharmacologic effect. Although drug therapeutic ranges are fairly consistent and relevant, they can vary from patient to patient based on a variety of physiological and pathological conditions. Providing the appropriate interpretation of drug concentrations and the observed or intended pharmacologic effect will aid in establishing the most optimal patient TDM plan.

7.5 SPECIFIC DRUG CLASSES

Specific drug classes are discussed next. Pharmacokinetic properties of various drugs are given in Table 7.1.

7.5.1 Antibiotics

Regarding TDM of antibiotics, most of them are not monitored, and management is based on empirical dosing [10]. Frequently monitored antibiotics are vancomycin and aminoglycosides [11–13]. Vancomycin is a glycopeptides antibiotic that inhibits bacterial growth by inhibiting bacterial cell wall synthesis of gram-positive bacteria, and it is commonly used in the treatment of gram-positive methicillin-resistant *Staphylococcus aureus* and some *Enterococcus* species infections. For systemic infections, vancomycin is given intravenously because it does not cross intestinal membranes due to its high polarity and large molecular weight. These properties make vancomycin the drug of choice for the treatment of pseudomembranous colitis caused by *Clostridium difficile*. For vancomycin TDM, trough levels are recommended, although peak levels may be helpful for pharmacokinetic studies in select patients. Trough levels are drawn at steady state and just before the next dose. Peak levels are drawn 30 min after completion of 60- or 120-min intravenous infusion. Protein binding of vancomycin in neonates is significantly higher (~75%) than in adults (~30%), and the drug is mostly (80–90%) excreted in the neonatal urine.

Aminoglycosides are used alone or, more commonly, in combination with other antibiotics (β-lactams) in the treatment of aerobic gram-negative bacteria such as *Enterobacter species, Escherichia coli, Klebsiella pneumoniae, Proteus mirabilis,* and *Pseudomonas aeruginosa*. They include amikacin, gentamicin, neomycin, streptomycin, and tobramycin. Like vancomycin, aminoglycosides are polar molecules, do not cross intestinal membranes, and are administered

Table 7.1 Pharmacokinetic Properties of Various Drugs [4,6]			
Drug	**Approximate Half-Life (h)**	**Average Protein Binding (%)**	**Therapeutic Range**
Antibiotics			
Amikacin	2–3	<10	Peak, 20–30; trough, <10 μg/mL
Gentamicin	2–3	<10	Peak, 5–10; trough, <2 μg/mL
Tobramycin	2–4	<10	Peak, 5–10; trough, <2 μg/mL
Vancomycin	3–10	55	Peak, 20–40; trough, 5–15 μg/mL
Anticonvulsants			
Carbamazepine	10–20	85	4–12 μg/mL
Ethosuximide	25–40	<5	40–100 μg/mL
Felbamate	14–21	25	40–120 μg/mL
Gabapentin	4–6	<5	12–20 μg/mL
Lamotrigine	30	55	3–15 μg/mL
Levetiracetam	5–7	<10	10–60 μg/mL
Oxcarbazapine	5–15	40	12–35 μg/mL
Phenobarbital	40–120	50	15–40 μg/mL
Phenytoin	7–30	90	10–20 μg/mL
Primidone	4–6	25	5–12 μg/mL
Topiramate	18–23	15	2–12 μg/mL
Valproic acid	8–12	90	50–100 μg/mL
Zonisamide	50–70	40–60	10–40 μg/mL
Immunosuppressive Drugs			
Cyclosporine A	6–24	95	100–400 ng/mL, variable with type and duration of transplant
Everolimus	18–35	75	3–10 ng/mL
Mycophenolic acid	8–18	95	1.0–3.5 μg/mL
Sirolimus	8–20	90	5–20 ng/mL
Tacrolimus	8–16	95	5–15 ng/mL
Antidepressants			
Amitriptyline[a]	21	95	120–250[b] ng/mL
Amoxapine	8	80	200–600 ng/mL
Citalopram	30	50	40–100 ng/mL
Desipramine	20	80	75–300 ng/mL
Fluoxetine[a]	60	94	300–1000[b] ng/mL
Fluvoxamine	23	77	50–900 ng/mL
Imipramine[a]	12	90	150–250[b] ng/mL
Maprotiline	40	90	200–600 ng/mL
Mirtazapine	30	85	4–40 ng/mL
Paroxetine	22	95	20–200 ng/mL
Sertraline	28	98	30–200 ng/mL
Trazodone	9	90	800–1600 ng/mL
Venlafaxine[a]	5	27	250–500[b] ng/mL

Continued...

Table 7.1 Pharmacokinetic Properties of Various Drugs [4,6] *Continued*

Drug	Approximate Half-Life (h)	Average Protein Binding (%)	Therapeutic Range
Cardiac Drugs			
Digoxin	12–48	25	0.5–2.0 ng/mL
Digitoxin	120–168	95	10–30 ng/mL
Disopyramide	2.5–3.8	45	2.0–6.0 μg/mL
Lidocaine	1–3	55	1.5–5.0 μg/mL
Procainamide and *N*-acetylprocainamide	1.5–2.0	20	4.0–8.0 μg/mL
Propranolol	3.9–6.4	90	50–100 ng/mL
Quinidine	2.5–6.7	85	2–7 μg/mL
Other Drugs			
Caffeine	40–100 (neonates); 3–5 (older children)	35	5–20 μg/mL
Methotrexate	8–15 (high dose)3–10 (low dose)	40	Variable, depends on therapeutic approach
Theophylline	1–8	55	10–20 μg/mL

[a]*Metabolized to active* N*-demethylated metabolite.*
[b]*Total concentrations of parent drug and active metabolite.*

intravenously. The mechanism of action is inhibition of bacterial protein synthesis by binding to the 30S ribosomal subunit of bacterial mRNA [14]. TDM of aminoglycosides is important due to their nephrotoxic and ototoxic effect. Aminoglycosides can result in auditory and vestibular dysfunction that may be irreversible. Both trough and peak levels are monitored. Samples for trough levels are drawn just before the next dose, and samples for peak levels are drawn 30–60 min after completion of intravenous infusion. In addition to measuring drug levels, renal functions should also be monitored. Audiology testing should be considered during long-term therapy with aminoglycosides.

Pharmacokinetic parameters and related TDM considerations in neonates and children are different for aminoglycosides and vancomycin [15]. These neonatal differences are mostly attributable to the larger volume of distribution (V_d) for these drugs, which in part results from immature renal function. For example, vancomycin distribution half-life is highly variable in neonates. It varies from 0.04 to 1.11 h compared to approximately 0.5–1.0 h in adults

[16]. Volume of distribution of gentamicin is approximately 0.70 L/kg in neonates with gestational age of 32 weeks compared to 0.32 L/kg in children aged 11–18 years [16]. Moreover, the V_d and plasma clearance for aminoglycosides are higher in patients with expressed cystic fibrosis, thus requiring higher dose treatment in this patient population [16].

Vancomycin and aminoglycosides are commonly assayed by immunoassays available on automated analyzers. For the most part, immunoassays provide fast turnaround time with satisfactory results. However, interferences in immunoassays have been reported. Inactive crystalline degradation products of vancomycin (CDP-1) are known to interfere in certain immunoassays. These products accumulate and give falsely high results in patients with renal failure and newborns younger than 30 weeks of age [17]. Interference from IgM paraprotein with a turbidimetric gentamicin assay has been reported [18]. In addition, vancomycin can be inactivated by heparin in high concentrations [19].

7.5.2 Anticonvulsants

Managing epilepsy in infants and children is challenging [20]. Despite anticonvulsant therapy, as many as 30% of children continue to experience epilepsy. Many anticonvulsants, including phenytoin, phenobarbital, carbamazepine, oxcarbazepine, primidone, ethosuximide, valproic acid, gabapentin, felbamate, lamotrigine, levetiracetam, oxcarbazepine, and zonisamide, are commonly assayed for TDM [21–23].

Special attention is required for TDM of anticonvulsants in infants and children. Generally, total concentrations are measured for TDM of anticonvulsants. However, certain anticonvulsants, such as phenytoin, carbamazepine, and valproate, are highly protein bound and may need measurement of free drug concentrations, particularly in neonates who have a higher free fraction of these drugs due to their lower plasma protein concentrations and reduced protein binding. For these reasons, infants require lower anticonvulsant dosing to reach the same therapeutic effect compared to adult dosing. On the other hand, young children have higher metabolic capacity and may require higher doses. For example, young children may require an up to four times higher dose of phenytoin (milligrams/kilogram) compared to adults [24]. Infants exposed in utero to phenytoin, phenobarbital, and carbamazepine may need higher doses of these drugs and certain other drugs due to increased induction of drug-metabolizing hepatic enzymes [25]. The newer anticonvulsants, however, are relatively free of this P450 enzyme induction effect [26].

Immunoassays are available and most commonly used for the assays of phenobarbital, phenytoin, carbamazepine, oxcarbazepine, primidone, ethosuximide, levetiracetam, and valproic acid. For the majority of applications, immunoassays provide fast and accurate results. However, immunoassays are prone to interferences due to cross-reactivity of antibody with drug metabolites or other structurally related drugs [9,27,28]. For example, phenytoin metabolite, 5-(*p*-hydroxyphenyl)-5-phenylhydantoin (HPPH), is known to cross-react with many commercial immunoassays. HPPH accumulates significantly in uremic patients and can result in falsely high levels of phenytoin by certain immunoassays [9,29]. Oxaprozin, a nonsteroidal anti-inflammatory drug, exhibits known cross-reactivity with most TDx phenytoin assays. The phenytoin precursor fosphenytoin and antihistamines such as hydroxyzine and cetirizine are known to interfere with a particle-enhanced turbidimetric inhibition immunoassay [27]. Carbamazepine's active metabolite, carbamazepine-10,11-epoxide, is known to interfere with various carbamazepine immunoassays. It is important to note that the cross-reactivity of carbamazepine-10,11-epoxide varies significantly among different immunoassays, ranging from less than 1% to greater than 95%. Immunoassays for phenobarbital have been shown to cross-react with other barbiturates, such as amobarbital, secobarbital, and butabarbital. Significant cross-reactivity of these barbiturates occurs only at toxic but not therapeutic concentrations. Immunoassays are not available for many newer anticonvulsants, such as gabapentin, felbamate, lamotrigine, and zonisamide.

7.5.3 Immunosuppressive Drugs

Organ transplantation in infants and children has become standard of care. Its success, in part, is due to availability of superior antirejection drugs and effective TDM. Commonly used antirejection drugs are steroids and other immunosuppressants, such as cyclosporine, tacrolimus, sirolimus, everolimus, and mycophenolic acid [30,31]. Steroids are not monitored and are managed mostly through empirical dosing. However, the aforementioned immunosuppressants are frequently monitored due to their narrow therapeutic windows and their significant intra- and interindividual variability.

Cyclosporine A, tacrolimus, and mycophenolic acid are given orally or intravenously, whereas sirolimus is administered only orally. Whole blood is the specimen of choice for TDM of cyclosporine A, tacrolimus, sirolimus, and everolimus because significant amounts of these drugs reside in red blood cells (RBCs). Plasma is a specimen of choice for the assay of mycophenolic acid. With each immunosuppressant listed previously, clinical outcome correlates best with area under the curve (AUC) when using the time–drug concentration curve. Trough levels are generally measured for the concentration curve because they are easier to obtain and are not greatly affected by the small time inaccuracies that can easily occur with sample collection. However, trough levels may not correlate very well with AUC for all drugs. With cyclosporine A, for example, samples collected 2 h post-dose (C_2) have been found to correlate better with total drug exposure than trough levels. With C_2 sampling, it is important that samples are collected within each 15-min time window of a 2-h total time interval. Many renal pediatric transplant centers now prefer C_2 levels over trough levels [32,33]. However, the appropriate time interval should be chosen for each drug. With tacrolimus, for example, a 12-h trough level correlates better with AUC and clinical outcome. Following collection, immunoassays are commonly used to quantitatively assay cyclosporine A and tacrolimus. Other immunosuppressants are generally measured by tandem mass spectrometry due to high cross-reactivities of metabolites with immunoassays [34–36]. Immunoassays are fast and more convenient than tandem mass spectrometry, but unlike tandem mass spectrometry, immunoassays are prone to cross-reactivity with the drug metabolites [31,36,37].

7.5.4 Antidepressants

Antidepressants can be divided into two broad categories: TCAs and non-TCAs. TCAs include amitriptyline, nortriptyline, imipramine, and desipramine. TCAs are well absorbed in the intestine and reach peak plasma levels in 2–12 h. They are extensively metabolized to hydroxy and demethylated metabolites that are pharmacologically active. Although metabolism of TCAs in children is not well characterized, it is usually faster compared to that in adults. In children, TCAs are additionally used for the treatment of enuresis and obsessive–compulsive disorder [38–41].

Non-TCAs include monoamine reuptake inhibitors (amoxapine and maprotiline), serotonin reuptake inhibitors (trazodone), inhibitors of norepinephrine and serotonin reuptake (venlafaxine and mirtazapine), and selective serotonin reuptake inhibitors (SSRIs; citalopram, escitalopram, fluoxetine, fluvoxamine, paroxetine, and sertraline). SSRIs are preferred over other antidepressants because they have fewer side effects. SSRIs are also used in the treatment of obsessive–compulsive disorder, panic disorder, bulimia, and many other conditions [38–40]. Immunoassays and chromatographic methods are used in measurement of various immunoassays [41–43].

7.5.5 Cardiovascular Drugs

Digoxin, a glycoside derived from the leaves of a digitalis plant, is the most commonly used and monitored cardiovascular drug. It is administered orally or parenterally for the treatment of congestive heart failure, chronic atrial fibrillation, and supraventricular arrhythmias. Digoxin acts by increasing the force of myocardial contraction by inhibiting the associated sodium–potassium pump.

Gastrointestinal bioavailability varies significantly from patient to patient. Many factors influence digoxin absorption, including P-glycoprotein, intestinal lumen pH, bacterial flora, and the presence of other drugs [44,45]. Digoxin is excreted unchanged in the urine, and its clearance correlates with renal function. Renal clearance of digoxin is low in the first few months of life and correlates with renal expression of P-glycoprotein. Digoxin toxicity is fairly common, particularly in infants during loading doses. Symptoms of toxicity include nausea, vomiting, anorexia, and visual disturbances. Hypokalemia, hypocalcemia, and hypomagnesemia are known to increase digoxin toxicity. Therefore, in addition to measuring digoxin concentration, potassium, calcium, and magnesium levels should also be monitored. Quinidine, another antiarrhythmic drug, can cause a significant increase in digoxin concentration when coadministered with digoxin due to decreased renal clearance.

Automated immunoassays are commonly used for the determination of digoxin. In several clinical conditions such as renal insufficiency, complicated third-trimester pregnancy, and infancy, there is an accumulation of digoxin-like endogenous substances. These compounds cross-react with many digoxin assays and give falsely high digoxin concentrations [28]. However, newer monoclonal antibody-based immunoassays are free from such interferences. An effective antidote, Fab fragment of antidigoxin antibodies, is available for the treatment of digoxin toxicity. The Fab antibody fragments interfere with most immunoassays and provide misleading results [46,47]. If needed, free digoxin can be measured after ultrafiltration of a sample obtained from a patient treated with Fab fragment of antidigoxin antibodies.

Less commonly used and measured cardiac drugs in children include lidocaine, disopyramide, mexiletine, flecainide, procainamide, quinidine, and verapamil [48]. Lidocaine, which is used in the treatment of ventricular arrhythmias and ventricular fibrillation, is administered intravenously due to significant first bypass metabolism. Procainamide is metabolized to active metabolite *N*-acetyl procainamide (NAPA). NAPA can accumulate to significant concentrations in renal dysfunction and fast acetylators. Therefore, both procainamide and NAPA should be monitored. Rarely monitored cardioactive drugs include angiotensin-converting enzyme inhibitors, calcium channel blockers, β-blockers, and catecholamines.

7.5.6 Bronchodilators

In general, β-adrenergic agonists, caffeine, and theophylline are commonly used bronchodilators. β-Adrenergic agonists are frequently used in the treatment of asthma. They are short-acting and used as inhalers. These drugs have rapid action and do not have significant systemic effects. TDM of β-adrenergic agonists is not warranted. Caffeine and theophylline are methylxanthines and frequently monitored. Toxic effects of these drugs include nausea, vomiting, tachycardia, hypotension, and seizures. Caffeine is preferred over theophylline for the treatment of newborn apnea because it has a longer half-life (40–100 h) and consistent intestinal absorption. Immunoassays and chromatographic methods are available for TDM of caffeine and theophylline.

7.5.7 Antineoplastic Drugs

Antineoplastic drugs are very toxic and have very low therapeutic indexes compared to most other drug classes. With the exception of a few drugs, including methotrexate and busulfan, the majority of antineoplastic drugs are usually not monitored [49]. Methotrexate, a nonmetabolizable folic acid antagonist, is used to treat a range of cancers, including acute lymphocytic leukemia, choriocarcinoma, non-Hodgkin lymphoma, osteogenic sarcoma, and various other carcinomas. It inhibits the formation of tetrahydrofolate by inhibiting dihydrofolate reductase. Tetrahydrofolate deficiency results in inhibition of DNA, RNA, and protein synthesis of rapidly proliferating cells. In addition, methotrexate in low doses is used in the management of autoimmune diseases such as rheumatoid arthritis, Crohn's disease, and psoriasis.

Methotrexate is administered orally, intramuscularly, intravenously, or intrathecally. TDM of methotrexate is most useful in patients receiving high-dose methotrexate exceeding 50 mg/m^2. For TDM, methotrexate concentrations are measured every 24 h for 3 days following high-dose administration. Methotrexate concentrations greater than 5 μmol/L after 24 h, greater than 0.5 μmol/L after 48 h, or greater than 0.05 μmol/L after 72 h are considered toxic [49]. If the drug concentrations exceed the toxic concentrations, folinic acid (leucovorin) rescue is initiated. Toxic effects of methotrexate include agranulocytosis, anemia, thrombocytopenia, neutropenia, hyperbilirubinemia, nausea, vomiting, and diarrhea. Extreme toxicity results in erosion of the oral cavity, gastrointestinal bleeding, and renal insufficiency. Methotrexate is generally analyzed by immunoassays [49]; however, chromatographic methods are also available.

Busulfan is an antileukemic DNA-alkylating agent that inhibits DNA replication and transcription of RNA by reacting with the N-7 position of guanosine. It is generally used in combination with cyclophosphamide in the conditioning regimen prior to allogeneic hematopoietic progenitor cell transplantation [50–53]. It binds to albumin and RBCs and is extensively metabolized in the liver. Busulfan has a narrow therapeutic range. High systemic exposure may result in nausea, vomiting, anorexia, anemia, hyperpigmentation, seizures, and infertility. Low exposure confers risk of incomplete myeloablative and graft rejection. Therefore, TDM of busulfan is warranted for dose adjustment and optimal drug exposure. Therapeutic effects correlate better with area under the plasma concentration–time curve or the average plasma concentrations at steady state rather than peak or trough levels. Various analytical methods, including immunoassays, GC, and liquid chromatography–mass spectrometry, are available [54–56]. Due to better specificity, chromatographic methods are preferred.

7.6 CONCLUSIONS

TDM considerations in children, particularly neonates, can vary significantly compared to adults due to differences in pharmacokinetic parameters such as drug absorption, distribution, metabolism, and elimination. Therefore, greater caution must be exercised in interpreting drug concentrations during TDM in infants and children. Moreover, immunoassays may be subjected to interference. As a result, if a drug level does not correlate with the clinical picture, blood should be sent to a reference laboratory to obtain a more accurate result using a chromatographic method. This is particularly important for TDM of immunosuppressants.

References

[1] Kauffman RE. Drug action and therapy in the infant and child. In: Yaffe SJ, Aranda JV, editors. Neonatal and pediatric pharmacology: therapeutic principles in practice. 3rd ed. Philadelphia, PA: Lippincott Williams and Wilkins; 2005. p. 20–31.

[2] Koren G. Therapeutic drug monitoring principles in the neonate. National Academy of Clinical Biochemistry. Clin Chem 1997;43:222–7.

[3] Capparelli EV. Clinical pharmacokinetics in infants and children. In: Yaffe SJ, Aranda JV, editors. Neonatal and pediatric pharmacology: therapeutic principles in practice. 3rd ed. Philadelphia, PA: Lippincott Williams and Wilkins; 2005. p. 9–19.

[4] Moyer TP, Shaw LM. Therapeutic drugs and their management. In: Burtis CA, Ashwood ER, Bruns DE, editors. Tietz textbook of clinical chemistry and molecular diagnostics. 4th ed. St. Louis, MO: Elsevier-Saunders; 2006. p. 1237–85.

[5] Rane A. Drug metabolism and disposition in infants and children. In: Yaffe SJ, Aranda JV, editors. Neonatal and pediatric pharmacology: therapeutic principles in practice. 4th ed. New York, NY: Lippincott Williams & Wilkins; 2011. p. 31–45.

[6] Garg U, Sandritter TL, Leeder JS. Pediatric therapeutic drug monitoring, toxicology and pharmacogenetics. In: Dietzen DJ, Bennett MJ, Wong ECC, editors. Biochemical and molecular basis of pediatric disease. Washington, DC: AACC Press; 2010. p. 531–60.

[7] Jones PM. Pediatric clinical biochemistry: why is it different. In: Dietzen DJ, Bennett MJ, Wong ECC, editors. Biochemical and molecular basis of pediatric disease. Washington, DC: AACC Press; 2010. p. 1–9.

[8] Dasgupta A. Effect of bilirubin, lipemia, hemolysis, paraproteins and heterophilic antibodies on immunoassays for therapeutic drug monitoring. In: Hammett-Stabler CA, Dasgupta A, editors. Therapeutic drug monitoring data.. 3rd ed. Washington, DC: AACC Press; 2007. p. 27–33.

[9] Dasgupta A. Introduction to therapeutic drug monitoring and chromatography. In: Dasgupta A, editor. Advances in chromatographic techniques for therapeutic drug monitoring. Boca Raton, FL: CRC Press; 2010. p. 1–38.

[10] Roberts JA, Norris R, Paterson DL, Martin JH. Therapeutic drug monitoring of antimicrobials. Br J Clin Pharmacol 2012;73:27–36.

[11] Dasgupta A, Hammett-Stabler CA, Broussard LA. Therapeutic drug monitoring of antimicrobial and antiviral agents. In: Hammett-Stabler CA, Dasgupta A, editors. Therapeutic drug monitoring data.. 3rd ed. Washington, DC: AACC Press; 2007. p. 163–92.

[12] Ye ZK, Tang HL, Zhai SD. Benefits of therapeutic drug monitoring of vancomycin: a systematic review and meta-analysis. PLoS One 2013;8:e77169.

[13] Ito H, Shime N, Kosaka T. Pharmacokinetics of glycopeptide antibiotics in children. J Infect Chemother 2013;19:352–5.

[14] Nagai J, Takano M. Molecular aspects of renal handling of aminoglycosides and strategies for preventing the nephrotoxicity. Drug Metab Pharmacokinet 2004;19:159–70.

[15] van den Anker JN, Allegaert K. Pharmacokinetics of aminoglycosides in the newborn. Curr Pharm Des 2012;18:3114–18.

[16] Hoog MD, vandenAnker JN. Aminoglycosides and glycopeptides. In: Yaffe SJ, Aranda JV, editors. Neonatal and pediatric pharmacology: therapeutic principles in practice. 4th ed. Philadelphia, PA: Lippincott Williams and Wilkins; 2005. p. 412–35.

[17] Sym D, Smith C, Meenan G, Lehrer M. Fluorescence polarization immunoassay: can it result in an overestimation of vancomycin in patients not suffering from renal failure? Ther Drug Monit 2001;23:441–4.

[18] Dimeski G, Bassett K, Brown N. Paraprotein interference with turbidimetric gentamicin assay. Biochem Med (Zagreb) 2015;25:117–24.

[19] Barg NL, Supena RB, Fekety R. Persistent staphylococcal bacteremia in an intravenous drug abuser. Antimicrob Agents Chemother 1986;29:209–11.

[20] Tolaymat A, Nayak A, Geyer JD, Geyer SK, Carney PR. Diagnosis and management of childhood epilepsy. Curr Probl Pediatr Adolesc Health Care 2015;45:3–17.

[21] Hitiris N, Brodie MJ. Modern antiepileptic drugs: guidelines and beyond. Curr Opin Neurol 2006;19:175–80.

[22] Patsalos PN, Berry DJ, Bourgeois BF, Cloyd JC, Glauser TA, Johannessen SI, et al. Antiepileptic drugs—best practice guidelines for therapeutic drug monitoring: a position paper by the subcommission on therapeutic drug monitoring, ILAE Commission on Therapeutic Strategies. Epilepsia 2008;49:1239–76.

[23] Garg U, Jacobs DS, Grady HJ, Foxworth J, Gorodetzky CW. Therapeutic drug monitoring. In: Jacobs DS, DeMott WR, Oxley DK, editors. Jacobs and DeMott Laboratory test handbook. 5th ed. Hudson, OH: Lexi-Comp; 2001. p. 731–71.

[24] Broussard LA. Monitoring anticonvulsant concentrations—general considerations. In: Hammett-Stabler CA, Dasgupta A, editors. Therapeutic drug monitoring data. 3rd ed. Washington, DC: AACC Press; 2007. p. 41–75.

[25] Lehr VT, Mathew M, Chugani HT, Aranda JV. Anticonvulsants. In: Yaffe SJ, Aranda JV, editors. Neonatal and pediatric pharmacology: therapeutic principles in practice. 4th ed. New York, NY: Lippincott Williams & Wilkins; 2011. p. 533–51.

[26] Perucca E. Clinically relevant drug interactions with antiepileptic drugs. Br J Clin Pharmacol 2006;61:246–55.

[27] Parant F, Moulsma M, Gagnieu MC, Lardet G. Hydroxyzine and metabolites as a source of interference in carbamazepine particle-enhanced turbidimetric inhibition immunoassay (PETINIA). Ther Drug Monit 2005;27:457–62.

[28] Datta P, Dasgupta A. Immunoassays for therapeutic drug monitoring: pitfalls and limitations. In: Dasgupta A, editor. Advances in chromatographic techniques for therapeutic drug monitoring. Boca Raton, FL: CRC Press; 2010. p. 53–68.

[29] Roberts WL, Rainey PM. Interference in immunoassay measurements of total and free phenytoin in uremic patients: a reappraisal. Clin Chem 1993;39:1872–7.

[30] Gummert JF, Ikonen T, Morris RE. Newer immunosuppressive drugs: a review. J Am Soc Nephrol 1999;10:1366–80.

[31] Butch AW. Introduction to immunosuppressive drug monitoring. In: Hammett-Stabler CA, Dasgupta A, editors. Therapeutic drug monitoring data.. 3rd ed. Washington, DC: AACC Press; 2007. p. 129–61.

[32] Ferraresso M, Ghio L, Zacchello G, Murer L, Ginevri F, Perfumo F, et al. Pharmacokinetic of cyclosporine microemulsion in pediatric kidney recipients receiving a quadruple immunosuppressive regimen: the value of C2 blood levels. Transplantation 2005;79:1164–8.

[33] Dello Strologo L, Pontesilli C, Rizzoni G, Tozzi AE. C2 monitoring: a reliable tool in pediatric renal transplant recipients. Transplantation 2003;76:444–5.

[34] Taylor PJ. Therapeutic drug monitoring of immunosuppressant drugs by high-performance liquid chromatography-mass spectrometry. Ther Drug Monit 2004;26:215–19.

[35] Deters M, Kirchner G, Resch K, Kaever V. Simultaneous quantification of sirolimus, everolimus, tacrolimus and cyclosporine by liquid chromatography-mass spectrometry (LC-MS). Clin Chem Lab Med 2002;40:285–92.

[36] Lee YW. Comparison between ultra-performance liquid chromatography with tandem mass spectrometry and a chemiluminescence immunoassay in the determination of cyclosporin A and tacrolimus levels in whole blood. Exp Ther Med 2013;6:1535–9.

[37] Kelly KA. Pharmacokinetics and therapeutic drug monitoring of immunosuppressants. In: Dasgupta A, editor. Advances in chromatographic techniques for therapeutic drug monitoring. Boca Raton, FL: CRC Press; 2010. p. 209–38.

[38] Ables AZ, Baughman III OL. Antidepressants: update on new agents and indications. Am Fam Physician 2003;67:547–54.

[39] Pacher P, Kecskemeti V. Trends in the development of new antidepressants. Is there a light at the end of the tunnel? Curr Med Chem 2004;11:925–43.

[40] Schatzberg AF. New indications for antidepressants. J Clin Psychiatry 2000;61(Suppl. 11):9–17.

[41] Garg U. Therapeutic drug monitoring of antidepressants. In: Hammett-Stabler CA, Dasgupta A, editors. Therapeutic drug monitoring data.. 3rd ed. Washington, DC: AACC Press; 2007. p. 107–28.

[42] Garg U. Chromatographic techniques for the analysis of antidepressants. In: Dasgupta A, editor. Advances in chromatographic techniques for therapeutic drug monitoring. Boca Raton, FL: CRC Press; 2010. p. 191–207.

[43] Krieg AK, Gauglitz G. Ultrasensitive label-free immunoassay for optical determination of amitriptyline and related tricyclic antidepressants in human serum. Anal Chem 2015;87:8845–50.

[44] Nakamura T, Kakumoto M, Yamashita K, Takara K, Tanigawara Y, Sakaeda T, et al. Factors influencing the prediction of steady state concentrations of digoxin. Biol Pharm Bull 2001;24:403–8.

[45] Al-Khazaali A, Arora R. P-glycoprotein: a focus on characterizing variability in cardiovascular pharmacotherapeutics. Am J Ther 2014;21:2–9.

[46] McMillin GA, Owen WE, Lambert TL, De BK, Frank EL, Bach PR, et al. Comparable effects of DIGIBIND and DigiFab in thirteen digoxin immunoassays. Clin Chem 2002;48:1580–4.

[47] Dasgupta A, McCudden CR. Therapeutic drug monitoring of cardiac drugs. In: Hammett-Stabler CA, Dasgupta A, editors. Therapeutic drug monitoring data.. 3rd ed. Washington, DC: AACC Press; 2007. p. 77–106.

[48] Zalzstein E, Gorodischer R. Cardiovascular drugs. In: Yaffe SJ, Aranda JV, editors. Neonatal and pediatric pharmacology: therapeutic principles in practice. 3rd ed. Philadelphia, PA: Lippincott Williams and Wilkins; 2005. p. 574–94.

[49] Dasgupta A, Hammett-Stabler CA, McCudden CR. Therapeutic drug monitoring of antineoplastic drugs. In: Hammett-Stabler CA, Dasgupta A, editors. Therapeutic drug monitoring data.. 3rd ed. Washington, DC: AACC Press; 2007. p. 209–20.

[50] Cavo M, Bandini G, Benni M, Gozzetti A, Ronconi S, Rosti G, et al. High-dose busulfan and cyclophosphamide are an effective conditioning regimen for allogeneic bone marrow transplantation in chemosensitive multiple myeloma. Bone Marrow Transplant 1998;22:27–32.

[51] Shah AJ, Lenarsky C, Kapoor N, Crooks GM, Kohn DB, Parkman R, et al. Busulfan and cyclophosphamide as a conditioning regimen for pediatric acute lymphoblastic leukemia patients undergoing bone marrow transplantation. J Pediatr Hematol Oncol 2004;26:91–7.

[52] Zao JH, Schechter T, Liu WJ, Gerges S, Gassas A, Egeler RM, et al. Performance of busulfan dosing guidelines for pediatric hematopoietic stem cell transplant conditioning. Biol Blood Marrow Transplant 2015;21:1471–8.

[53] Tesfaye H, Branova R, Klapkova E, Prusa R, Janeckova D, Riha P, et al. The importance of therapeutic drug monitoring (TDM) for parenteral busulfan dosing in conditioning regimen for hematopoietic stem cell transplantation (HSCT) in children. Ann Transplant 2014;19:214–24.

[54] Athanasiadou I, Angelis YS, Lyris E, Archontaki H, Georgakopoulos C, Valsami G. Gas chromatographic-mass spectrometric quantitation of busulfan in human plasma for therapeutic drug monitoring: a new on-line derivatization procedure for the conversion of busulfan to 1,4-diiodobutane. J Pharm Biomed Anal 2014;90:207–14.

[55] Danso D, Jannetto PJ, Enger R, Langman LJ. High-throughput validated method for the quantitation of busulfan in plasma using ultrafast SPE-MS/MS. Ther Drug Monit 2015;37:319–24.

[56] French D, Sujishi KK, Long-Boyle JR, Ritchie JC. Development and validation of a liquid chromatography-tandem mass spectrometry assay to quantify plasma busulfan. Ther Drug Monit 2014;36:169–74.

CHAPTER 8

Therapeutic Drug Monitoring in Pregnancy

Sarah C. Campbell[1], Laura M. Salisbury[2], Jessica K. Roberts[1], Manijeh Kamyar[3], Jeunesse Fredrickson[3], Maged M. Costantine[4] and Catherine M.T. Sherwin[1]

[1]Division of Clinical Pharmacology, Department of Pediatrics, University of Utah, Salt Lake City, UT, United States

[2]Department of Nursing, Clinical Pharmacology and Pediatric Nursing, Eagle Gate College, Salt Lake City, UT, United States

[3]Department of Obstetrics and Gynecology, Maternal–Fetal Medicine, University of Utah School of Medicine, Salt Lake City, UT, United States

[4]Department of Obstetrics and Gynecology, University of Texas Medical Branch, Galveston, TX, United States

CONTENTS

8.1 Introduction 185

8.2 TDM in Pregnancy....................186

8.3 What Drugs Warrant TDM?...........................186

8.4 Steps Involved in the TDM Process..............187

8.4.1 Decision to Request a Drug Concentration........188

8.4.2 Collection of the Sample............................188

8.4.3 Laboratory Measurement of Drug Concentrations in Specimens.......................190

8.4.4 Communication of Lab Result.......................191

8.4.5 Clinical Interpretation of Drug Levels.................191

8.4.6 Implementation and Therapeutic Management....................192

8.5 Pregnancy-Induced Physiologic, Pharmacokinetic, and Pharmacodynamic Changes in Drug Disposition..................192

8.5.1 Pregnancy-Induced Physiologic Changes.......193

8.1 INTRODUCTION

Many women take pharmaceutical agents for acute or chronic medical conditions during pregnancy [1–9]. The risks associated with not taking these medications can be more serious than the risks associated with taking them for the mother and the infant [10–18]. Therapeutic drug monitoring (TDM) has been suggested as a way to mitigate these potential risks and prevent under- or overdosing. TDM may also increase therapeutic success for certain medications used during pregnancy [19,20].

TDM is generally defined as the clinical laboratory measurement of drug concentrations by which appropriate medical interpretation can influence the individualization of a drug's dosage to attain a desired clinical outcome and improve safety. In doing so, TDM can play a valuable role in guiding therapeutic management and aiding in the diagnosis of clinical toxicity. TDM can also be used to establish an individual's therapeutic drug concentration range so that drug therapy can be assessed at subsequent times for potential changes in drug response.

Much of the pharmacological research in pregnant women has focused on the teratogenicity of drugs and remains very limited regarding the safety and

W. Clarke & A. Dasgupta (Eds): Clinical Challenges in Therapeutic Drug Monitoring. DOI: http://dx.doi.org/10.1016/B978-0-12-802025-8.00008-8

8.5.2 Pregnancy-Induced Changes in Drug Metabolism.....195

8.6 Drugs Used in Pregnancy With Recommended TDM ..196

8.6.1 Antiarrhythmics....197

8.6.2 Antibiotics.............198

8.6.3 Anticoagulants......198

8.6.4 Antiepileptics........198

8.6.5 Antidepressants/ Antimanics/ Antipsychotics.................201

8.6.6 Antiretrovirals 201

8.6.7 Bronchodilators....202

8.6.8 Immune Modulators.....................203

8.7 Limitations of TDM in Pregnancy..............203

8.8 Conclusions........206

References.................206

efficacy of drug therapy during pregnancy. Although TDM is not commonly applied for the majority of drugs used in practice today, it may aid in filling gaps in the current knowledge and assist clinicians in making very important decisions on dose adjustments in pregnant women. Currently, no professional society in the field of obstetrics or maternal–fetal medicine has published recommendations or guidelines on the use of TDM in pregnancy.

It is well known that vast physiological changes occur during pregnancy. These changes can lead to altered pharmacokinetics and pharmacodynamics that can lead to changes in a drug's efficacy or toxicity. TDM can provide insight to these changes during a pregnancy and is recommended for certain drugs in the antidepressant, antiepileptic, and antivirals/anti-infectives drug classes to optimize therapeutic drug concentrations and prevent toxicity [21]. This chapter outlines the practice of TDM in pregnancy and reviews these and other drug classes that warrant TDM in pregnant women.

8.2 TDM IN PREGNANCY

Pregnancy-induced changes in a drug's pharmacokinetics, pharmacodynamics, and metabolism can be altered during the course of a pregnancy [22–30]. Therefore, the need for TDM during pregnancy can be much more frequent than in the nonpregnant state [19,31–33]. The schedule for performing TDM during pregnancy will depend on the drug(s) given. However, it is recommended that most drugs have a blood level checked pre-pregnancy, at the beginning of each trimester, and every month in the third trimester. In addition, serum levels should be checked weekly to biweekly in the postpartum period, particularly if dose adjustments were necessary during pregnancy [19,34,35]. If applicable and whenever possible, it is recommended that pre-pregnancy levels be taken at a time when the individual is clinically asymptomatic in order to make TDM more meaningful and more representative of the individual [34]. These pre-pregnancy levels may then serve as a baseline for dosing adjustments during pregnancy to maintain symptomatic control and prevent toxicity. Using individualized therapeutic ranges helps to serve the goal of minimizing maternal and fetal drug exposure, preventing toxicity and other complications, and maintaining asymptomatic control of the given disease or disorder [11,36].

8.3 WHAT DRUGS WARRANT TDM?

The majority of drugs on the market today do not require TDM because they are considered safe and have a low incidence of toxicity. With many concerns over growing health care costs, it is important that TDM be used appropriately and only for drugs that warrant its use. The pharmacoeconomic impact of TDM is beyond the scope of this chapter and is reviewed elsewhere [37–40]. Note that this impact remains largely understudied in pregnant populations.

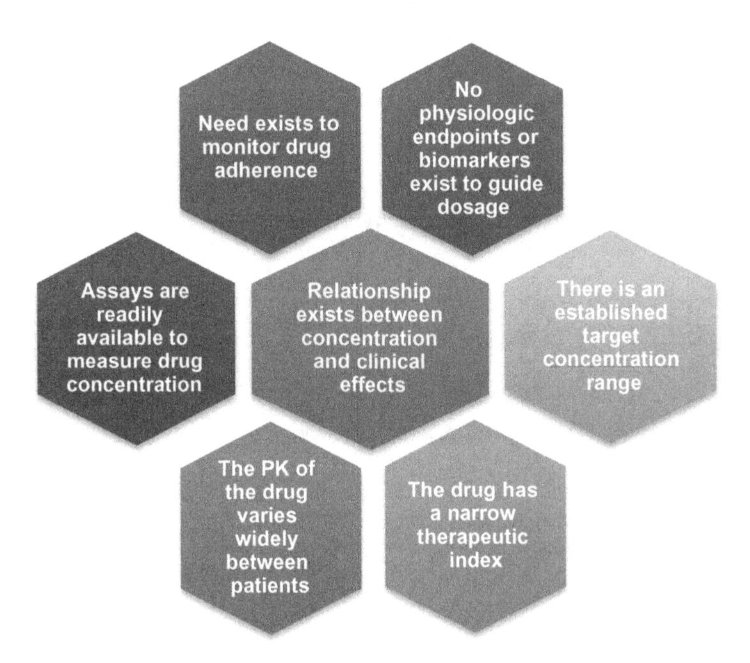

FIGURE 8.1
Criteria of drugs and conditions that warrant TDM. PK, pharmacokinetics.

A number of criteria have been identified that make a drug suitable for TDM (Fig. 8.1) [37,41,42]. These criteria are as follows: (1) There must be a readily available assay to measure the drug (and/or active or toxic metabolites) in blood, (2) there must be an established therapeutic concentration range, and (3) a relationship between blood concentrations and clinical effects must exist. A number of conditions also exist that warrant TDM (Fig. 8.1). These conditions are as follows: (1) A drug has a narrow therapeutic index (ie, there is little difference between toxic and therapeutic dose), (2) drug adherence needs to be monitored, (3) there are no physiologic endpoints (ie, blood pressure, blood glucose, lipid concentrations, etc.), and (4) the pharmacokinetics of the drug vary widely between patients (therefore, predicting blood concentrations that may be subtherapeutic or toxic from a given dose would be difficult).

8.4 STEPS INVOLVED IN THE TDM PROCESS

Several steps are required in the TDM process. First, the clinician must make the decision to request a drug concentration. Second, there must be a collection of the sample at the appropriate time and in the appropriate collection tube/device. Third, the sample must be measured by an accredited laboratory. Fourth, the results must be communicated back to the clinician. Fifth, the clinical interpretation of the results must be made. Finally, there

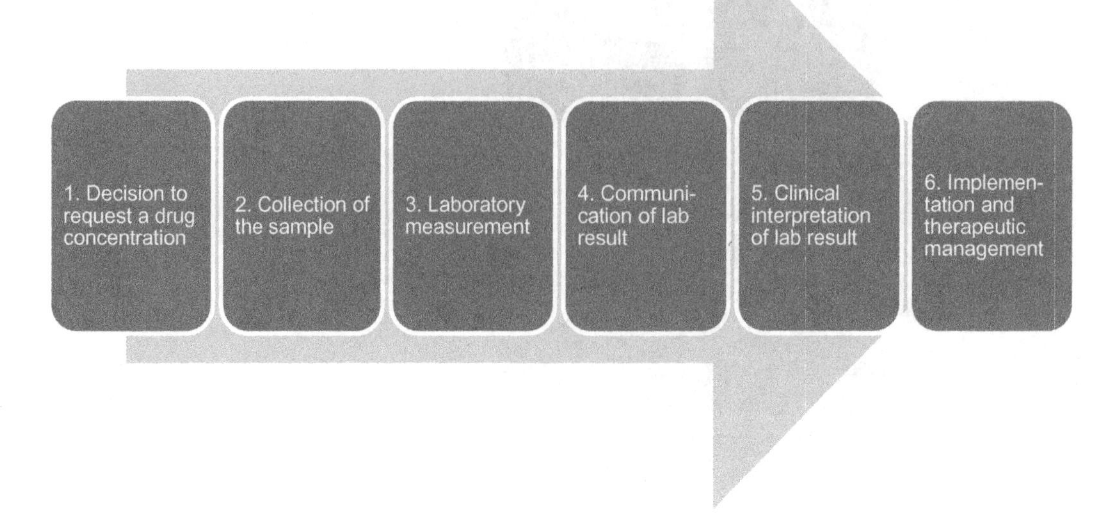

FIGURE 8.2
Overview of the steps involved in the TDM process.

must be implementation of the results for therapeutic management of the patient (Fig. 8.2).

8.4.1 Decision to Request a Drug Concentration

There are several indications/circumstances that could prompt a clinician to make a request to obtain a patient's drug concentration in blood, including the following: (1) A patient's drug dosage or regimen has been changed; (2) the patient is at risk for toxicity, or toxicity is suspected (eg, decreased renal function for renally excreted drugs); (3) there is a need to monitor adherence; (4) it is suspected that the patient may be in a subtherapeutic range (eg, morbidly obese patients who may require larger doses); (5) a comedication has been added; (6) a potentially clinically relevant drug–drug interaction may occur; or (7) there is a need to individualize therapy for the patient.

8.4.2 Collection of the Sample

Once the decision has been made to perform TDM, the sample must be collected at a specific time during the dosing interval for a clinically meaningful measurement. It is important to record and include the time the blood sample was collected, the time of the last dose (or start time of infusion and flow rate), and current dosage regimen for the patient in order for the correct clinical interpretation. Because the drug concentration for the patient

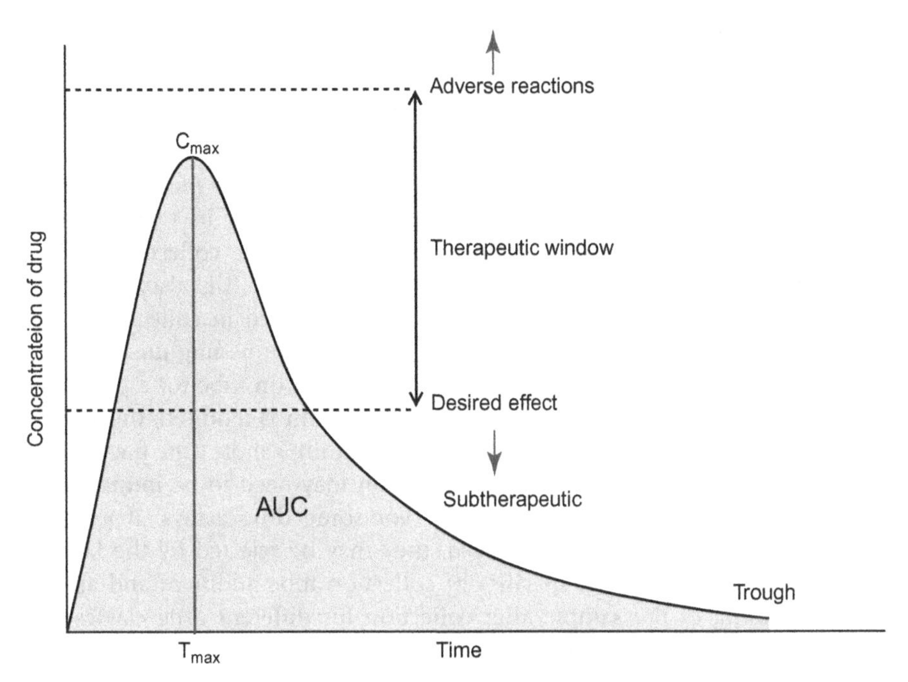

FIGURE 8.3
Drug concentration versus time profile.

provides the information to critically evaluate and individualize the dosage of the drug, the timing of when the blood sample is collected is critically important.

An example of a pharmacokinetic profile of a drug is represented in Fig. 8.3. The least variable point in the interval tends to be just before the next dose (ie, trough or pre-dose concentration) and less likely to be influenced by drug absorption and distribution issues [37,43]. Trough concentrations are measured for many antibiotics, antiepileptic drugs (AEDs), and antidepressants and can indicate efficacy and/or toxicity. However, for most drugs, a sample should be collected at steady state, which occurs four or five half-lives after starting the drug (earlier if a loading dose was given) [36,41,42]. When a drug is at steady state, it is assumed that 95% of the drug has accumulated and the blood concentration is proportional to its receptor concentration. For drugs such as the aminoglycosides, the timing of blood collection will be determined by the dosing interval used and type of monitoring being requested. For example, when the area under the concentration–time curve (AUC) is desired, the collection of blood will need to occur at least twice within the dosing interval. When a nomogram is used, then collection during a specific time window may be necessary. In addition, it may be necessary to collect a sample when side effects are at their greatest in order to aid in detecting toxicity related to high drug concentrations. On the other hand, for drugs with long half-lives, samples may be collected at any time. These are only general guidelines, and suggestions for precise collection times should be reviewed for each drug of interest receiving TDM.

Drug concentrations obtained for TDM typically involve the measurement of active or toxic forms of the drugs (and/or metabolites) in whole blood, plasma, or serum. Therefore, careful consideration of the collection tube is critical in order to obtain the correct matrix for analysis. It is also important to note that some anticoagulants and preservatives used in collection tubes can interfere with certain drug assays. The laboratory providing the TDM service should provide advice on the appropriate collection tube for a particular drug's measurement. For example, if plasma or serum is required, the separation from the cells may need to be performed within a short time frame (typically 1 or 2 h), and the serum/plasma portion may need to be immediately refrigerated or frozen for drug stability. For some drug assays, if samples become severely hemolyzed or thawed, they may be rejected by the laboratory and not analyzed. The specifics of collection tube additives and appropriate handling of the sample after collection for different drug classes can generally be found on the manufacturer and laboratory websites.

Although some assays can measure drug concentrations in saliva, this will usually require a specific type of collection device that can vary depending on the drug of interest. Note that therapeutic drug concentration ranges in saliva are still under evaluation for most drugs and therefore may not be currently defined. Saliva assays for TDM are reviewed elsewhere [44,45].

8.4.3 Laboratory Measurement of Drug Concentrations in Specimens

Routine measurement of drug concentrations in blood typically involves the measurement of total drug concentrations (ie, protein-bound and -unbound "free" drug concentrations). Although the measurement of the total drug concentration will generally require a smaller volume of sample, take less time to prepare, and cost less, the monitoring of the free drug fraction for some drugs may provide greater clinical utility. The free drug fraction is the only fraction that is pharmacologically active and available for metabolism, renal excretion, and transport across the placenta. However, assays that measure free drug are generally not available, and results from these assays may vary widely between laboratories depending on the techniques used to separate the fractions (eg, dialysis, protein precipitation, ultracentrifugation, and gel filtrations) and the type of assay and instrumentation used (eg, immunoassays, liquid chromatography, and flame ion absorption chromatography) [37,46]. Moreover, therapeutic ranges for free drug concentrations may not be well-defined for the drug of interest and, therefore, results may be more difficult to interpret.

There are numerous instances in which protein binding of drugs can change, and pregnancy is one of them. Due to decrease in serum albumin concentrations, the protein binding capacity for certain drugs has been demonstrated to be significantly lower in the second and third trimesters compared to that of nonpregnant women [47]. The serum albumin binding of drugs then becomes significantly greater for women postpartum [47]. Several of the antiepileptic, antidepressant, antipsychotic, and immunosuppressant drugs are highly protein bound (≥80%) and therefore may show variations in protein binding during pregnancy. In addition, for pregnant patients receiving polytherapy with highly protein-bound comedications, or for those with renal insufficiency or hepatic disease, the total drug concentration may inaccurately reflect the therapeutically active concentration. This supports the theory that regular TDM of free drug concentration during pregnancy could provide some clinical utility; however, only a few of these drugs have an assay commercially available to measure free drug concentrations. Nevertheless, for most drugs, the total drug concentration has become standard practice to provide an index of the therapeutically active drug concentration.

8.4.4 Communication of Lab Result

The time it takes the laboratory to return the results of a patient's drug concentration to the clinician is referred to as "turnaround time." Turnaround times vary greatly depending on the laboratory and the type of assay that needs to be performed to obtain the drug concentration. Occasionally, results can be almost immediate, as with bedside tests (eg, immunoassay-based tests used to test for toxic concentrations of drugs), or they may require a week or longer when sophisticated preparation and instrumentation are required. The results from drug measurements in whole blood/plasma/serum are typically reported in mass or molar units, and reports may not contain formulas to assist with conversion to other units. To relate the concentration measured back to the dose, it is preferred for the results to be in mass units. The therapeutic range for the drug assayed should be included with the report, as well as toxic ranges for the drug. However, the therapeutic ranges listed are usually not for pregnant women and may differ between the types of assay performed or the specific laboratories used to measure the drug.

8.4.5 Clinical Interpretation of Drug Levels

Obtaining a precise drug concentration is only one piece that makes up TDM. The correct interpretation of that result is equally important. Ideally, an interpretation service would be available with every report. Being able to adequately interpret the drug concentration requires a comprehensive understanding of pharmacology, pharmacokinetics, pharmacodynamics, pharmaceutics, and analytical sciences. Some large hospital systems have teams that provide support on the clinical interpretation of TDM results. To thoroughly evaluate the drug concentration, the details of the dosage regimen (dose and duration), time the sample was drawn, and patient demographics and comorbidities must be known. Before any dose adjustments are made on the TDM result, it is important to consider many factors: Was the sample collected at the correct time (ie, steady state, trough, etc.)? Are there other important considerations for the patient (eg, patient demographics, renal and hepatic function, and the patient's comedications)? Is the result for total or free drug concentration? TDM interpretation services may also offer dosage prediction techniques to help with individualized drug dosage regimens for a given patient. This involves Bayesian estimations to predict the pharmacokinetic parameters of a drug in a given patient. Although not used routinely in pregnant populations, Bayesian estimations may better predict appropriate dosage regimens to attain defined target concentrations and efficacy and safety in pregnant women.

8.4.6 Implementation and Therapeutic Management

The goal of TDM to individualize dosing regimens requires implementation and therapeutic management. The implementation of the recommendations set forth by the TDM interpretation service (when available) is central to the therapeutic management success of the patient. Additional or routine monitoring may be necessary to maintain drug concentrations in a therapeutic range during a pregnancy. A clinician's clinical judgment may always override the recommendations from a TMD interpretation service to modify a drug's dosage, and it is common for obstetric providers to continue patients on the lowest effective dose if their clinical situation is stable or disease status is unchanged despite recommendations to increase dose.

8.5 PREGNANCY-INDUCED PHYSIOLOGIC, PHARMACOKINETIC, AND PHARMACODYNAMIC CHANGES IN DRUG DISPOSITION

In order to obtain clinically relevant data from TDM during pregnancy, it is important to understand the physiological, pharmacokinetic/pharmacodynamic, and metabolic changes that occur throughout a pregnancy. These changes can be dynamic and lead to variations in drug absorption, distribution, metabolism, and excretion (ADME). These changes may be clinically relevant and require dose adjustments throughout the pregnancy [19,35]. When these changes are not taken into consideration, a general underdosing may occur, leading to subtherapeutic drug concentrations [19,35]. In pregnant women who are adherent to their medications, the dose requirements may increase with the changing maternal physiology and the drug's ADME [19,35]. Conversely, there are some drugs that will require a lower dose during pregnancy. A broad overview of these maternal changes is provided below. A more detailed description of obstetric pharmacology and pregnancy-induced pharmacokinetic changes can be found elsewhere [19,22,48].

8.5.1 Pregnancy-Induced Physiologic Changes

Pregnancy is associated with a multitude of physiological changes for which the causative mechanisms are still poorly understood; however, they are believed to be regulated by pregnancy-related hormones [19,49,50]. A summary of some of the major changes is provided in Table 8.1.

These maternal physiological changes can have a direct effect on a drug's ADME. The absorption of some drugs may be affected by the reduction in intestinal motility [19,35,48]. This reduction results in a 30–50% reduction in gastric and intestinal emptying time [49,50]. This could lead to an increase in systemic drug exposure because more drug can be absorbed while in the intestinal tract due to decreased motility. In addition, gastric acid secretions are reduced by 40% and mucus secretions increase [20]. This can lead to an increase in gastric pH and buffering capacity, altering the ionization of weak acids and bases, and possible decreased drug absorption [56]. It is important to note that pregnant women who experience nausea and vomiting

Table 8.1 Pregnancy-Induced Physiologic Changes

Physiologic Change	Overall Effect in Pregnancy	References
Gastric pH	Increased	[20,51,52]
Gastric and intestinal emptying time	Increased	[20]
Glomerular filtration rate	Increased	[52]
Protein binding capacity	Decreased	[20]
Intestinal motility	Decreased	[20,51–53]
Blood volume	Increased	[20]
Plasma volume	Increased	[53]
Total body water	Increased	[53]
Cardiac output	Increased	[20]
Tidal volume	Increased	[20,53]
Respiratory rate	Increased	[54]
Renal plasma blood flow	Increased	[20]
Hepatic blood flow	Increased	[51]
Uterine blood flow	Increased	[55]

can also have a decrease in gastrointestinal absorption [19,20]. For drugs that rely on pulmonary absorption, the increased cardiac output, pulmonary blood flow, respiratory rate, and tidal volumes in pregnant women may have important considerations for drug therapy for drugs administered by inhalation, such as a decreased requirement for inhaled anesthetic agents [57,58].

During pregnancy, the distribution of drugs is generally altered as a result of the increase in plasma volume (up to 50%) [59]. Furthermore, there is a total mean increase of total body water of 8 L (25% increase from nonpregnant state) [50,60]. These factors combined can lead to a decrease in peak serum concentrations (C_{max}), especially for drugs with relatively small volumes of distribution (V_d) [20,48]. Drugs that mainly distribute to water compartments will have the greatest decrease in C_{max}. For these drugs, the dosage requirements are expected to be higher to achieve the same therapeutic effect during pregnancy.

Plasma volume continues to increase as the pregnancy advances. The rate of albumin production does not match this increase, resulting in a dilution of albumin and hypoalbuminemia as a consequence [50,61,62]. In addition, protein binding sites are occupied with placental and steroid hormones, leading to an overall decrease in drug binding capacity for albumin and an increase in the free drug fraction [61]. Although it is thought that pregnant women would then have increased drug effects due to an increase in free, pharmacologically active drug concentrations, this is generally not the case. The increase in free fraction leads to more drug being subject to the enhanced renal and hepatic drug elimination during pregnancy, which is further exemplified by a lower amount of drug exposure per kilogram of body weight [20,48]. This has an overall result of decreased blood concentrations for most drugs during pregnancy.

Due to the increased hepatic and renal blood flow and function, the overall drug elimination rate (clearance) in pregnancy is thought to be increased for most drugs. The increase in estrogen and progesterone in pregnancy can affect hepatic drug elimination in different ways: (1) For some drugs, an increase in hepatic enzyme activity induced by progesterone can lead to increased drug elimination; (2) for other drugs, a decreased elimination occurs due to competitive inhibition of metabolic enzymes by progesterone and estradiol; or (3) the cholestatic effect of estrogen may interfere with drugs that are excreted into the biliary system [63,64]. For renal elimination, increases in the glomerular filtration rate and renal secretion (via transporters) have been shown for drugs that are renally excreted unchanged during pregnancy [51]. However, unless renal secretion is a predominant route of elimination for the drug, these increases may not translate to overall increases in drug elimination [65].

Through the changes in V_d and clearance mentioned previously, the terminal elimination half-life of drugs may be decreased, increased, or remain unchanged. Therefore, evaluation of half-life needs to be on a drug-by-drug basis (Table 8.2).

Table 8.2 Pregnancy-Induced Pharmacokinetic/Pharmacodynamic Changes

Pharmacokinetic Change	Overall Effect in Pregnancy	References
Absorption	Increased, decreased, or unchanged	[51–53]
Volume of distribution	Increased	[20,51,52]
Clearance	Increased for most drugs	[20,51]
Terminal elimination half-life	Increased, decreased, or unchanged	[52,66]
Peak serum concentrations	Decreased for many drugs	[66,67]
Steady-state plasma concentration	Decreased	[20,53]
Renal drug elimination	Increased	[51]
Hepatic drug elimination	Increased, decreased, or unchanged	[51]

Table 8.3 Pregnancy-Induced Metabolic Changes

Metabolizing Enzymes and Transporters	Overall Effect in Pregnancy	References
CYP1A2	Decreased	[23,51,53,68]
CYP2A6	Increased	[51,53]
CYP2C9	Increased	[51,53]
CYP2C19	Decreased	[51,53]
CYP2D6	Increased	[23,51,53]
CYP34A	Increased	[23,27,51,53]
UGT1A4	Increased	[51]
UGT2B7	Increased	[51]
NAT2	Decreased	[51,68]
P-gp	Increased	[27,53]

8.5.2 Pregnancy-Induced Changes in Drug Metabolism

The metabolism of drugs (phase I oxidation and phase II conjugation) occurs primarily in the liver. Other organs, such as the intestine or kidney, may also contribute to drug metabolism and elimination (drug clearance). The drug metabolites that are formed are usually more polar than the parent drug, which makes them more readily excreted by the kidney or into bile. Much of the research on drug metabolism and pregnancy has focused on the cytochrome P450 (CYP) enzymes, with limited research being done on other drug-metabolizing enzymes. Changes induced by pregnancy in these drug-metabolizing enzymes are isoform specific and are shown in Table 8.3.

The majority of CYP enzymes are upregulated during pregnancy. As such, drug concentrations generally decline due to an increase in drug clearance during pregnancy. Significant increases in clearance during pregnancy have been noted for many drugs metabolized by CYPs, including carbamazepine, fluoxetine, and phenytoin [25,33]. This induction has also been shown for lamotrigine, an AED that is metabolized outside the cytochrome 450 system by uridine diphosphate glucuronosyltransferase (UGT) isoforms 1A4 and 2B7. As such, lamotrigine clearance has been demonstrated to be as much as 300% greater than pre-pregnancy levels [30,69–71]. Therefore, the dosages of drugs predominantly metabolized by these isoenzymes may need to be increased during pregnancy in order to avoid loss of efficacy. In contrast, two CYP450 isoforms, CYP1A2 and CYP2C19, have been shown to decrease in activity during pregnancy [20,23,27,51,68]. Compared to 1 month postpartum, CYP1A2 metabolism has been shown to decrease 32%, 50%, and 52% when using caffeine as a probe drug during the first, second, and third trimester, respectively [68]. A decrease in activity during pregnancy has also been demonstrated for *N*-acetyltransferase 2 (NAT2) using caffeine as a probe drug [68]. As such, a decrease in dose for drugs metabolized through these pathways may be necessary.

Pregnancy has also been shown to affect drug transporters, which are responsible for the movement of drug and/or metabolites in or out of an organ. Very limited research has been conducted on pregnancy-induced changes in transporters across the trimesters. However, one study during the third trimester demonstrated that metformin renal secretion clearance mediated by organic cation transporters and P-glycoprotein increased by 50% and 120%, respectively, requiring increased doses of metformin [72]. Data are lacking for pregnancy-related effects for drugs not eliminated by CYP, UGT, or NAT pathways and for drugs for which multiple pathways are involved.

8.6 DRUGS USED IN PREGNANCY WITH RECOMMENDED TDM

The focus of this section is on drug classes used in pregnancy for which TDM is currently recommended for all populations. The drugs discussed in this section meet the criteria for TDM and may be used in pregnancy if the benefit outweighs the risks. Although not an exhaustive list, these drugs represent some of the most thoroughly evaluated classes of drugs for which a relationship between clinical effects and blood concentrations has been evaluated and for which optimal ranges have been defined. As more research is done on pregnant populations, this list will certainly evolve. For therapeutic ranges of these drugs, one should refer to the report provided with the test results because ranges may vary depending on the type of assay, instruments used, and laboratory performing the analysis. Because no professional obstetric or maternal–fetal organization has published guidelines on TDM during pregnancy, the following recommendations were pulled from multiple sources and are referenced with each drug in their respective table.

The recommendations in the following tables are denoted as follows:

Crucial: Regular TDM has high clinical relevance.

Strongly recommended: Regular TDM may have high clinical relevance for certain patients.

Recommended: Regular monitoring has been advocated, including testing before pregnancy, each trimester, and soon after delivery.

Considered useful: Although TDM is considered useful, the frequency is undetermined.

Still being evaluated: Current knowledge on the clinical utility of TDM is incomplete. Therefore, the recommended frequency of TDM is not established.

Insufficient data: Some information is available that supports the potential for TDM use, but insufficient evidence is available to support widespread use.

8.6.1 Antiarrhythmics

Maternal and fetal arrhythmias occurring during pregnancy may threaten the lives of the mother and the fetus. For the treatment of fetal tachyarrhythmias, drugs may be administered to the fetus via maternal ingestion and rely on transplacental delivery [73,74]. Antiarrhythmic therapy should be offered with careful fetal and maternal monitoring and an appreciation of the pharmacology and pharmacokinetics of the maternal and placental–fetal systems. Drugs in this class include digoxin, disopyramide, and procainamide. Digoxin has long been used in the treatment of both maternal and fetal arrhythmias and is considered safe during pregnancy [73]. Close and regular TDM of antiarrhythmics in pregnancy is recommended to prevent toxicity and complications for both the mother and the fetus [73]. TDM of antiarrhythmics is summarized in Table 8.4.

Table 8.4 Antiarrhythmics that should be Monitored in Pregnancy

Drug Name	Recommendation	Rationale	References
Digoxin	Recommended	Regularly monitoring to prevent toxicity and complications	[36,73]
Disopyramide	Recommended	Regularly monitoring to prevent toxicity and complications	[73]
Procainamide	Recommended	Regularly monitoring to prevent toxicity and complications	[73]

Table 8.5 TDM of Antibiotics in Pregnancy

Drug Name	Recommendation	Rationale	References
Amikacin	Recommended—considered useful	Dosing adjustments may be required.	[19,36]
Gentamicin	Recommended—considered useful	Dosing adjustments may be required.	[19,36,80]
Tobramycin	Recommended—considered useful	Dosing adjustments may be required.	[19,36,80]
Vancomycin	Recommended—considered useful	Dosing adjustments may be required.	[19,36]

8.6.2 Antibiotics

Many antibiotics given in pregnancy do not have TDM recommendations, except for the following: amikacin, gentamicin, tobramycin, and vancomycin. Dosing adjustments may be required because total clearance increases in pregnant versus nonpregnant patients have been demonstrated for many drugs in this class [19,20,75–79]. An increase in dose may be required to achieve and maintain effective therapeutic drug levels [66,78–80]. TDM requirements of certain antibiotics in pregnant woman are summarized in Table 8.5.

8.6.3 Anticoagulants

Drugs in this class include low-molecular-weight heparin (LMWH), unfractionated heparin (UFH), and warfarin. Whereas warfarin is able to cross the placenta, LMWH and UFH are unable to do so, therefore limiting fetal exposure concerns [81,82]. LMWH has been shown to be more effective than UFH during pregnancy with less fetal risks to maintain adequate anticoagulation [81,82]. Both LMWH and UFH are recommended over warfarin during pregnancy, except in very rare cases such as anticoagulation to prevent valve thrombosis [81,82]. Warfarin is not recommended for use during pregnancy, and TDM is strongly recommended if a patient requires it for anticoagulation [82]. Although drug assays do exist for these drugs, it is rare for the actual drug concentration to be monitored, and biomarkers (eg, internal normalized ratio and anti-Xa) are typically monitored instead. TDM of anticoagulants in pregnancy are summarized in Table 8.6.

8.6.4 Antiepileptics

Certain older AEDs are recommended over the newer AEDs due to the lack of research demonstrating safety of the use of newer AEDs during pregnancy [83]. Individual therapeutic levels should be identified when the patient is experiencing adequate seizure control. These individual baseline levels can then be used during pregnancy with clinical correlation to decrease seizure activity and improve overall seizure control and management of symptoms [34]. TDM is recommended at the beginning of each trimester, monthly

Table 8.6 Monitoring of Anticoagulant Therapy in Pregnancy

Drug Name	Recommendation	Rationale	References
Low-molecular-weight heparin (LMWH)	Recommended	Individualized anticoagulation strategy is helpful for increasing effectiveness. LMWH is more effective than unfractionated heparin with decreased risks to mother and infant; however, more treatment failures have been reported.	[81,82]
Unfractionated heparin	Recommended	Heparin decreases the risk for fetal toxicity during pregnancy; however, there is an increased risk for valve thrombosis even with TDM and dose adjustments during pregnancy.	[81,82]
Warfarin	Strongly recommended	Warfarin is *not* recommended for use during pregnancy, and heparin and LMWH have been shown to be more effective with less fetal risks to maintain adequate anticoagulation. TDM is strongly recommended during pregnancy if the patient requires warfarin for anticoagulation.	[81,82]

during the third trimester, and weekly to biweekly in the postpartum period [84,85]. Drugs in this class include carbamazepine, felbamate, gabapentin, lamotrigine, levetiracetam, oxcarbazepine, phenobarbital, phenytoin, valproic acid, and vigabatrin.

The clearance of phenytoin, an older AED primarily metabolized by CYP2C9, has been shown to increase by 117%, leading to a 61% decrease in total serum blood concentrations compared to baseline [86]. Free phenytoin serum concentrations were also observed to decrease slightly [86]. Lamotrigine is the most extensively studied of the newer AEDs in pregnancy. The apparent clearance of lamotrigine has been reported to increase progressively throughout pregnancy. In one prospective study of lamotrigine monotherapy, the median clearance was elevated 197%, 236%, and 248% during the first, second, and third trimester, respectively [30]. Another study demonstrated an increase greater than 300% from baseline in late pregnancy [87]. This effect is thought to be most likely due to enhanced metabolism by CYP450s [87,88] and results in significant dose increases (up to threefold) in order to obtain therapeutic serum levels [30]. Felbamate use is associated with an increased incidence of liver failure and aplastic anemia; therefore, monitoring for toxic levels of drug is strongly advised [88]. Phenobarbital is not commonly used in pregnancy, but TDM is recommended to minimize potential risks of complications to the mother and fetus [34]. If the patient requires valproic acid during pregnancy, a dose increase during late pregnancy with a decrease in the postpartum period may be required [19,34,36]. For vigabatrin, TDM has been suggested to be useful as a means of checking compliance with medication but not as a guide for dose adjustment and therapy [88]. TDM of antiepileptics during pregnancy is summarized in Table 8.7.

Table 8.7 TDM of Antiepileptics in Pregnancy

Drug Name	Recommendation	Rationale	References
Carbamazepine	Strongly recommended	Reduction in plasma albumin levels may lead to an increase of the active drug and dose may need to be adjusted.	[19,21,25,34,36,86]
Felbamate	Still being evaluated	Insufficient data; pharmacokinetics vary widely, particularly with comedications, age, and/or compromised renal function. Felbamate use is associated with an increased incidence of liver failure and aplastic anemia.	[88]
Lamotrigine (for seizures)	Strongly recommended	Lamotrigine is the most commonly prescribed AED in pregnancy. Close monitoring with dose adjustments in the antenatal, prenatal, and postpartum periods are strongly recommended. Clearance increases through the 32nd week of pregnancy, and an increased dose may be required to sustain therapeutic drug levels across pregnancy in women with epilepsy. There is a rapid drop in the postpartum period, and dose will need to be decreased to prevent toxic levels. This is a common problem in the postpartum period for women who required a dose increase of lamotrigine during pregnancy.	[11,21,30,31,33,34,36,86,88]
Lamotrigine (for mood disorders)	May be useful—insufficient data	For use in mood disorders, the use of TDM may be useful in conjunction with clinical features monitoring because the dosing is more typically driven by clinical response.	
Levetiracetam	Still being evaluated	Pharmacokinetic studies have demonstrated an increase in clearance, particularly in the third trimester.	[12,34,88–90]
Oxcarbazepine	Recommended	Due to pharmacokinetic changes and competition for metabolic pathways during pregnancy, a dose adjustment may be required.	[31,34,88]
Phenobarbital	Recommended	Not commonly used in pregnancy, but TDM is recommended to minimize potential risks of complications to the mother and fetus.	[34]
Phenytoin	Strongly recommended	Strongly recommended because this medication is routinely monitored in all patients.	[19,25,34,36,86]
Valproic acid	Recommended	Recommended because a dose increase may be required during late pregnancy with a decrease in the postpartum period.	[19,21,34,36]
Vigabatrin	Still being evaluated	Insufficient data—may be useful as a means of checking compliance with medication, but not as a guide for dose adjustment and therapy.	[88]

8.6.5 Antidepressants/Antimanics/Antipsychotics

The literature suggests that one-third of all pregnant women take at least one psychoactive drug [32]. There are several drugs in these classes with sufficient evidence to support TDM during pregnancy [21,32,33]. In studies in which TDM has been done, lower blood concentrations are often reported during pregnancy [19–21,33]. Drugs in these classes commonly used during pregnancy include bupropion, citalopram, clomipramine, clozapine, duloxetine, fluoxetine, haloperidol, imipramine, lithium, nortriptyline, olanzapine, paroxetine, quetiapine, risperidone, sertraline, and venlafaxine.

There are established therapeutic ranges for many of these drugs. Individualized baseline drug levels should be determined prior to pregnancy at a time when the patient is clinically asymptomatic. These pre-pregnancy, asymptomatic baseline levels can be used as a target level during pregnancy and can help ensure that the patient is not treated with higher doses than necessary to maintain symptomatic control [21,32]. Clinical monitoring is indicated for mood disorders because a dose increase may be indicated [21,33]. Dosing adjustment and treatment changes should be based on therapeutic drug levels and clinical features monitoring [21].

Increased gonadal steroid activity during pregnancy may lower serum plasma levels of selective serotonin reuptake inhibitors (SSRIs), which may require dose adjustments [33]. Due to inadequate data and an insufficient relationship between blood SSRI levels and clinical response, dosage adjustments should also be correlated clinically with symptoms to achieve the desired therapeutic effects [33,88].

For lithium therapy, a lower circulating volume has been noted due to dehydration from hyperemesis gravidarum and vomiting in late pregnancy [33]. A lower circulating volume may increase serum levels, leading to higher risk for toxic levels [19,21,32,33]. In the postpartum period, serum levels increase, which can also lead to toxicity [33]. TDM of antidepressants, antimanics, and antipsychotic drugs in pregnancy is summarized in Table 8.8.

8.6.6 Antiretrovirals

Lower concentrations have been noted during the second and third trimesters and the postpartum period, and increased dosing may be necessary to maintain virological control [10,91]. TDM has been shown to provide inconsistent results with antiretroviral therapy based on the individual's clinical course of treatment and previous virological control [91]. Further studies are needed to correlate blood concentrations of these drugs to clinical findings, decreased viral loads, and a patient response to treatment [10]. It is more common to use immunologic and virologic (viral load) monitoring strategies for

Table 8.8 TDM of Antidepressants/Antimanics/Antipsychotics in Pregnancy

Drug Name	Recommendation	Rationale	References
Amitriptyline	Recommended	Dose increase may be indicated.	[33]
Bupropion	Recommended	Dose increase may be indicated.	[21]
Citalopram	Recommended	Dose increase may be indicated, especially in late pregnancy.	[21,33]
Clomipramine	Recommended	Dose increase may be indicated.	[21]
Clozapine	Strongly recommended	Dose increase may be indicated, correlate clinically with symptoms.	[21]
Duloxetine	Recommended	Dose increase may be indicated.	[21]
Fluoxetine	Recommended	Dose increase may be indicated.	[21,33]
Haloperidol	Strongly recommended	Dose increase may be indicated, correlate clinically with symptoms.	[33]
Imipramine	Recommended	Dose increase may be indicated.	[21]
Lithium	Crucial	Lower serum concentrations are often reported during pregnancy. Later in pregnancy, a lower circulating volume may be noted due to dehydration from hyperemesis gravidarum and vomiting. A lower circulating volume may increase serum levels, leading to a higher risk for toxic levels. In the postpartum period, serum levels can also increase leading to toxicity. Close monitoring during pregnancy and the postpartum period should be completed to minimize the risk of toxicity and harmful effects to mother and infant.	[19,21,32,33]
Nortriptyline	Recommended	Dose increase may be indicated.	[21]
Olanzapine	Strongly recommended	Dose increase may be indicated, correlate clinically with symptoms.	[21]
Paroxetine	Recommended	Dose increase is very often indicated, correlate clinically with symptoms.	[21,33]
Quetiapine	Recommended	Not recommended in pregnant patients, although the risks versus the benefits to mother and infant should be considered. TDM is recommended and should be correlated clinically with symptoms.	[21]
Risperidone	Recommended	Correlate dosage adjustments with clinical symptoms.	[32]
Sertraline	Recommended	Dose increase may be indicated.	[21,32,33]
Venlafaxine	Recommended	Dose increase may be indicated.	[21]

individuals on these therapies than actual blood concentrations of these drugs [92]. Drugs in this class include atazanavir, indinavir, nelfinavir, lopinavir, ritonavir, and saquinavir. The TDM of antiretrovirals in pregnancy are summarized in Table 8.9.

8.6.7 Bronchodilators

The only bronchodilator that has TDM recommendations is theophylline. Bronchodilation occurs over a specific theophylline concentration range [93]. Clinically important improvement in symptom control has been found to

Table 8.9 TDM of Antiretrovirals in Pregnancy

Drug Name	Recommendation	Rationale	References
Atazanavir	Recommended	Dosage adjustments may be indicated, especially late in pregnancy.	[10,91]
Indinavir	Recommended	Dosage adjustments may be indicated, especially late in pregnancy.	[91]
Nelfinavir	Considered useful	Dosage adjustments may be indicated, especially late in pregnancy.	[91]
Nevirapine	Recommended	Dosage adjustments may be indicated, especially late in pregnancy.	[91]
Lopinavir	Considered useful	Dosage adjustments may be indicated, especially late in pregnancy.	[91]
Ritonavir	Recommended	Dosage adjustments may be indicated, especially late in pregnancy.	[91]
Saquinavir	Recommended	Dosage adjustments may be indicated, especially late in pregnancy.	[91]

correlate with peak serum concentrations of theophylline [30]. TDM is recommended (1) when initiating therapy to guide final dosage adjustment after titration, (2) before making a dose increase to determine whether the serum concentration is subtherapeutic in a patient who continues to be symptomatic, (3) whenever there are signs or symptoms of theophylline toxicity, (4) whenever a new illness or worsening of a chronic illness occurs, or (5) whenever the patient's treatment regimen is altered [30]. The half-life of theophylline has been demonstrated to increase modestly throughout pregnancy, indicating a possible decrease in clearance [30].

8.6.8 Immune Modulators

The optimal therapeutic range for a given patient may differ from the suggested range based on the type of organ transplanted, the treatment phase (initiation or maintenance), whether an immune-modulating drug is used in combination with other drugs, and/or the therapeutic approach of the transplant center. In patients with lupus, TDM is recommended to maintain adequate control and prevent flare-ups during pregnancy [94]. Drugs in this class include azathioprine, cyclosporine, hydroxychloroquine, sirolimus, and tacrolimus. The TDM of immune modulators in pregnancy are summarized in Table 8.10.

8.7 LIMITATIONS OF TDM IN PREGNANCY

Although there is sufficient evidence to support the need for TDM for several drugs used in pregnancy, it is not without its limitations. In general, therapeutic ranges have been derived in nonpregnant populations (which are frequently male) and have not been validated for most drugs during pregnancy. In addition, the clinical relevance of TDM has not been specifically studied in pregnant women, nor has the benefit for TDM in pregnancy been well

Table 8.10 TDM of Immune Modulators in Pregnancy

Drug Name	Recommendation	Rationale	References
Azathioprine	Recommended	TDM is recommended in patients with lupus to maintain adequate control and prevent flare-ups during pregnancy.	[94]
Cyclosporine	Recommended	Clinical monitoring is considered useful in pregnancy to maintain appropriate immune suppression.	[36,95]
Hydroxychloroquine	Recommended	TDM is recommended in patients with lupus to maintain adequate control and prevent flare-ups during pregnancy.	[94]
Sirolimus	Recommended	Clinical monitoring is considered useful in pregnancy to maintain appropriate immune suppression.	[36,95]
Tacrolimus	Recommended	Clinical monitoring is considered useful in pregnancy to maintain appropriate immune suppression.	[36,95]
Corticosteroids, (glucocorticoids)	Considered useful	It is considered useful to monitor steroid levels during pregnancy, particularly in patients with lupus; however, there are insufficient data to support widespread monitoring.	[94]

established. Furthermore, the cost-effectiveness of TDM in pregnancy still needs to be investigated for most drugs. Assays to determine blood concentrations of many drugs (and/or metabolites) may not be readily available in some settings, and not all hospitals/clinics have TDM interpretation services to help guide the clinician. In addition, many different methods exist to prepare samples for analysis, and there are several different types of analytical instruments, all of which have varying costs. These include several different types of analytical techniques with immunoassays, high-performance liquid chromatography, gas chromatography, liquid chromatography mass spectrometry, spectrophotometry/colorimetry/fluorometry, electrochemical luminescence, and flame photometry being the most common [46]. Use of immunoassays can be associated with lower cost; however, they are subject to interferences due to the ability of the antibodies to cross-react with the metabolites of the target drug and possibly other drugs that the patient is taking. Immunoassays can therefore lack specificity, and drug concentrations may be reported to be higher than they are in reality. Conversely, methods that provide more specificity, such as liquid chromatography mass spectrometry, will lead to more accurate drug concentration measurements but tend to cost more.

Therapeutic drug concentration windows are not always universal, and specific ranges can vary depending on the instrumentation and laboratory used. The resultant drug concentrations measured may vary significantly, and caution is warranted when interpreting results in these settings. Caution is also warranted when obtaining results from less invasive methods of TDM, such as measuring saliva or blood spot concentrations, because therapeutic ranges in these matrices are still being evaluated for most drugs.

There is an overall lack of results from randomized controlled trials for TDM in pregnancy. Much of the data come from the use of only two drug classes—antiretrovirals to prevent HIV transmission to the fetus and AEDs to control seizures. For some of the AEDs, there can be wide interindividual variations in serum/plasma drug concentrations, often with seizure control poorly correlating with a given therapeutic range. Furthermore, therapeutic ranges have not been well defined for the newer AEDs, and active metabolites (eg, carbamazepine-10,11-epoxide) that may contribute to therapeutic response are generally not measured.

For drugs that are highly protein bound, the measurement of free drug concentrations may more accurately reflect drug availability during pregnancy than total drug concentrations, but assays that measure free drug are not always available and may be unreliable due to complications of laboratory separation techniques. For drugs in which free and total drug are available, it is useful prior to pregnancy to establish the total and free drug concentrations required to achieve optimal symptom relief for a given individual. It is also important to note that reported therapeutic ranges may be based on a specific brand, such as enoxaparin brand for LMWH [82,96].

Another limitation of TMD in pregnant populations is drug adherence. In pregnant women, drug adherence is less than optimal due to fear of teratogenic risk or persistent nausea and vomiting, especially of concern in patients with hyperemesis gravidarum [19,20]. Low drug concentrations may therefore not be reflective of the altered pharmacokinetics (discussed previously) but, rather, due to decreased drug exposure and low gastrointestinal absorption. Another possibility leading to low or toxic blood concentrations in the mother that should be considered for certain drugs is pharmacogenomics (how an individual's genes affect his or her response to drugs). For instance, an individual could have an ultra-rapid metabolizing phenotype (which can cause an increase in drug clearance) or a poor metabolizing phenotype (which can cause a decrease in drug clearance) contributing to drug effects that occur independent of pregnancy-induced changes. Without knowing the individual's pharmacogenomics profile, one may not know whether it is pregnancy or a polymorphism in a drug-metabolizing enzyme or transporter that is contributing to the overall drug concentration measured. Currently, much remains unknown about pharmacogenomics applications in maternal–fetal medicine, and no recommendations or guidelines specific to pregnant populations have been put into place.

A branch of science called pharmacometrics is currently being used to address some of these limitations. Modeling approaches such as physiologically based pharmacokinetics and population pharmacokinetics can integrate physiological data, preclinical data, and clinical data to quantify the anticipated changes in the pharmacokinetics of drugs during pregnancy. These approaches may identify which drugs' pharmacokinetics are likely to be altered in pregnancy and support dose adjustments for pregnant women in order to maintain therapeutic objectives [22,97–99].

8.8 CONCLUSIONS

Understanding of the basic physiologic, pharmacokinetic/pharmacodynamic, and metabolic changes that occur during pregnancy, in conjunction with TDM, can aid clinicians in individualizing drug therapy and improving drug safety for pregnant women. There is also theoretical potential for TDM to identify maternal drug concentrations that may be safer or potentially toxic to the fetus. To date, very few drugs have been studied adequately in pregnancy, and many opportunities exist to better understand pharmacotherapy in pregnant women. The literature currently reflects great interindividual variability regarding the effects of pregnancy on the pharmacokinetics/pharmacodynamics and metabolism of drugs. Further research is required to identify biomarkers that would specifically inform which women are at risk for clinically significant pharmacokinetic changes in their medications across pregnancy and the postpartum period [21]. More studies are needed to determine whether the implementation of TDM for individual drugs is cost-effective and can lead to improved clinical outcomes beyond clinical judgment alone.

References

[1] Irvine L, Flynn RW, Libby G, Crombie IK, Evans JM. Drugs dispensed in primary care during pregnancy: a record-linkage analysis in Tayside, Scotland. Drug Saf 2010;33(7):593–604.

[2] Crespin S, Bourrel R, Hurault-Delarue C, Lapeyre-Mestre M, Montastruc JL, Damase-Michel C. Drug prescribing before and during pregnancy in south west France: a retrolective study. Drug Saf 2011;34(7):595–604.

[3] Medication during pregnancy: an intercontinental cooperative study. Collaborative Group on Drug Use in Pregnancy (C.G.D.U.P.). Int J Gynaecol Obstet 1992;39(3):185–96.

[4] Egen-Lappe V, Hasford J. Drug prescription in pregnancy: analysis of a large statutory sickness fund population. Eur J Clin Pharmacol 2004;60(9):659–66.

[5] Drug use in pregnancy: a preliminary report of the International co-operative Drug Utilization Study. Collaborative Group on Drug Use in Pregnancy. Pharm Weekbl Sci 1990;12(2):75–8.

[6] Bonati M, Bortolus R, Marchetti F, Romero M, Tognoni G. Drugs and pregnancy. Medicina (Firenze) 1989;9(3):265–70.

[7] Brocklebank JC, Ray WA, Federspiel CF, Schaffner W. Drug prescribing during pregnancy. A controlled study of Tennessee Medicaid recipients. Am J Obstet Gynecol 1978;132(3):235–44.

[8] Piper JM, Baum C, Kennedy DL. Prescription drug use before and during pregnancy in a Medicaid population. Am J Obstet Gynecol 1987;157(1):148–56.

[9] Bonati M, Bortolus R, Marchetti F, Romero M, Tognoni G. Drug use in pregnancy: an overview of epidemiological (drug utilization) studies. Eur J Clin Pharmacol 1990;38(4):325–8.

[10] Else L, Jackson V, Brennan M, Back D, Khoo S, Coulter-Smith S, et al. Therapeutic drug monitoring of atazanavir/ritonavir in pregnancy. HIV Med 2014;15(10):604–10.

[11] Pirie DA, Al Wattar BH, Pirie AM, Houston V, Siddiqua A, Doug M, et al. Effects of monitoring strategies on seizures in pregnant women on lamotrigine: a meta-analysis. Eur J Obstet Gynecol Reprod Biol 2014;172:26–31.

[12] Reisinger TL, Newman M, Loring DW, Pennell PB, Meador KJ. Antiepileptic drug clearance and seizure frequency during pregnancy in women with epilepsy. Epilepsy Behav 2013;29(1):13–18.

[13] Eberhard-Gran M, Eskild A, Opjordsmoen S. Treating mood disorders during pregnancy: safety considerations. Drug Saf 2005;28(8):695–706.

[14] Schou M. Treating recurrent affective disorders during and after pregnancy. What can be taken safely? Drug Saf 1998;18(2):143–52.

[15] Fishell A. Depression and anxiety in pregnancy. J Popul Ther Clin Pharmacol 2010;Fall 17(3):e363–9.

[16] Marcus SM. Depression during pregnancy: rates, risks and consequences—Motherisk Update 2008. Can J Clin Pharmacol 2009;Winter 16(1):e15–22.

[17] Kelly RH, Russo J, Holt VL, Danielsen BH, Zatzick DF, Walker E, et al. Psychiatric and substance use disorders as risk factors for low birth weight and preterm delivery. Obstet Gynecol 2002;100(2):297–304.

[18] Steer RA, Scholl TO, Hediger ML, Fischer RL. Self-reported depression and negative pregnancy outcomes. J Clin Epidemiol 1992;5(10):1093–9.

[19] Loebstein R, Koren G. Clinical relevance of therapeutic drug monitoring during pregnancy. Ther Drug Monit 2002;24(1):15–22.

[20] Lobstein RLA, Koren G. Pharmacokinetic changes during pregnancy and their clinical relevance. 3rd ed. New York, NY: Marcel Dekker Inc.; 2001.

[21] Deligiannidis KM, Byatt N, Freeman MP. Pharmacotherapy for mood disorders in pregnancy: a review of pharmacokinetic changes and clinical recommendations for therapeutic drug monitoring. J Clin Psychopharmacol 2014;34(2):244–55.

[22] Ke AB, Rostami-Hodjegan A, Zhao P, Unadkat JD. Pharmacometrics in pregnancy: an unmet need. Annu Rev Pharmacol Toxicol 2014;54:53–69.

[23] Tracy TS, Venkataramanan R, Glover DD, Caritis SN. Temporal changes in drug metabolism (CYP1A2, CYP2D6 and CYP3A Activity) during pregnancy. Am J Obstet Gynecol 2005;192(2):633–9.

[24] Gardner MJ, Schatz M, Cousins L, Zeiger R, Middleton E, Jusko WJ. Longitudinal effects of pregnancy on the pharmacokinetics of theophylline. Eur J Clin Pharmacol 1987;32(3):289–95.

[25] Tomson T, Lindbom U, Ekqvist B, Sundqvist A. Disposition of carbamazepine and phenytoin in pregnancy. Epilepsia 1994;35(1):131–5.

[26] Hogstedt S, Lindberg B, Peng DR, Regardh CG, Rane A. Pregnancy-induced increase in metoprolol metabolism. Clin Pharmacol Ther 1985;37(6):688–92.

[27] Hebert MF, Easterling TR, Kirby B, Carr DB, Buchanan ML, Rutherford T, et al. Effects of pregnancy on CYP3A and P-glycoprotein activities as measured by disposition of midazolam and digoxin: a University of Washington specialized center of research study. Clin Pharmacol Ther 2008;84(2):248–53.

[28] Unadkat JD, Wara DW, Hughes MD, Mathias AA, Holland DT, Paul ME, et al. Pharmacokinetics and safety of indinavir in human immunodeficiency virus-infected pregnant women. Antimicrob Agents Chemother 2007;51(2):783–6.

[29] Hebert MF, Ma X, Naraharisetti SB, Krudys KM, Umans JG, Hankins GD, et al. Are we optimizing gestational diabetes treatment with glyburide? The pharmacologic basis for better clinical practice. Clin Pharmacol Ther 2009;85(6):607–14.

[30] Fotopoulou C, Kretz R, Bauer S, Schefold JC, Schmitz B, Dudenhausen JW, et al. Prospectively assessed changes in lamotrigine-concentration in women with epilepsy during pregnancy, lactation and the neonatal period. Epilepsy Res 2009;85(1):60–4.

[31] Tomson T, Battino D. Pharmacokinetics and therapeutic drug monitoring of newer antiepileptic drugs during pregnancy and the puerperium. Clin Pharmacokinet 2007;46(3):209–19.

[32] DeVane CL, Stowe ZN, Donovan JL, Newport DJ, Pennell PB, Ritchie JC, et al. Therapeutic drug monitoring of psychoactive drugs during pregnancy in the genomic era: challenges and opportunities. J Psychopharmacol 2006;20(Suppl. 4):54–9.

[33] Deligiannidis KM. Therapeutic drug monitoring in pregnant and postpartum women: recommendations for SSRIs, lamotrigine, and lithium. J Clin Psychiatry 2010;71(5):649–50.

[34] Adab N. Therapeutic monitoring of antiepileptic drugs during pregnancy and in the postpartum period: is it useful? CNS Drugs 2006;20(10):791–800.

[35] Knott C, Reynolds F. Therapeutic drug monitoring in pregnancy: rationale and current status. Clin Pharmacokinet 1990;19(6):425–33.

[36] Ghiculescu R. Therapeutic drug monitoring: which drugs, why and when, and how to do it. Aust Prescr 2008;31(2):42.

[37] Kang JS, Lee MH. Overview of therapeutic drug monitoring. Korean J Intern Med 2009;24(1):1–10.

[38] Potter JM. Pharmacoeconomics of therapeutic drug monitoring in transplantation. Ther Drug Monit 2000;22(1):36–9.

[39] Schumacher GE, Barr JT. Economic and outcome issues for therapeutic drug monitoring in medicine. Ther Drug Monit 1998;20(5):539–42.

[40] Destache CJ. Use of therapeutic drug monitoring in pharmacoeconomics. Ther Drug Monit 1993;15(6):608–10.

[41] Gross AS. Best practice in therapeutic drug monitoring. Br J Clin Pharmacol 2001;52(Suppl. 1):5S–10S.

[42] Birkett D. Therapeutic drug monitoring. Aust Prescr 1997;20:9–11.

[43] Winter M. Basic clinical pharmacokinetics. Philadelphia, PA: Lippincott Williams & Wilkins; 2004.

[44] Raju KS, Taneja I, Singh SP, Wahajuddin M. Utility of noninvasive biomatrices in pharmacokinetic studies. Biomed Chromatogr 2013;27(10):1354–66.

[45] Pichini S, Altieri I, Zuccaro P, Pacifici R. Drug monitoring in nonconventional biological fluids and matrices. Clin Pharmacokinet 1996;30(3):211–28.

[46] Norris RL, Martin JH, Thompson E, Ray JE, Fullinfaw RO, Joyce D, et al. Current status of therapeutic drug monitoring in Australia and New Zealand: a need for improved assay evaluation, best practice guidelines, and professional development. Ther Drug Monit 2010;32(5):615–23.

[47] Echizen H, Nakura M, Saotome T, Minoura S, Ishizaki T. Plasma protein binding of disopyramide in pregnant and postpartum women, and in neonates and their mothers. Br J Clin Pharmacol 1990;29(4):423–30.

[48] Zhao Y, Hebert MF, Venkataramanan R. Basic obstetric pharmacology. Semin Perinatol 2014;38(8):475–86.

[49] Parry E, Shields R, Turnbull AC. Transit time in the small intestine in pregnancy. J Obstet Gynaecol Br Commonw 1970;77(10):900–1.

[50] Hytten FELT. The physiology of pregnancy. Oxford: Blackwell; 1971.

[51] Anderson GD. Pregnancy-induced changes in pharmacokinetics: a mechanistic-based approach. Clin Pharmacokinet 2005;44(10):989–1008.

[52] Pavek P, Ceckova M, Staud F. Variation of drug kinetics in pregnancy. Curr Drug Metab 2009;10(5):520–9.

[53] Koren G. Pharmacokinetics in pregnancy; clinical significance. J Popul Ther Clin Pharmacol 2011;18(3):e523–7.

[54] Grindheim G, Toska K, Estensen ME, Rosseland LA. Changes in pulmonary function during pregnancy: a longitudinal cohort study. BJOG 2012;119(1):94–101.

[55] Thaler I, Manor D, Itskovitz J, Rottem S, Levit N, Timor-Tritsch I, et al. Changes in uterine blood flow during human pregnancy. Am J Obstet Gynecol 1990;162(1):121–5.

[56] Koren G, Pastuszak A, Ito S. Drugs in pregnancy. N Engl J Med 1998;338(16):1128–37.

[57] Palahniuk RJ, Shnider SM, Eger II EI. Pregnancy decreases the requirement for inhaled anesthetic agents. Anesthesiology 1974;41(1):82–3.

[58] Kerr MG. Cardiovascular dynamics in pregnancy and labour. Br Med Bull 1968;24:19.

[59] Walters Waw LY. Blood volume and haemodynamics in pregnancy. Obstet Gynecol 1975;2:301–2.

[60] Assali NS, Rauramo L, Peltonen T. Measurement of uterine blood flow and uterine metabolism. VIII. Uterine and fetal blood flow and oxygen consumption in early human pregnancy. Am J Obstet Gynecol 1960;79:86–98.

[61] Parker WA. Effects of pregnancy on pharmacokineticsIn: Benet LZ, editor. New York, NY: Raven Press; 1984. p. 249–68.

[62] Quilligan EJKI. Maternal physiology. 3rd ed. New York, NY: Harper and Rox; 1982.

[63] Juchau MR, Mirkin DL, Zachariah PK. Interactions of various 19-nor steroids with human placental microsomal cytochrome P-450 (P-450 hpm). Chem Biol Interact 1976;15(4):337–47.

[64] Harrison LI, Gibaldi M. Influence of cholestasis on drug elimination: pharmacokinetics. J Pharm Sci 1976;65(9):1346–8.

[65] Zamek-Gliszczynski MJ, Kalvass JC, Pollack GM, Brouwer KL. Relationship between drug/metabolite exposure and impairment of excretory transport function. Drug Metab Dispos 2009;37(2):386–90.

[66] Philipson A. Pharmacokinetics of antibiotics in pregnancy and labour. Clin Pharmacokinet 1979;4(4):297–309.

[67] Yerby MS, Friel PN, McCormick K. Antiepileptic drug disposition during pregnancy. Neurology 1992;42(4 Suppl. 5):12–16.

[68] Tsutsumi K, Kotegawa T, Matsuki S, Tanaka Y, Ishii Y, Kodama Y, et al. The effect of pregnancy on cytochrome P4501A2, xanthine oxidase, and N-acetyltransferase activities in humans. Clin Pharmacol Ther 2001;70(2):121–5.

[69] Pennell PB, Newport DJ, Stowe ZN, Helmers SL, Montgomery JQ, Henry TR. The impact of pregnancy and childbirth on the metabolism of lamotrigine. Neurology 2004;62(2):292–5.

[70] Ohman I, Beck O, Vitols S, Tomson T. Plasma concentrations of lamotrigine and its 2-N-glucuronide metabolite during pregnancy in women with epilepsy. Epilepsia 2008;49 (6):1075–80.

[71] Ohman I, Luef G, Tomson T. Effects of pregnancy and contraception on lamotrigine disposition: new insights through analysis of lamotrigine metabolites. Seizure 2008;17 (2):199–202.

[72] Eyal S, Easterling TR, Carr D, Umans JG, Miodovnik M, Hankins GD, et al. Pharmacokinetics of metformin during pregnancy. Drug Metab Dispos 2010;38 (5):833–40.

[73] Joglar JA, Page RL. Treatment of cardiac arrhythmias during pregnancy: safety considerations. Drug Saf 1999;20(1):85–94.

[74] Abele H, Meyer-Wittkopf M. Diagnosis and antenatal treatment of fetal arrhythmias. Ther Umsch 2005;62(1):43–51.

[75] Schentag JJ, Smith IL, Swanson DJ, DeAngelis C, Fracasso JE, Vari A, et al. Role for dual individualization with cefmenoxime. Am J Med 1984;77(6A):43–50.

[76] Chamberlain A, White S, Bawdon R, Thomas S, Larsen B. Pharmacokinetics of ampicillin and sulbactam in pregnancy. Am J Obstet Gynecol 1993;168(2):667–73.

[77] Philipson A. Pharmacokinetics of ampicillin during pregnancy. J Infect Dis 1977;136 (3):370–6.

[78] Heikkila AM. Antibiotics in pregnancy—a prospective cohort study on the policy of antibiotic prescription. Ann Med 1993;25(5):467–71.

[79] Heikkila AM, Erkkola RU. The need for adjustment of dosage regimen of penicillin V during pregnancy. Obstet Gynecol 1993;81(6):919–21.

[80] Fernandez H, Bourget P, Delouis C, Demirdjian S. The administration of tobramycin in the 2nd and 3rd trimester of pregnancy: contribution to a pharmacokinetic study for the adaptation of posology. J Gynecol Obstet Biol Reprod (Paris) 1991;20(1):107–15.

[81] Castellano JM, Narayan RL, Vaishnava P, Fuster V. Anticoagulation during pregnancy in patients with a prosthetic heart valve. Nat Rev Cardiol 2012;9(7):415–24.

[82] Biasiutti FD, Strebel JK. Anticoagulation and antiaggregation during pregnancy. Ther Umsch 2003;60(1):54–8.

[83] Adab N, Smith CT, Vinten J, Williamson PR, Winterbottom JB. Common antiepileptc drugs in pregnancy in women with epilepsy (Review). Cochrane Database of Systematic Reviews 2004. Issue 6. Art No.: CD004848. Available from: http://dx.doi.org/10.1002/14651858.

[84] ACOG educational bulletin. Seizure disorders in pregnancy. Number 231, December 1996. Committee on Educational Bulletins of the American College of Obstetricians and Gynecologists. Int J Gynaecol Obstet 1997;56(3):279–86.

[85] Practice parameter: management issues for women with epilepsy (summary statement). Report of the Quality Standards Subcommittee of the American Academy of Neurology. Epilepsia 1998;39(11):1226–31.

[86] Harden CL, Pennell PB, Koppel BS, Hovinga CA, Gidal B, Meador KJ, et al. Management issues for women with epilepsy—focus on pregnancy (an evidence-based review): III. Vitamin K, folic acid, blood levels, and breast-feeding: report of the quality standards

subcommittee and therapeutics and technology assessment subcommittee of the American Academy of Neurology and the American Epilepsy Society. Epilepsia 2009;50(5):1247–55.

[87] de Haan GJ, Edelbroek P, Segers J, Engelsman M, Lindhout D, Devile-Notschaele M, et al. Gestation-induced changes in lamotrigine pharmacokinetics: a monotherapy study. Neurology 2004;63(3):571–3.

[88] Johannessen SI, Tomson T. Pharmacokinetic variability of newer antiepileptic drugs: when is monitoring needed? Clin Pharmacokinet 2006;45(11):1061–75.

[89] Tomson T, Palm R, Kallen K, Ben-Menachem E, Soderfeldt B, Danielsson B, et al. Pharmacokinetics of levetiracetam during pregnancy, delivery, in the neonatal period, and lactation. Epilepsia 2007;48(6):1111–16.

[90] Longo B, Forinash AB, Murphy JA. Levetiracetam use in pregnancy. Ann Pharmacother 2009;43(10):1692–5.

[91] Roustit M, Jlaiel M, Leclercq P, Stanke-Labesque F. Pharmacokinetics and therapeutic drug monitoring of antiretrovirals in pregnant women. Br J Clin Pharmacol 2008;66(2):179–95.

[92] Tucker JD, Bien CH, Easterbrook PJ, Doherty MC, Penazzato M, Vitoria M, et al. Optimal strategies for monitoring response to antiretroviral therapy in HIV-infected adults, adolescents, children and pregnant women: a systematic review. Aids 2014;28(Suppl. 2): S151–60.

[93] Hendeles L, Weinberger M. Theophylline. A "state of the art" review. Pharmacotherapy 1983;3(1):2–44.

[94] Lateef A, Petri M. Managing lupus patients during pregnancy. Best Pract Res Clin Rheumatol 2013;27(3):435–47.

[95] Paziana K, Del Monaco M, Cardonick E, Moritz M, Keller M, Smith B, et al. Ciclosporin use during pregnancy. Drug Saf 2013;36(5):279–94.

[96] Salt Lake City: ARUP Laboratories; [11.12.2014]; Available from: <http://ltd.aruplab.com/Tests/Pub/0030144>.

[97] Ke AB, Nallani SC, Zhao P, Rostami-Hodjegan A, Isoherranen N, Unadkat JD. A physiologically based pharmacokinetic model to predict disposition of CYP2D6 and CYP1A2 metabolized drugs in pregnant women. Drug Metab Dispos 2013;41(4):801–13.

[98] Ke AB, Nallani SC, Zhao P, Rostami-Hodjegan A, Unadkat JD. Expansion of a PBPK model to predict disposition in pregnant women of drugs cleared via multiple CYP enzymes, including CYP2B6, CYP2C9 and CYP2C19. Br J Clin Pharmacol 2014;77(3):554–70.

[99] Ke AB, Nallani SC, Zhao P, Rostami-Hodjegan A, Unadkat JD. A PBPK model to predict disposition of CYP3A-metabolized drugs in pregnant women: verification and discerning the site of CYP3A induction. CPT Pharmacometrics Syst Pharmacol 2012;1:e3.

CHAPTER 9

Therapeutic Drug Monitoring in Older People

Andrew J. McLachlan[1,2]

[1]Faculty of Pharmacy and Centre for Education and Research on Ageing, University of Sydney and Concord Hospital, Sydney, Australia

[2]National Health and Medical Research Council, Centre of Research Excellence in Medicines and Ageing, University of Sydney, Sydney, Australia

CONTENTS

9.1 Introduction 213

9.2 Clinical Pharmacology of Medicines in Older People: Implications for TDM 214

9.2.1 Age-Related Changes in Pharmacokinetics 215

9.2.2 Age-Related Changes in Pharmacodynamics 217

9.3 Which Drugs and When Should TDM Be Used in Older People? 218

9.4 Studies Investigating TDM in Older People .220

9.5 Pharmacodynamic Monitoring to Guide Dosing in Older People 223

9.6 Cost-Effectiveness of TDM in Older People 223

9.7 Big Data and Best Evidence to Support TDM in Older People 224

9.8 Conclusions 225

Acknowledgments......225

References 225

9.1 INTRODUCTION

The International Association of Therapeutic Drug Monitoring and Clinical Toxicology defines therapeutic drug monitoring (TDM) as a

> multidisciplinary clinical specialty aimed at improving patient care by individually adjusting the dose of drugs for which clinical experience or clinical trials have shown it improved outcome in the general or special populations. It can be based on a priori pharmacogenetic, demographic and clinical information, and/or on the posteriori measurement of blood concentrations of drugs (pharmacokinetic monitoring) and/or biomarkers (pharmacodynamic monitoring). [1]

TDM has also been positively reframed into an active process and described as the "target concentration intervention" (TCI) [2,3]. TCI is an affirmative approach to optimizing medicine dosing using all of the available information, including the known clinical pharmacology properties of the drug, factors that affect variability in response, the clinical and demographic characteristics (age, weight, and serum creatinine) of the patient, and dosing strategies to "target" a specific drug or metabolite concentration to achieve therapeutic benefits while avoiding medication-related harms. As a strategy, TCI also combines modeling and simulation approaches to guide and refine drug dosing regimens [2] based on drug or metabolite concentration observations and changing metrics of organ function or patient clinical status. This is an iterative process, beyond simply "monitoring," that is, a positive intervention to optimize therapeutic outcomes. Both of these working definitions

W. Clarke & A. Dasgupta (Eds): Clinical Challenges in Therapeutic Drug Monitoring. DOI: http://dx.doi.org/10.1016/B978-0-12-802025-8.00009-X

of TDM reflect an active approach of translating information to guide and refine pharmacotherapy. Importantly in geriatrics, the definitions closely align with many of the hallmarks of comprehensive geriatric assessment, which has been recognized as a cornerstone of geriatric care [4]. Like TDM, comprehensive geriatric assessment includes a systematic multidimensional interdisciplinary process with careful planning focused on integrating information and monitoring to guide treatment and long-term follow-up. Comprehensive geriatric assessment, like TDM and TCI, includes both diagnostic and therapeutic elements seeking to identify problems, quantify the response, and manage the care of the older person appropriately [4].

The rationale for using TDM in older people is clearly supported by the elements of comprehensive geriatric assessment and embedded within the important ethical dimensions of quality care for older people. Medical ethics provides an overarching framework that guides prescribing of medicines in older people [5,6]. This framework is dominated by four key principles—beneficence, nonmaleficence, autonomy, and justice [7]—and these can be applied meaningfully to any decisions to commence (prescribe), adjust, or discontinue a medicine [5,6]. Beneficence is equivalent to the effectiveness of a medicine that takes into account the evidence of benefit in older people and individualizing the dose regimen of a medicine to achieve optimal outcomes. Nonmaleficence is the need to avoid harms that can be more frequent and serious in older people. TDM can have a role in avoiding harms by dose optimization and appropriate monitoring. Autonomy relates to informed consent prior to commencing a medicine and embraces shared decision making about the benefits and harms that a medicine might have and how this aligns with priorities for the older person. Justice is the important aspect of avoiding decisions simply based on a person's age or frailty status [5]. TDM is firmly embedded within this ethical meta-framework [5] and is a helpful tool for optimizing outcomes and avoiding harms of medicines in older people [8].

This chapter explores the utility of TDM in guiding ethical decisions related to the optimal use of medicines in older people. It also explores the factors that are unique to the use of TDM in the care of older people.

9.2 CLINICAL PHARMACOLOGY OF MEDICINES IN OLDER PEOPLE: IMPLICATIONS FOR TDM

The application of TDM requires the integration of information about the characteristics and clinical needs of the patient and an understanding of the likely pharmacokinetic (PK) behavior of the drug in the patient cohort and the nature of the concentration–effect relationship or target concentration.

Table 9.1 Summary of the Impact of Pharmacokinetic Changes in Older People and Frailty

Pharmacokinetic Parameter	Aging	Frailty
Absorption	↔	?
Distribution	↓ Total body water, ↑ adipose tissue, ↓ serum albumin concentration	↓↓ Total body water, ↑↑ adipose tissue, ↓↓ serum albumin concentration
Metabolism	↓ Phase 1 metabolism, ? phase 2 metabolism	?↓ Phase 1 metabolism, ↓ phase 2 metabolism
Excretion	↓	↓↓

Note: ↔, no change; ↓, reduced; ↓↓, major decrease; ↑, increased; ↑↑, major increase; ?, unclear based on current data.
Adapted from Refs. [9–12].

9.2.1 Age-Related Changes in Pharmacokinetics

There is a growing body of evidence on the impact of aging on the PK of drugs and metabolites that is summarized in Table 9.1. It is now recognized that there are differences between robust older people and frail older people, with more significant changes in PK of drugs expected in the latter group of patients [9]. Frailty in older people is a multifactorial biological syndrome characterized by the cumulative dysregulation of physiological processes and can be associated with changes in both pharmacodynamics (PD) and PK [9,10].

The rate and extent of drug absorption do not appear to be substantially affected by the aging process. Although there is some suggestion that the rate of absorption of drugs may be slower due to age-related changes in gastrointestinal motility, the extent of drug absorption appears unaffected in robust older people but may be reduced in frail older people [10,11]. Drug distribution is significantly influenced by body composition and the molecular and physicochemical characteristics of the drug or metabolite. Age-related changes in body composition, such as decreased total body water and an increase in the portion of adipose tissue in the body, are likely to impact on the distribution and tissue uptake of aqueous-soluble and lipid-soluble drugs and metabolites [11]. Frailty-associated sarcopenia results in an exaggeration of these changes in body composition (Table 9.1) that is more likely to have an impact on drug PK [10].

The concentration of human serum albumin (and other soluble proteins) is significantly lower (or altered) in older people [11], which has possible implications for the protein binding of drugs and metabolites [13]. Changes in protein binding of drugs and metabolites generally have limited clinical significance because protein binding is typically in equilibrium and rapidly reversible [14]. However, Butler and Begg [15], in an insightful commentary and analysis, identified that age-related changes in the extent of protein binding influences the interpretation of data on hepatic clearance of medicines in older people. These authors highlighted that many studies suggest that hepatic clearance is not influenced by aging, but a reconsideration of drugs with a lower hepatic extraction ratio, which are influenced by the fraction unbound in plasma and the intrinsic activity of the hepatic enzymes, does show a marked decline in the unbound hepatic clearance of these drugs in older people [15]. These findings align with the body of evidence regarding the aging liver and the substantial changes that occur in older people compared to younger robust people.

Age-related changes in the liver include a reduction in liver blood flow (with an effect on the hepatic clearance of high-extraction-ratio drugs and metabolites), decreased liver volume, impaired transfer of solutes across the hepatic vascular endothelium, and substantial changes to hepatic phase 1 and phase 2 drug-metabolizing enzymes [11,12]. The changes in the cytochrome P450 superfamily of enzymes suggest that both the amount and the activity of these isoenzymes are decreased to some extent in robust older people and that these changes are exacerbated in frail older people [9,12]. Current data suggest that conjugation drug metabolism pathways (phase 2 metabolism) appear to be relatively unaffected in robust older people but substantially lower in frail older people [12,16]. A recent commentary speculated on the reasons for the substantial changes in drug metabolism observed in frail older people and attributed this to the impact of frailty as a proinflammatory state that leads to the downregulation of the expression of key drug-metabolizing enzymes [17]. Taken together, these findings have (at least theoretical) implications for prodrugs that require the metabolic generation of an active metabolite. Conversely, the impact of frailty on the aging liver may have a protective effect in circumstances in which a hepatotoxic metabolite is formed. Mitchell et al. [18] investigated the impact of frailty on the hepatotoxicity of acetaminophen (paracetamol) and found that although frail older people achieved higher plasma concentrations of the drug, this was not associated with changes in the activity of liver enzymes, suggesting that frail older people did not generate the highly reactive hepatotoxic metabolite of paracetamol.

The appropriate assessment of renal function is a key aspect of TDM of drugs that are predominately renally excreted in older people. Although increasing age is generally considered to be associated with a decline in renal function, the extent of renal impairment shows considerable interpatient variability in older people [19]. Glomerular filtration rate is reported to decline at an average of approximately 1 mL/min/year as a person ages, but it does not change substantially in healthy robust older people [11,19]. A recent study investigated the impact of frailty on gentamicin pharmacokinetics and found that frailty independently predicted the reduced clearance of gentamicin in older patients (over and above the impact of renal function) [20]. These authors suggested that frailty be considered as an influential covariate when developing dosing guidelines for medicines in older people.

9.2.2 Age-Related Changes in Pharmacodynamics

Understanding the exposure–response relationship and the factors that affect this is a critical component to TDM but often more difficult to establish [21–23], yet this is essential information for guiding dosing recommendations. It is generally accepted that older people are more sensitive to the effects of drugs [11,24]. Older people demonstrate greater sensitivity to drugs that act on the central nervous system, including sedatives, antipsychotics, antidepressants, and opioid analgesics [24,25]. There is evidence that selected cardiovascular drugs also have altered PD in older people [24]. Typically, this means that therapeutic benefit and also harm occur at lower concentrations in older people [11]. However, this increased PD sensitivity is not universal for all drugs and all older patients, highlighting the need for careful clinical monitoring when commencing drugs or altering the dose regimen. The diminished physiological reserve and homeostatic control observed in frail older people also has implications for dose adjustment due to enhanced PD sensitivity [9,10]. The key PK and PD factors that impact the relationships between the administered dose, blood concentration, pharmacological response, and clinical outcome of drugs related to TDM are summarized in Fig. 9.1.

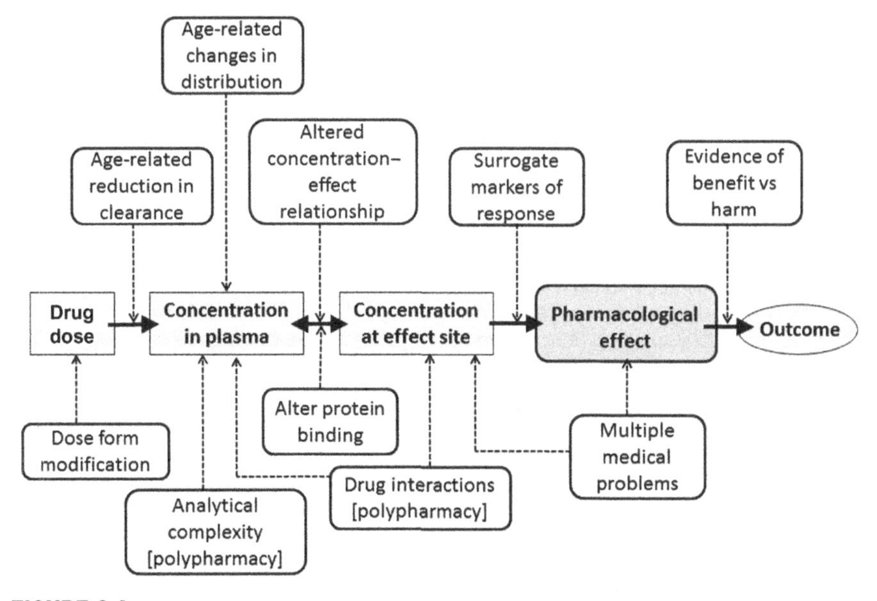

FIGURE 9.1
Factors influencing TDM in older people.

9.3 WHICH DRUGS AND WHEN SHOULD TDM BE USED IN OLDER PEOPLE?

The main indications for TDM in older people are summarized in Table 9.2. A feature of older people as a cohort is significant interpatient variability in response to drugs [9]. In part, this can be attributed to variability in age-related changes in organ function leading to interpatient variability in PK and PD and therefore provides an indication for TDM to guide drug and doses selection. Adverse effects and drug-related problems are often difficult to diagnose in older people with multiple medical problems receiving multiple drugs [26,33,34]. TDM is a tool that can be used to investigate suspected adverse effects or toxicity related to drugs in older people by detecting and confirming changes in drug exposure, especially when drug-related adverse effects may present in an atypical manner [8,9]. In a similar manner, inadequate response to drugs can be a

Table 9.2 Indications for Therapeutic Drug Monitoring in Older Patients

Indication for TDM	Comment
Changes in organ function	Age-related changes in drug elimination (renal and hepatic) pathways require careful assessment of clearance and dose adjustment [9,10].
Investigating suspected adverse effects or toxicity related to drugs	Adverse events and drug-related problems can present in a complex and atypical manner in older people, especially in those who are frail. Most drug-related problems are concentration dependent; hence, the judicious use of concentration or PD monitoring can guide appropriate management for new signs and symptoms when a drug is commenced [26].
Administration of dose forms that have been modified	Dose form modification is common in older people who have swallowing difficulties. This has the potential to substantially alter PK behavior and the need to carefully monitor changes in systemic exposure and clinical response [27].
Inadequate response to drugs	Where a drug has been commenced and there is a clear plan for review of that treatment, it is often difficult to investigate why a person may not have responded adequately. TDM has a useful role in confirming that drug (and metabolite) exposure is adequate to achieve a pharmacological response [28].
Frailty	Sarcopenia can result in significant changes in drug distribution and some drug metabolism/excretion pathways [12,20]. The diminished physiological reserve seen in frail older people can lead to changes in PD and predispose people to adverse events [25].
Understanding adherence to drugs	TDM has been identified as one tool (among many strategies) for assessing adherence to a prescribed dose regimen. A number of studies have highlighted the utility of this approach [29,30].
Drug–drug and drug–disease interactions	Polypharmacy is a key feature of drug use in older people [31] that can lead to potential drug–drug and drug–disease interactions [32]. TDM can be used to explore the impact of concomitant drug therapy or comorbidities on drug exposure and how this might guide drug and dose selection in older people.

challenge in the therapeutic decision-making process. Using TDM to establish that an older person has achieved an adequate systemic exposure is an essential first step before increasing the dose of a drug or switching to an alternative therapy [28]. Related to this is the role that TDM might have in establishing adherence to a prescribed medication regimen as part of a review of therapy due to inadequate response. There are a number of examples in which TDM has been used to identify that poor adherence has been the cause of treatment failure [29,30].

Older frail people more commonly experience swallowing difficulties necessitating the alteration of dose forms for oral administration. This might include crushing tablets, opening capsules, and administration via a nasogastric tube to minimize the risk of aspiration pneumonia in older people with dysphagia in residential aged care. Dose form modification has the potential to significantly alter systemic bioavailability, and TDM has a role in assessing the impact of changes to the methods administration [27].

Polypharmacy is a common feature of the care of older patients [31] and poses a number of challenges in managing older people. The potential for serious drug–drug and drug–disease interactions increases as the number of drugs and comorbid medical conditions increases for older people [32]. TDM has a role in managing this complexity and unmasking multiple drug and disease interactions to guide optimal therapy [35]. In this situation, TDM may also have a role in the safe and judicious cessation of selected drugs [36].

Polypharmacy also poses a substantial analytical challenge for the routine implementation of TDM in older people. For example, Koeber et al. [37] developed an assay for the routine measurement of donepezil in the treatment of older people living with Alzheimer's disease. As part of the analytical validation for this assay, the authors evaluated the potential analytical interference for 100 of the most commonly used concomitant medications in this patient population [37].

There is some debate about which drugs should be the focus of TDM in older people [8]. At least one rationale for using TDM relates to the safety margin of the drugs of interest, such that TDM is recommended for drugs with a narrow therapeutic index such as antiarrhythmic agents and anticonvulsants [28,38]. In older people, especially those who are frail, the harm-to-benefit ratio for many drugs is substantially shifted due to age-related changes in PK and PD. As discussed previously, this means that frail older people are more sensitive to the harmful effects of drugs and thus more drugs meet the criteria of having a narrow safety margin.

Dose adjustment is a critical part of TDM for renally excreted drugs (and their metabolites) in older people, such as aminoglycosides, metformin, β-lactams, digoxin, lithium, and allopurinol [39]. Most guidelines currently recommend the use of the Cockcroft–Gault equation, modification of diet in renal disease (MDRD) equation, or the chronic kidney disease epidemiology collaboration (CKD-EPI) equation, all of which use serum creatinine concentration and include age as a variable to estimate a person's renal function. Roberts et al. [40] highlighted that the MDRD equation overestimates renal function in some older people, which would lead to higher than appropriate doses for renally excreted drugs. A detailed analysis by Chin et al. [41] highlighted that the CKD-EPI equation (corrected for body surface area) and the Cockcroft–Gault equation provided the best estimates of renal function for dose adjustment of aminoglycosides.

The introduction of the nonvitamin K antagonist oral anticoagulants (NOACs), including dabigatran, apixaban, and rivaroxaban, represented a major development in the management of venous thromboembolism and the prevention of complications in older people with atrial fibrillation. Although the initial premise of using these agents (compared to warfarin) assumed no need for routine monitoring, recent studies have focused on how monitoring coagulation markers and drug concentration [42,43] have utility in individualizing dosing and reducing the risk of bleeding complications in older people [44]. Taken together, there is now compelling evidence for the need to routinely monitor renal function [39,45] in older people receiving NOACs and to monitor for the clinical signs of bleeding. This complexity in the clinical pharmacology of NOACs coupled with the comorbidity and frailty associated with many older people has led some authors to conclude that NOACs have limited utility in stroke prevention in older people [46], but if these drugs are used, they should be used with careful monitoring [47].

9.4 STUDIES INVESTIGATING TDM IN OLDER PEOPLE

Selected PK studies with relevance to TDM in older people are summarized in Table 9.3. Most published studies are observational in nature and investigate routinely collected TDM data for older people in both hospital and community settings. Some studies utilize these data to provide insights into the causes and consequences of variability in PK using pharmacometrics analyses. Other studies investigate the real-world aspects of treatment, such as the impact of drug interactions and medical comorbidity. In a series of well-designed studies, Berkovitch, Leibovitz, and Segal (Table 9.2) investigated the influence of oral dose form modification on the PK of drugs in frail older people and the importance of using a TDM approach.

Table 9.3 Examples of Pharmacokinetic Studies in Older People That Have Implications for Therapeutic Drug Monitoring and Individualized Dosing

Drug(s) of Interest	Study Cohort	Study Design	Comments
Theophylline [48]	32 frail older people (age: 63–98 years); 17 people with swallowing difficulties	Prospective parallel group PK study; modified-release (MR) tablet versus opening a capsule of MR granules via nasogastric tube	Opening a theophylline capsule of MR granules and administration via a nasogastric tube led to theophylline concentrations not reaching therapeutic targets (especially compared to theophylline MR tablets). This has implications for administration of drugs in frail older people with swallowing difficulties.
Valproic acid [49]	146 nursing home residents (age: 65–99 years); 405 observations	Retrospective observational study from routine TDM and pharmacometric analysis	Apparent clearance not affected by age or weight but was lower in female residents (27%), greater in people also receiving carbamazepine or phenytoin (41%), and higher (25%) in people receiving the liquid formulation.
Digoxin [50]	119 older people of Chinese ancestry (age: 60–88 years); 173 observations	Retrospective study of routinely collected TDM data and pharmacometric analysis	Concomitant administration of spironolactone, body weight, and creatinine concentration influenced clearance of digoxin in older patients.
Cyclosporine [51]	36 stable older renal allograft recipients (age: 65–77 years)	Observational study of PK and PK (residual nuclear factor of activated T cell-regulated gene expression); opportunistic infection rate	Level of immunosuppression correlated with more opportunistic infections. PK and PD endpoints can be used to individualize cyclosporine dosing in older renal transplant recipients to improve health outcomes.
Antidepressants [52]	17,930 patients with 15% aged >65 years; 32,126 observations	Retrospective observational study from a routine TDM database	Older patients (>65 years) had an exposure of 1.5- to 2-fold higher for most antidepressants compared to people younger than 40 years at the same dose; implications for adverse effects in older people.
Carbamazepine [53]	121 community-dwelling patients (age: 60 years or older); 555 observations	Prospective pharmacometric study investigating factors affecting clearance	Concomitant phenytoin administration significantly reduced clearance (23%); no effect of body size, age, or hepatic or renal function. Dosing to a targeted concentration is critical in older patients.

Continued...

Table 9.3 Examples of Pharmacokinetic Studies in Older People That Have Implications for Therapeutic Drug Monitoring and Individualized Dosing *Continued*

Drug(s) of Interest	Study Cohort	Study Design	Comments
Lithium [54]	759 in- and outpatients from three teaching hospitals (288 aged 60 years or older); 19,392 observations	Retrospective observational study from a routine TDM; consider time in target range	Age was not identified as a factor that influenced instability in lithium serum concentrations. The time within the target concentration range was comparative across adult age ranges, including those >70 years.
Anticonvulsants [55]	1050 patients, 82 older people (age: 65–93 years)	Retrospective observational study of routinely collected TDM data from the National Center for Epilepsy, Norway	Age affected the apparent clearance of levetiracetam (40% reduction in older people). Concomitant administration of inducers and inhibitors affected lamotrigine (± 70%) apparent clearance. Inducers increased topiramate (25%), levetiracetam (25%), and oxcarbazepine (75%) apparent clearance. Valproic acid inhibited topiramate (25%) apparent clearance.
Ciprofloxacin [56]	20 frail older people (mean age: 80 years); 10 people with swallowing difficulties	Prospective parallel group PK study; tablet versus crushed tablet via nasogastric tube	Ciprofloxacin concentrations (C_{max} and AUC) were not different in frail older people when the tablet dose form was crushed and administered through a nasogastric tube compared to an oral tablet.
Donepezil [57]	106 patients with Alzheimer's dementia (age: 72 ± 9 years); 206 observations	Retrospective observational study Clinical Global Impression (CGI) scores to assess dementia response	High variability in donepezil serum concentrations at standard dosing. Concentrations at or above 50 ng/mL were associated with an improved response based on CGI. TDM may be used to individualize donepezil treatment.
Selective serotonin reuptake inhibitors [58]	6333 patients (age: 60 years or older); 8457 observations	Retrospective observational study from routine TDM of inpatients and outpatients	TDM in older patients could aid in the early detection of drug-related problems. TDM is less frequent in older people compared to younger patients. People older than 90 years had the lowest rate of TDM.
Roxithromycin [27]	20 frail older people (mean age: 80 years); 10 people with swallowing difficulties	Prospective parallel group PK study; tablet versus crushed tablet via nasogastric tube	Administration as crushed tablet via a nasogastric tube led to substantially higher roxithromycin concentrations compared with oral administration as a tablet.

9.5 PHARMACODYNAMIC MONITORING TO GUIDE DOSING IN OLDER PEOPLE

TDM is not just about trough concentration measurements at steady state but also increasingly about using smart strategies to measure and monitor drug response as part of routine patient care. There is now well-established evidence for the efficacy and safety of point-of-care self-monitoring of the international normalized ratio for older people receiving warfarin [59], and there is growing evidence of the utility of this TDM approach in more vulnerable older people residing in residential aged care facilities [60]. The integration of monitoring systems with technology platforms has led to significant improvements in patient outcomes. For example, remote PD telemonitoring to guide treatment adjustment and individualization can lead to improved outcomes in people with chronic heart failure [61].

There is a growing understanding of monitoring surrogate endpoints of clinical outcomes with portable and wearable [62,63] technologies to guide dose optimization. These innovative approaches to real-time monitoring of response to drugs will have increased importance in the future, especially in older and more vulnerable patients. For example, point-of-care (portable) PD testing could allow for optimal dosing in assessing cardiac rhythm in older people receiving antiarrhythmic agents. In this context, antiarrhythmic agents are used to prevent the occurrence of potentially fatal cardiac events. Real-time monitoring of heart rhythm using a minimally invasive approach makes perfect sense. A study by Lowres et al. [64] demonstrated the utility and cost-effectiveness of using a smartphone (iPhone) electrocardiogram (iECG) in community pharmacy to screen for atrial fibrillation but also investigated heart rhythm in people receiving antiarrhythmic or anticoagulant drugs for atrial fibrillation prevention. This study identified a new paradigm in patient-controlled screening of heart rhythm that includes features such as algorithms to assess and report on cardiac rhythm on the device or uploading the iECG using cloud-based computing to allow independent external review and advice [64].

9.6 COST-EFFECTIVENESS OF TDM IN OLDER PEOPLE

Any health care intervention needs to establish that it is not only safe and effective in practice but also cost-effective. Touw et al. [65] have made a compelling case for the need to consider the cost-effectiveness of TDM by presenting a systematic review of available studies investigating the value of using TDM as a therapeutic intervention. Although a number of studies have established the cost-effectiveness of TDM as a health intervention [66], few studies have specifically focused on this aspect of TDM in older people. The study by Ostad Haji et al. [67] indicated that TDM is cost-effective in managing hospitalized older people undergoing treatment with citalopram. This benefit relied on the use of TDM to optimize citalopram treatment resulting in a shorter duration of hospitalization. It is clear that further evidence of the cost-effectiveness of TDM in older people is needed. Although the value of TDM may seem obvious in avoiding harm, it is essential to quantify this, as would be expected for any health intervention in a resource-constrained environment.

9.7 BIG DATA AND BEST EVIDENCE TO SUPPORT TDM IN OLDER PEOPLE

One of the challenges of understanding the role and impact of TDM (or TCI) in improving health outcomes for older people is the limited number of rigorously designed studies linked to meaningful outcomes (Table 9.3). Indeed, systematic reviews (eg, Kredo et al. [68]) that have attempted to quantify the benefit (and harms) of TDM in different settings of care have reported that studies tend to be underpowered (with small sample sizes), have relatively short duration of follow-up to allow assessment of meaningful outcomes, and there is generally poor uptake of TDM recommendations. Although there are an increasing number of well-designed studies that rigorously investigate the impact of targeting a specific systemic exposure [69] and the use of TDM for drugs such as theophylline [70], anticancer drugs [71], antimicrobial agents [72], and antifungal agents [73], the direct application to TDM in older people is lacking. Generating strong evidence to inform the optimal use of TDM is difficult and typically involves controlled trials that have logistic and funding challenges. However, there are some excellent examples of randomized controlled trials investigating TDM in the setting of infectious disease [73,74] and oncology [71].

An alternative approach is to consider using linked health data sets to connect health outcomes of clinical interest (eg, hospitalization) to measures of drug exposure such as dispensed drugs and monitoring to investigate the real-world impact of TDM in older people. This would require electronic health records [75], including clinical chemistry data that could be linked to health outcomes. Such use of pharmacoepidemiologic methods has already demonstrated the ability of this approach to investigate side effects [76] and provide insight into the real-world impact of drug interactions [77].

9.8 CONCLUSIONS

The need to individualize drug dosing is a critical component of quality care for older vulnerable patients and forms an essential part of comprehensive geriatric assessment. TDM in older people is complicated by age-related changes in PK and PD. Despite this, there is a key role for the judicious and appropriate use of TDM in older people to ensure optimal clinical outcomes from drugs. Innovations in analytical and emerging technologies for monitoring provide a key opportunity for guiding the outcomes of pharmacotherapy in older people who are vulnerable to the adverse effects of drugs.

Acknowledgments

Professor David Le Couteur and Professor Sarah Hilmer are acknowledged for their fruitful discussions and insights.

References

[1] International Association of Therapeutic Drug Monitoring and Clinical Toxicology, therapeutic drug monitoring definition, <http://www.iatdmct.org/about-us/about-association/about-definitions-tdm-ct.html>; 2015 [accessed 12.11.15].

[2] Holford NGH. Target concentration intervention: beyond Y2K. Br J Clin Pharmacol 1999;48:9–13.

[3] Saleem M, Dimeski G, Kirkpatrick CM, Taylor PJ, Martin JH. Target concentration intervention in oncology: where are we at? Ther Drug Monit 2012;34(3):257–65.

[4] Ellis G, Whitehead MA, O'Neill D, Langhorne P, Robinson D. Comprehensive geriatric assessment for older adults admitted to hospital. Cochrane Database Syst Rev 2011;(7):CD006211.

[5] Le Couteur DG, Ford GA, McLachlan AJ. Evidence, ethics and medication management in older people. J Pharm Pract Res 2010;40:148–52.

[6] Kouladjian L, Chen TF, Hilmer SN. First do no harm: a real need to deprescribe in older patients. Med J Aust 2015;202(4):179.

[7] Gillon R. Ethics needs principles—four can encompass the rest—and respect for autonomy should be "first among equals". J Med Ethics 2003;29(5):307–12.

[8] Ruiz JG, Array S, Lowenthal DT. Therapeutic drug monitoring in the elderly. Am J Ther 1996;3(12):839–60.

[9] McLachlan AJ, Hilmer SN, Le Couteur DG. Variability in response to drugs in older people: phenotypic and genotypic factors. Clin Pharmacol Ther 2009;85(4):431–3.

[10] Hilmer SN, McLachlan AJ, Le Couteur DG. Clinical pharmacology in the geriatric patient. Fundam Clin Pharmacol 2007;21(3):217–30.

[11] McLean AJ, Le Couteur DG. Aging biology and geriatric clinical pharmacology. Pharmacol Rev 2004;56(2):163–84.

[12] McLachlan AJ, Pont LG. Drug metabolism in older people—a key consideration in achieving optimal outcomes with drugs. J Gerontol A Biol Sci Med Sci 2012;67(2):175–80.

[13] Grandison MK, Boudinot FD. Age-related changes in protein binding of drugs: implications for therapy. Clin Pharmacokinet 2000;38(3):271–90.

[14] Benet LZ, Hoener BA. Changes in plasma protein binding have little clinical relevance. Clin Pharmacol Ther 2002;71(3):115–21.

[15] Butler JM, Begg EJ. Free drug metabolic clearance in elderly people. Clin Pharmacokinet 2008;47(5):297–321.

[16] Wynne HA, Cope LH, Herd B, Rawlins MD, James OF, Woodhouse KW. The association of age and frailty with paracetamol conjugation in man. Age Ageing 1990;19(6):419–24.

[17] Tan JL, Eastment JG, Poudel A, Hubbard RE. Age-related changes in hepatic function: an update on implications for drug therapy. Drugs Aging 2015; Nov 7 (in press)

[18] Mitchell SJ, Hilmer SN, Murnion BP, Matthews S. Hepatotoxicity of therapeutic short-course paracetamol in hospital inpatients: impact of ageing and frailty. J Clin Pharm Ther 2011;36(3):327–35.

[19] Aymanns C, Keller F, Maus S, Hartmann B, Czock D. Review on pharmacokinetics and pharmacodynamics and the aging kidney. Clin J Am Soc Nephrol 2010;5(2):314–27.

[20] Johnston C, Hilmer SN, McLachlan AJ, Matthews ST, Carroll PR, Kirkpatrick CM. The impact of frailty on pharmacokinetics in older people: using gentamicin population pharmacokinetic modeling to investigate changes in renal drug clearance by glomerular filtration. Eur J Clin Pharmacol 2014;70(5):549–55.

[21] Dolton MJ, McLachlan AJ. Optimizing azole antifungal therapy in the prophylaxis and treatment of fungal infections. Curr Opin Infect Dis 2014;27(6):493–500.

[22] Dolton MJ, Brüggemann RJ, Burger DM, McLachlan AJ. Understanding variability in posaconazole exposure using an integrated population pharmacokinetic analysis. Antimicrob Agents Chemother 2014;58(11):6879–85.

[23] Dolton MJ, McLachlan AJ. Voriconazole pharmacokinetics and exposure-response relationships: assessing the links between exposure, efficacy and toxicity. Int J Antimicrob Agents 2014;44(3):183–93.

[24] Bowie MW, Slattum PW. Pharmacodynamics in older adults: a review. Am J Geriatr Pharmacother 2007;5(3):263–303.

[25] McLachlan AJ, Bath S, Naganathan V, Hilmer SN, Le Couteur DG, Gibson SJ, et al. Clinical pharmacology of analgesic drugs in older people: impact of frailty and cognitive impairment. Br J Clin Pharmacol 2011;71(3):351–64.

[26] Rochon PA, Gurwitz JH. Optimising drug treatment for elderly people: the prescribing cascade. Br Med J 1997;315(7115):1096–9.

[27] Britzi M, Berkovitch M, Soback S, Leibovitz A, Segal R, Smagarinsky M, et al. Roxithromycin pharmacokinetics in hospitalized geriatric patients: oral administration of whole versus crushed tablets. Ther Drug Monit 2015;37(4):512–15.

[28] Gross AS. Best practice in therapeutic drug monitoring. Br J Clin Pharmacol 1998;46 (2):95–9.

[29] Chung O, Vongpatanasin W, Bonaventura K, Lotan Y, Sohns C, Haverkamp W, et al. Potential cost-effectiveness of therapeutic drug monitoring in patients with resistant hypertension. J Hypertens 2014;32(12):2411–21.

[30] Samsonsen C, Reimers A, Bråthen G, Helde G, Brodtkorb E. Nonadherence to treatment causing acute hospitalizations in people with epilepsy: an observational, prospective study. Epilepsia 2014;55(11):e125–8.

[31] Hilmer SN, Gnjidic D. The effects of polypharmacy in older adults. Clin Pharmacol Ther 2009;85(1):86–8.

[32] Dumbreck S, Flynn A, Nairn M, Wilson M, Treweek S, Mercer SW, et al. Drug-disease and drug–drug interactions: systematic examination of recommendations in 12 UK national clinical guidelines. Br Med J 2015;350:h949.

[33] Handler SM, Wright RM, Ruby CM, Hanlon JT. Epidemiology of medication-related adverse events in nursing homes. Am J Geriatr Pharmacother 2006;4(3):264–72.

[34] Spinewine A, Schmader KE, Barber N, Hughes C, Lapane KL, Swine C, et al. Appropriate prescribing in elderly people: how well can it be measured and optimised? Lancet 2007;370 (9582):173–84.

[35] Mannheimer B, von Bahr C, Pettersson H, Eliasson E. Impact of multiple inhibitors or substrates of cytochrome P450 2D6 on plasma risperidone levels in patients on polypharmacy. Ther Drug Monit 2008;30(5):565–9.

[36] Scott IA, Hilmer SN, Reeve E, Potter K, Le Couteur D, Rigby D, et al. Reducing inappropriate polypharmacy: the process of deprescribing. JAMA Intern Med 2015;175(5):827–34.

[37] Koeber R, Kluenemann HH, Waimer R, Koestlbacher A, Wittmann M, Brandl R, et al. Implementation of a cost-effective HPLC/UV-approach for medical routine quantification of donepezil in human serum. J Chromatogr B Analyt Technol Biomed Life Sci 2012;881–882:1–11.

[38] Ghiculescu RA. Therapeutic drug monitoring: which drugs, why, when and how to do it. Aust Prescr 2008;31:42–4.

[39] Helldén A, Odar-Cederlöf I, Nilsson G, Sjöviker S, Söderström A, Euler Mv, et al. Renal function estimations and dose recommendations for dabigatran, gabapentin and valaciclovir: a data simulation study focused on the elderly. BMJ Open 2013;3(4) pii: e002686

[40] Roberts GW, Ibsen PM, Schioler CT. Modified diet in renal disease method overestimates renal function in selected elderly patients. Age Ageing 2009;38:698–703.

[41] Chin PK, Florkowski CM, Begg EJ. The performances of the Cockcroft-Gault, modification of diet in renal disease study and chronic kidney disease epidemiology collaboration equations in predicting gentamicin clearance. Ann Clin Biochem 2013;50(Pt 6):546–57.

[42] Owada S, Tomita H, Kinjo T, Ishida Y, Itoh T, Sasaki K, et al. CHA2DS2-VASc and HAS-BLED scores and activated partial thromboplastin time for prediction of high plasma concentration of dabigatran at trough. Thromb Res 2015;135(1):62–7.

[43] Šinigoj P, Malmström RE, Vene N, Rönquist-Nii Y, Božič-Mijovski M, Pohanka A, et al. Dabigatran concentration: variability and potential bleeding prediction in "Real-Life" patients with atrial fibrillation. Basic Clin Pharmacol Toxicol 2015;117(5):323–9.

[44] Kim D, Barna R, Bridgeman MB, Brunetti L. Novel oral anticoagulants for stroke prevention in the geriatric population. Am J Cardiovasc Drugs 2014;14(1):15–29.

[45] Chin PK, Wright DF, Zhang M, Wallace MC, Roberts RL, Patterson DM, et al. Correlation between trough plasma dabigatran concentrations and estimates of glomerular filtration rate based on creatinine and cystatin C. Drugs R D 2014;14(2):113–23.

[46] Turagam MK, Velagapudi P, Flaker GC. Stroke prevention in the elderly atrial fibrillation patient with comorbid conditions: focus on non-vitamin K antagonist oral anticoagulants. Clin Interv Aging 2015;10:1431–44.

[47] Skeppholm M, Hjemdahl P, Antovic JP, Muhrbeck J, Eintrei J, Rönquist-Nii Y, et al. On the monitoring of dabigatran treatment in "real life" patients with atrial fibrillation. Thromb Res 2014;134(4):783–9.

[48] Berkovitch M, Dafni O, Leiboviz A, Mayan H, Habut B, Segal R. Therapeutic drug monitoring of theophylline in frail elderly patients: oral compared with nasogastric tube administration. Ther Drug Monit 2002;24(5):594–7.

[49] Birnbaum AK, Ahn JE, Brundage RC, Hardie NA, Conway JM, Leppik IE. Population pharmaco-kinetics of valproic acid concentrations in elderly nursing home residents. Ther Drug Monit 2007;29(5):571–5.

[50] Zhou XD, Gao Y, Guan Z, Li ZD, Li J. Population pharmacokinetic model of digoxin in older Chinese patients and its application in clinical practice. Acta Pharmacol Sin 2010;31(6):753–8.

[51] Sommerer C, Schnitzler P, Meuer S, Zeier M, Giese T. Pharmacodynamic monitoring of cyclosporin A reveals risk of opportunistic infections and malignancies in renal transplant recipients 65 years and older. Ther Drug Monit 2011;33(6):694–8.

[52] Waade RB, Molden E, Refsum H, Hermann M. Serum concentrations of antidepressants in the elderly. Ther Drug Monit 2012;34(1):25–30.

[53] Punyawudho B, Ramsay ER, Brundage RC, Macias FM, Collins JF, Birnbaum AK. Population pharmacokinetics of carbamazepine in elderly patients. Ther Drug Monit 2012;34(2):176–81.

[54] van Melick EJ, Souverein PC, den Breeijen JH, Tusveld CE, Egberts TC, Wilting I. Age as a determinant of instability of serum lithium concentrations. Ther Drug Monit 2013;35(5):643–8.

[55] Johannessen Landmark C, Baftiu A, Tysse I, Valsø B, Larsson PG, Rytter E, et al. Pharmacokinetic variability of four newer antiepileptic drugs, lamotrigine, levetiracetam, oxcarbazepine, and topiramate: a comparison of the impact of age and comedication. Ther Drug Monit 2012;34(4):440–5.

[56] Lubart E, Berkovitch M, Leibovitz A, Britzi M, Soback S, Bukasov Y, et al. Pharmacokinetics of ciprofloxacin in hospitalized geriatric patients: comparison between nasogastric tube and oral administration. Ther Drug Monit 2013;35(5):653–6.

[57] Hefner G, Brueckner A, Hiemke C, Fellgiebel A. Therapeutic drug monitoring for patients with Alzheimer dementia to improve treatment with donepezil. Ther Drug Monit 2015;37(3):353–61.

[58] Hermann M, Waade RB, Molden E. Therapeutic drug monitoring of selective serotonin reuptake inhibitors in elderly patients. Ther Drug Monit 2015;37(4):546–9.

[59] Heneghan C, Alonso-Coello P, Garcia-Alamino JM, Perera R, Meats E, Glasziou P. Self-monitoring of oral anticoagulation: a systematic review and meta-analysis. Lancet 2006;367(9508):404–11.

[60] Bereznicki LR, Jackson SL, Kromdijk W, Gee P, Fitzmaurice K, Bereznicki BJ, et al. Improving the management of warfarin in aged-care facilities utilising innovative technology: a proof-of-concept study. Int J Pharm Pract 2014;22(1):84–91.

[61] Kitsiou S, Paré G, Jaana M. Effects of home telemonitoring interventions on patients with chronic heart failure: an overview of systematic reviews. J Med Internet Res 2015;17(3):e63.

[62] Redmond SJ, Lovell NH, Yang GZ, Horsch A, Lukowicz P, Murrugarra L, et al. What does big data mean for wearable sensor systems? Contribution of the IMIA wearable sensors in healthcare WG. Yearb Med Inform 2014;9:135–42.

[63] Appelboom G, Camacho E, Abraham ME, Bruce SS, Dumont EL, Zacharia BE, et al. Smart wearable body sensors for patient self-assessment and monitoring. Arch Public Health 2014;72(1):28.

[64] Lowres N, Neubeck L, Salkeld G, Krass I, McLachlan AJ, Redfern J, et al. Feasibility and cost-effectiveness of stroke prevention through community screening for atrial fibrillation using iPhone ECG in pharmacies. The SEARCH-AF study. Thromb Haemost 2014;111(6):1167–76.

[65] Touw DJ, Neef C, Thomson AH, Vinks AA. Cost-effectiveness of therapeutic drug monitoring committee of the international association for therapeutic drug monitoring and clinical toxicology. Cost-effectiveness of therapeutic drug monitoring: a systematic review. Ther Drug Monit 2005;27(1):10–17.

[66] van Lent-Evers NA, Mathôt RA, Geus WP, van Hout BA, Vinks AA. Impact of goal-oriented and model-based clinical pharmacokinetic dosing of aminoglycosides on clinical outcome: a cost-effectiveness analysis. Ther Drug Monit 1999;21(1):63–73.

[67] Ostad Haji E, Mann K, Dragicevic A, Müller MJ, Boland K, Rao ML, et al. Potential cost-effectiveness of therapeutic drug monitoring for depressed patients treated with citalopram. Ther Drug Monit 2013;35(3):396–401.

[68] Kredo T, Van der Walt JS, Siegfried N, Cohen K. Therapeutic drug monitoring of antiretrovirals for people with HIV. Cochrane Database Syst Rev 2009;(3):CD007268.

[69] Grahnén A, Karlsson MO. Concentration-controlled or effect-controlled trials: useful alternatives to conventional dose-controlled trials? Clin Pharmacokinet 2001;40(5):317–25.

[70] Holford N, Black P, Couch R, Kennedy J, Briant R. Theophylline target concentration in severe airways obstruction—10 or 20 mg/L? A randomised concentration-controlled trial. Clin Pharmacokinet 1993;25(6):495–505.

[71] Gotta V, Widmer N, Decosterd LA, Chalandon Y, Heim D, Gregor M, et al. Clinical usefulness of therapeutic concentration monitoring for imatinib dosage individualization: results from a randomized controlled trial. Cancer Chemother Pharmacol 2014;74(6):1307–19.

[72] Sime FB, Roberts MS, Tiong IS, Gardner JH, Lehman S, Peake SL, et al. Can therapeutic drug monitoring optimize exposure to piperacillin in febrile neutropenic patients with haematological malignancies? A randomized controlled trial. J Antimicrob Chemother 2015;70(8):2369–75.

[73] Park WB, Kim NH, Kim KH, Lee SH, Nam WS, Yoon SH, et al. The effect of therapeutic drug monitoring on safety and efficacy of voriconazole in invasive fungal infections: a randomized controlled trial. Clin Infect Dis 2012;55(8):1080–7.

[74] De Waele JJ, Carrette S, Carlier M, Stove V, Boelens J, Claeys G, et al. Therapeutic drug monitoring-based dose optimisation of piperacillin and meropenem: a randomised controlled trial. Intensive Care Med 2014;40(3):380–7.

[75] Nebeker JR, Hurdle JF, Bair BD. Future history: medical informatics in geriatrics. J Gerontol A Biol Sci Med Sci 2003;58(9):M820–5.

[76] Pearce A, Haas M, Viney R, Haywood P, Pearson SA, van Gool K, et al. Can administrative data be used to measure chemotherapy side effects? Expert Rev Pharmacoecon Outcomes Res 2015;15(2):215–22.

[77] Vitry AI, Roughead EE, Ramsay EN, Preiss AK, Ryan P, Gilbert AL, et al. Major bleeding risk associated with warfarin and co-medications in the elderly population. Pharmacoepidemiol Drug Saf 2011;20(10):1057–63.

CHAPTER 10

Therapeutic Drug Monitoring in Obese Patients

Ventzislava Hristova and William Clarke
Department of Pathology, Johns Hopkins University School of Medicine, Baltimore, MD, United States

CONTENTS

10.1 Introduction 231

10.2 PK in Obesity 234
10.2.1 Absorption...........234
10.2.2 Distribution.........234
10.2.3 Clearance............235
10.2.4 Metabolism.........235
10.2.5 Renal Elimination 236

10.3 Chemotherapy in Obese Patients.......237

10.4 Antimicrobial Treatment in Obese Patients.......................239

10.5 PK After Bariatric Surgery.......................240

10.6 Conclusions 242

References 242

10.1 INTRODUCTION

Special populations, with respect to medical toxicology, represent individuals with a unique physiology that leads to altered pharmacokinetics (PK). Therapeutic drug monitoring (TDM) is essential in these groups because the absorbance, metabolism, and clearance of drugs do not follow a clinically predicted route by standard PK. Altered drug PK and pharmacodynamics are reported in obese patients due to increased blood volume, body mass, and cardiac output.

The rate of obesity is fast growing and represents an emerging special population with respect to medical toxicology, in which effective and safe dosing must be determined for individuals of varying age who belong to this category. The World Health Organization (WHO) defines obesity as a body mass index (BMI) greater than or equal to 30. WHO estimates that in 2014, more than 600 million adults (aged 18 years or older), close to 13% of the world population, were obese. The Centers for Disease Control and Prevention (CDC) indicates that in the United States, 34.9% of adults and 17% of children and adolescents (aged 2–19 years) are obese [1,2]. Obesity is associated with numerous health complications of varying severity that require treatment and often hospitalization. According to the CDC, in 2008 the medical cost of obesity amounted to a staggering $147 billion. These statistics emphasize the need to adjust patient treatment to prevent toxicity and optimize efficacy in the obese population. Obesity is linked to a number of health problems, including hypertension, cardiovascular disease, type II diabetes mellitus (T2DM), respiratory complications, hepatic steatosis, chronic pain, and depression. A high prevalence of serious health complications coupled with poorly understood dosing parameters in obese patients make treatment

W. Clarke & A. Dasgupta (Eds): Clinical Challenges in Therapeutic Drug Monitoring. DOI: http://dx.doi.org/10.1016/B978-0-12-802025-8.00010-6

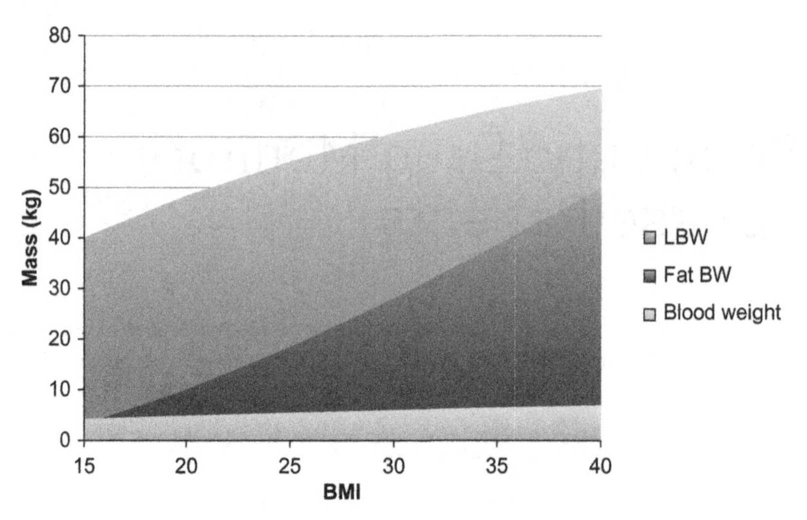

FIGURE 10.1
TBW in obesity can be attributed to an increase in LBW, fat body weight (fat BW), and blood weight. Despite a rise in all parameters, fat body weight increases at a greater rate with higher BMI. *Source: Data from Zuckerman et al. [3].*

very difficult, and TDM is essential to achieve and maintain required drug levels while preventing toxicity or suboptimal dosing.

Although BMI is used to define obesity, it does not differentiate adipose tissue from muscle mass, and it serves as a descriptor of body size but not composition. Obesity leads to high total body weight (TBW), which is attributed to increased adipose tissue mass as well as lean body weight (LBW) [3]. As an individual's weight increases, his or her body composition changes, resulting in higher LBW and disproportionately higher adipose content (Fig. 10.1). Large body mass causes increased blood volume and cardiac output in obese individuals, which is required for adequate oxygen and nutrient supply. These changes alter volume of distribution, as well as renal and hepatic clearance, which is dependent on perfusion and blood circulation. Increased blood volume and cardiac output are closely linked to hypertension and cardiomyopathy in obese individuals.

Volume of distribution (V_d) refers to the distribution of a drug to various body compartments for absorption and metabolism. Increased blood volume and cardiac output alter drug transport and deposition, leading to inaccurate pharmacokinetic modeling for the drug in a patient. Calculating optimal dosing in obese patients depends on the properties of the drug and the physiology of the individual patient; several parameters have been used to determine optimal concentrations and dosing intervals (Table 10.1) [4]. Body surface area (BSA) takes into account weight and height but not gender. It is often used to calculate

Table 10.1 Body Weight Parameters Routinely Used in Drug Dosing Calculations

Measurement	Calculation	Gender
BMI	Weight (kg)/height $(m)^2$	
BSA	TBW $(kg)^{0.425}$ × height $(cm)^{0.725}$ × 0.007184	
IBW	Devine's estimation = 45.4 + 0.89 × (height (cm)−152.4) + 4.5 (if male)	
LBW	James equations:	
	1.1 × TBW (kg)−0.0128 × BMI × TBW (kg)	Males
	1.07 × TBW (kg)−0.0148 × BMI × TBW (kg)	Females
	Janmahasatian formulae:	
	$(9.27 \times 10^3 \times$ weight (kg))/$(6.68 \times 10^3 + 216 \times$ BMI)	Males
	$(9.27 \times 10^3 \times$ weight (kg))/$(8.78 \times 10^3 + 244 \times$ BMI)	Females

Note: BMI, body mass index; BSA, body surface area; IBW, ideal body weight; LBW, lean body weight; TBW, total body weight.

dosing for chemotherapeutic drugs in order to decrease toxicity. Ideal body weight (IBW) is used for dosing drugs with weak lipophilicity. It is derived via Devine's estimation equation, which takes gender into consideration but assumes that individuals with the same weight and height have the same LBW. The LBW parameter is derived by subtracting the fat mass component from the TBW, providing a mass equal to the nonadipose weight composed of bones, muscles, and organs. Choosing the optimal body mass parameter for dosing also depends on the drug's molecular weight, lipid solubility, and protein binding, among other properties.

Increased volume of distribution is only one factor that requires dose adjustment in obese patients. Altered metabolism and clearance must also be considered when determining optimal drug treatment. As mentioned previously, increased blood volume and cardiac output alter volume of distribution, but they also play a role in clearance by the kidney and liver. Increased delivery of blood to the kidneys has been linked to higher glomerular filtration rate (GFR) and renal clearance of drugs and their metabolites [3,5,6]. However, obesity can also lead to renal dysfunction, resulting in reduced elimination [6]. The complex physiology of obese individuals does not fit a single model; in fact, due to the extent of health complications, obesity represents a special population with a need for individually based TDM. Therapeutic drug ranges are determined through clinical trials, as are dosing regimens; however, obese patients are rarely incorporated in clinical studies, resulting in very limited understanding of the drug PK in this special population.

10.2 PK IN OBESITY

Obesity can cause changes in pharmacokinetic parameters of a drug, including absorption, distribution, clearance, and metabolism.

10.2.1 Absorption

Obesity does not significantly alter oral absorption, but parenteral absorption can be affected due to increased subcutaneous tissue thickness and higher BMI. Cardiac output affects organ perfusion, which may alter drug delivery to target tissues. The protein binding properties of a drug affect its transport in blood and permeability into tissues. Changes in albumin and α_1 acidic glycoprotein reported in obese patients can affect the rate at which the drug is taken up from circulation. Absorption is dramatically altered in obese individuals following gastric bypass, which leads to decreased gastric mixing and mucosal absorption along a shortened surface area [7,8]. Cyclosporine, thyroxine, and rifampin are several examples of drugs with decreased absorption due to gastric bypass surgery [7].

10.2.2 Distribution

Volume of distribution (V_d) relates the amount of drug in circulation to the amount that is distributed to various compartments in the body. Drug properties such as hydrophilicity, lipophilicity, and protein binding play a role in tissue uptake; enhanced tissue uptake leads to increased V_d. Blood flow and body composition alterations affect V_d and thus render drug PK unreliable in obese individuals. Variability in V_d associated with obesity is especially pronounced when rapid onset of the drug effect is needed and the peak concentrations after a single dose are determined based on the clinically established V_d in normal-weight individuals [3].

Initial dosing is particularly affected for certain lipophilic drugs that are taken up in adipose tissue resulting in lower drug levels in circulation and at target compartments. The body mass parameters (TBW, IBW, BSA, etc.) used to calculate dosing are crucial for achieving the target loading level. For example, if IBW (which does not account for adipose tissue mass) is used to determine the loading dose of a lipophilic drug in obese patients, this may lead to underdosing and the drug will be deposited in the metabolically inactive adipose tissue. In such a case, the drug is stored in the adipose compartment without being metabolized or eliminated, and as drug levels in other tissues decline, the drug is redistributed from the adipose tissue, resulting in a prolonged clinical effect [3]. Unfortunately, not all lipophilic drugs exhibit similar distribution patterns in obese patients, and TBW cannot be used for all dosing calculations. Hydrophilic drugs such as lithium and protein binding drugs such as warfarin are not taken up rapidly and remain in circulation longer, resulting in smaller V_d [3]. Because lithium and warfarin distribution is not significantly altered by excess adipose tissue in obese individuals, IBW should be used for calculating optimal dosing. If TBW is used instead, serum drug levels will be higher than anticipated and may lead to toxicity in obese patients. This requires clinical studies that investigate individual drug metabolism and clearance in obese patients.

10.2.3 Clearance

Drug clearance by metabolism and elimination is essential in PK and dictates maintenance dosing concentrations and frequency. The liver and the kidneys are responsible for metabolism and elimination, respectively; blood flow and uptake in these organs dictate the extent of clearance. Liver function may be impaired in obesity due to hepatic steatosis. In the absence of renal impairment and kidney failure, obese patients have higher cardiac output and GFR resulting in increased renal clearance. Drugs with a high V_d are present at lower levels in circulation and undergo slower hepatic and renal clearance [3]. The elimination half-life of a drug can be calculated as $t_{1/2} = (\ln 2 \times V_d)/\mathrm{CL}$, where changes in V_d and clearance (CL) due to obesity have a significant effect [4].

10.2.4 Metabolism

Obesity results in altered drug metabolism and a state of low-grade chronic inflammation that can result in obesity-related metabolic disorders. Obesity-induced inflammation predominantly occurs in adipose tissue, liver, pancreas, and muscle, resulting in metabolic disruption and insulin resistance leading to T2DM [9]. Insulin resistance is triggered by the release of proinflammatory cytokines such as tumor necrosis factor-α (TNF-α), interleukin (IL)-6, and IL-1β, which reduce insulin sensitivity and lead to hyperglycemia [9,10]. Adipose tissue inflammation begins with enlargement of adipocytes and release of cytokines and chemokines that recruit macrophages, leukocytes, and mast cells along with other proinflammatory factors to the adipose tissue. This in turn results in increased secretion of TNF-α and IL-6, leading to insulin resistance and ultimately T2DM [9–11]. Obesity is also associated with inflammation of pancreatic islet cells due to increased release of cytokines and infiltration of macrophages; this reduces insulin secretion and leads to β-islet cell death [9].

Another organ subjected to obesity-induced chronic inflammation is the liver, in which macrophages and Kupffer cells experience enhanced activation with obesity and release proinflammatory factors that further contribute to insulin resistance [9]. Liver abnormalities linked to obesity include nonalcoholic fatty liver disease, which can result in steatosis and hepatic inflammation. Hepatic steatosis and inflammation alter enzyme activity required for drug metabolism. The liver is the primary site for drug metabolism, which occurs via phase I or phase II. Phase I consists of cytochrome P450 enzymes (CYPs) located at the endoplasmic reticulum of hepatocytes that modify drugs via oxidation, reduction, or hydrolysis [5]. CYP metabolism accounts for approximately 75% of all drug metabolism. CYP3A4 metabolic activity is reduced in obese individuals. Substrates for CYP3A4, including triazolam, carbamazepine, alfentanil, *N*-methylerythromycin, and midazolam, show reduced clearance in obese compared to nonobese patients [5]. Although CYP3A4 exhibits reduced metabolism with respect to the previously mentioned compounds, enzyme activity toward trazodone and docetaxel is not altered in obese individuals [5]. Following gastric bypass surgery, increased CYP3A4 drug metabolism is reported [5]. On the other hand, CYP2E1 metabolic activity is enhanced in obese individuals. Increased CYP2E1-mediated clearance of chlorzoxazone, enflurane, and sevoflurane was observed with obesity [5]. Another enzyme, CYP2C9, mediates metabolism of ibuprofen and phenytoin, which is increased with obesity. However, CYP2C9 clearance of glimepiride is not altered among obese individuals, emphasizing that enzyme activity in obesity is altered in a drug-specific manner [5].

Phase II metabolism relies on the conjugation of glucuronide, *N*-acetyl, methyl, glutathione, and sulfate groups to drug molecules for clearance [5]. Uridine diphosphate glucuronosyltransferase (UGT) enzymes are the predominant phase II enzymatic factors. UGT enzymes are primarily expressed in the liver, although they can also be present in the gastrointestinal tract, adipose tissue, and kidneys. UGT facilitates the conjugation of a glucuronosyl group to a substrate/drug, resulting in polarization and subsequent increased solubility and elimination efficiency. Liver enlargement in obesity accounts for enhanced UGT enzymatic activity, hence altering PK for UGT substrates [5]. The presence of UGT isoforms in adipose tissue, which also expands with obesity, further contributes to the rapid drug metabolism observed in obese patients. Increased UGT-mediated clearance of paracetamol, oxazepam, garenoxacin, and lorazepam is reported in obese patients compared to nonobese individuals [5].

10.2.5 Renal Elimination

Kidney function is responsible for drug and metabolite elimination. The rate of elimination depends on glomerular filtration, tubular secretion, and tubular reabsorption. The exact correlation between obesity and these parameters is not fully understood, but glomerular hyperfiltration is reported in obese individuals. Increased cardiac output and blood flow are attributed to glomerular hyperfiltration. Vancomycin clearance by GFR increases with TBW in obese compared to nonobese individuals [12]. Conversely, T2DM, hypertension, and a chronic state of inflammation associated with obesity can induce kidney damage and lead to renal failure, reducing GFR and clearance [6]. Advanced obesity induces structural changes in the kidney, resulting in glomerulopathy and eventually glomerulosclerosis and tubulointerstitial fibrosis with declining GFR and worsening renal function [6]. Obesity is associated with hypertension, sodium retention, and elevated renin levels. This increase in renin is attributed to stimulation of the sympathetic nervous system at the kidney due to obesity (potentially triggered by leptin release by adipose tissue) [11,13]. Activation of the renin–angiotensin pathway can alter tubular reabsorption in obese patients. Procainamide is eliminated by glomerular filtration and active tubular secretion. Comparing procainamide excretion between obese and nonobese individuals showed increased tubular secretion associated with obesity [5]. Cisplatin and ciprofloxacin elimination by tubular secretion is also higher in obesity [5]. There is very limited information regarding the effect of obesity on tubular reabsorption of drugs, with preliminary work suggesting lower reuptake for lithium. Renal drug elimination due to GFR and tubular secretion show a trend of increased clearance with obesity.

10.3 CHEMOTHERAPY IN OBESE PATIENTS

Obesity is associated with increased recurrence and higher mortality for breast cancer and other malignancies. Chemotherapeutic drugs exhibit dose-dependent cytotoxicity, and physicians try to balance toxic effects with therapeutic outcomes. This approach results in systemic underdosing of obese patients and poor clinical outcomes. Clinical trials are the basis for developing treatment guidelines and obese patients are significantly underrepresented in these studies, resulting in poor correlation between chemotherapy dosing and increased body mass. Currently, chemotherapy dosing is based on BSA, which is determined by height and weight measurements [14]. To prevent toxicity, arbitrary limits are used when calculating dosing for obese patients, and often physicians cap the maximum dose at a BSA of 2.0 m^2 or use IBW to determine dose concentration [14–16]. This potentially results in systemic underdosing of obese patients.

Griggs et al. [17] conducted a retrospective study examining 9672 breast cancer patients who received lower than standard doses for the first course of a doxorubicin hydrochloride and cyclophosphamide regimen. The cohort was composed of healthy weight, overweight, obese, and severely obese women. A first-cycle dose reduction of 10% or more was reported mostly for severely obese and obese patients. Of these patients, 16% received additional cycles of doxorubicin and cyclophosphamide, resulting in a larger overall dose of chemotherapy over time that was consistently below the standard therapeutic concentration [17]. This dose-reduction approach has been linked to poor outcomes for obese breast cancer patients.

Similar dose reductions are reported by Baillargeon et al. [18] in a pediatric cohort (aged 2–18 years) of 199 patients with β-precursor acute lymphoblastic leukemia. For children, obesity is defined as a BMI greater than or equal to the 95th percentile for age and sex; 33 patients (16.2%) in this cohort met the BMI criteria for obesity. Monitoring L-asparaginase treatment in this cohort showed that obese patients received a modified chemotherapy dose that was 7% lower than the protocol determined dose, whereas nonobese patients received L-asparaginase levels equivalent to the protocol-based, BSA-dependent dose [18,19]. Dose reduction was greater among obese patients aged 10–18 years compared to those younger than age 10 years. As in obese adults, chemotherapeutic dosing in obese pediatric patients is difficult to calculate with high accuracy. Often, instead of using TBW or BSA to determine the necessary dose concentration, physicians follow a modified dosing regimen based on IBW or reduce the TBW-based levels to prevent toxicity. This leads to underdosing, reduced drug efficacy, and increased morbidity and mortality. Guidelines for chemotherapeutic treatment of obese pediatric patients are not established, and there is very limited knowledge of dosing in this special population because it is underrepresented in clinical trials, even more so than the corresponding adult population.

Studies have shown that treatment with chemotherapy doses based on TBW does not result in adverse toxicity in obese compared to nonobese patients, but despite these findings, dosing is limited from the start in overweight individuals [14,20]. Rosner et al. [20] investigated the effects of chemotherapeutic toxicity on obese patients who received actual body weight-based dosing. Approximately 1600 women with stage II breast cancer (representing healthy weight, obese, and severely obese populations) received chemotherapy either within 5% of their weight-based dose or below 95% of their weight-based dose. Both nonobese and obese patients who received weight-based dosing showed similar rates of toxicity. The absence of enhanced toxicity in obese individuals who received 120% of the dose they would have received based on IBW supports the notion that toxicity in dose-intense therapy is similar between healthy-weight and obese patients [20]. This study also emphasized that patients who received dose-intense therapy calculated using actual body weight experienced improved recovery and survival.

Higher doses of chemotherapy, within the therapeutic range, have been correlated with increased remission rate and lower mortality. Mathematical models developed by Norton and Simon demonstrate that increasing chemotherapy dose intensity by shortening dosing intervals reduces the opportunity for tumor regrowth [14]. Despite accumulating evidence that chemotherapy requires sufficiently high dosing, clinicians continue cautious treatment of obese patients due to lack of specific guidelines for dosing. The repercussions of inadequate treatment are evident in studies demonstrating increased risk of mortality among obese patients with hormone receptor-positive and hormone receptor-negative breast cancer compared to healthy-weight patients.

10.4 ANTIMICROBIAL TREATMENT IN OBESE PATIENTS

Obesity has been associated with increased prevalence of inflammation, emphasizing the need for adequate antimicrobial dosing. Antimicrobial efficacy can depend on systemic maximum concentration (C_{max}); for these drugs, achieving a high peak concentration early on is necessary to achieve maximal killing of the pathogen. Other antimicrobial drugs exhibit a time-dependent effect, where the length of time the drug is present above the minimum inhibitory concentration (MIC) is directly related to the efficacy of the treatment [21]. Physiological properties such as V_d and clearance are key determinants of drug levels for both concentration-dependent and time-dependent antimicrobials. These parameters are altered in obese patients, so initial and maintenance dosing must be adjusted and blood levels have to be monitored especially for drugs with narrow therapeutic ranges.

V_d is the main parameter to consider for concentration-based antimicrobials; obese patients have increased V_d especially for lipophilic drugs. TBW-based dosing should not be used for all antimicrobials because it is more accurate for lipophilic than hydrophilic drugs. For example, aminoglycosides are not taken up by adipose tissue due to their polarity; hence, using TBW to calculate aminoglycoside dosing can lead to toxicity in obese individuals [22]. Schwartz et al. [23] demonstrated that in dosing gentamycin and tobramycin, V_d correlates best for obese and nonobese patients when 40% of the excess body weight is added to the IBW. Although variable correlation factors have been reported by numerous studies investigating gentamycin dosing in obese patients, the majority of investigations indicate 40% is optimal correction [22]. Bauer et al. [24] examined gentamycin, tobramycin, and amikacin dosing in normal-weight and obese patients and proposed the use of 40% as a correction factor for the aminoglycoside class. Traynor et al. [25] investigated gentamycin and tobramycin PK in a cohort of 1708 patients and calculated the optimal correction factor for obese patient dosing as 0.43 times the excess body weight. The potency of time-dependent antimicrobials is dependent on V_d as well as clearance because the rate at which the drug is eliminated determines the length of time the drug level is above the MIC. An increase in V_d results in a reduced C_{max}; however, it does not reduce the length of time the drug is present in circulation (in fact, translocation from adipose tissue storage may result in a prolonged half-life, especially in obese patients). On the other hand, increased renal excretion will reduce the length of time the antimicrobial is present in circulation, and higher GFR associated with obesity can accelerate drug clearance and reduce efficacy.

Vancomycin is a hydrophilic antibiotic, most often administered intravenously, with a fast elimination time. Both V_d and clearance of vancomycin increase with obesity, and current consensus guidelines recommend dosing vancomycin based on TBW with a maximum of 2 g per dose prior to dose adjustments based on TDM [22]. Dosing frequency is determined based on renal function and clearance. Bauer et al. [24] examined 24 morbidly obese patients matched with a normal-weight control population and showed a strong correlation between TBW and clearance as well as TBW and V_d, to a lesser extent. This work indicates that high daily doses are necessary for obese patients to achieve the desired steady-state level. The loading dose and subsequent maintenance dosing for antimicrobials depend on the properties of the drug as well as the V_d and clearance rate. To determine if the weight-based recommended dosing guideline was implemented in clinical practice, a multicenter study of 421 patients identified only 12.7% of obese patients as receiving greater than 10 mg/kg/dose, which is below the recommended 15 mg/kg/dose [26]. These findings indicate that vancomycin dosing remains inadequate in obese patients despite evidence that weight-based dosing does not lead to toxicity and is necessary for drug efficacy. Because TBW may not be the optimal size descriptor for dosing, TDM is crucial for antimicrobials with a narrow therapeutic range when trying to prevent toxicity while maintaining efficacy in obese patients.

10.5 PK AFTER BARIATRIC SURGERY

Bariatric surgery is performed in severely obese individuals, who often have comorbidities and are prescribed a range of therapeutic drugs. An overall decrease in drug absorption is reported for thyroxine, phenytoin, and rifampin in patients following bariatric surgery [7,8]. Changes in drug disintegration, solubility, and dissolution linked to reduced gastric mixing, changes in gastric pH, and reduced gastric emptying emphasize the need for dose adjustment and TDM postbariatric surgery [7]. Reduced gastric mixing results after several bariatric procedures, including gastric banding, sleeve gastrectomy, and gastric bypass, which reduces drug disintegration that is often the rate-limiting step for drug absorption. Procedures involving gastric restriction, such as gastric bypass, result in an elevated pH in the newly formed stomach pouch because acid-producing cells are partitioned from the compartment [7,27]. A high gastric pH enhances solubility of basic drugs but decreases solubility of acidic drugs and reduces disintegration.

Gastric bypass also results in reduced gastrointestinal length, which decreases mucosal exposure, thus limiting drug absorption. Roux-en-Y gastric bypass is one of the most common bariatric procedures; it alters the gastrointestinal anatomy, leading to changes in nutrient absorption and drug metabolism [27]. For example, cytochrome P450 and other enzymes found in the proximal small intestine are bypassed following this procedure [27]. This results in insufficient first-pass intestinal metabolism, and drugs such as metformin have increased bioavailability and absorption following Roux-en-Y bypass. There are limited studies investigating the impact of bariatric surgery on the PK of antidepressants, antimicrobials, and statins, although these are the most common classes of drugs prescribed to obese patients with other health complications.

Tricyclic antidepressants are highly lipophilic and absorbed through the gastrointestinal tract. Following bariatric surgery, reduced levels of tricyclic antidepressants are attributed to poor absorbance and lower V_d due to reduction in adipose tissue [8]. Lower postbariatric levels of these antidepressants require close psychiatric monitoring and dose adjustments in patients. Selective serotonin reuptake inhibitors (SSRIs) are another class of lipophilic antidepressants, which have been shown to have decreased serum levels post-bariatric surgery due to reduced solubility and intestinal surface area available for drug absorption. Hamad et al. [27] reported reduced SSRI bioavailability 1 month post-Roux-en-Y gastric bypass in 12 patients (11 female and 1 male). Area under the curve (AUC) values for the SSRIs (venlafaxine, citalopram, escitalopram, sertraline, and duloxetine) had decreased to as low as 54% of the preoperative levels. Six months postsurgery, the AUC for most patients had returned to the baseline levels observed with SSRI dosing before the procedure [27]. Marzinke et al. [28] described reduced escitalopram serum concentrations following Roux-en-Y gastric bypass in four patients, who exhibited a 33% decrease in circulating drug levels compared to presurgical concentrations. Higher pH in the newly formed stomach and reduced intestinal surface area altered drug solubility and absorption, thus hindering therapeutic efficacy and requiring postsurgery dose adjustment. Physiological changes implemented through bariatric surgery render patients susceptible to underdosing or toxicity in the absence of adequate TDM and dose adjustment.

10.6 CONCLUSIONS

The prevalence of obesity among adults and children in the United States and throughout the world is increasing at a staggering rate. Obesity is associated with a range of comorbidities, including insulin resistance and T2DM, hypertension, cardiac complications, and renal disease. These health complications are attributed largely to physiological changes induced by obesity. Increased body fat is the predominant factor linked to obesity; however, the gain in adipose tissue is accompanied by an increase in blood volume, cardiac output, and lean body mass, as well as organ enlargement. PK is altered in obese patients due to increased V_d and clearance, which affect drug distribution, metabolism, and elimination. Dosing is based on the properties of the drug, as well as parameters such as TBW, BSA, and IBW. The lipophilicity or hydrophilicity of a drug determines its distribution and translocation to adipose tissue, which significantly affects V_d. Kidney hyperfiltration is observed early on in obesity, which enhances drug elimination and results in suboptimal drug levels. On the other hand, chronic obesity can lead to renal disease or failure and reduced clearance, leading to elevated drug levels in circulation and potential toxicity. Uncertainties with dosing in the obese population stem from fluctuations in distribution and clearance that are not conserved among therapeutics and patients, emphasizing the need for TDM to prevent toxicity.

Chemotherapeutic and antimicrobial PK in obesity has received the most attention in clinical studies due to the high prevalence of malignancies and chronic inflammation in this population. However, antidepressants and anesthetics, among other classes of drugs, are frequently utilized in the treatment of obese patients, indicating a need for more extensive clinical trials in an obese population across a variety of drug classes. The extent and variability of physiological alterations and comorbidities associated with obesity make treatment of obese patients a challenging and individualized process in which TBW, IBW, and/or BSA should be used as a guideline for dosing, but drug concentrations should be monitored to ensure optimal treatment and prevent adverse effects.

References

[1] World Health Organization [WHO]. Obesity general information [online], <http://www.who.int/mediacentre/factsheets/fs311/en/>; 2015 accessed 11/25/15.

[2] Centers for Disease Control and Prevention [CDC]. Obesity data & statistics [online], <http://www.cdc.gov/obesity/data/index.html>; 2015 accessed 11/25/15.

[3] Zuckerman M, et al. A review of the toxicologic implications of obesity. J Med Toxicol 2015;11:342–54.

[4] Hanley MJ, et al. Effect of obesity on the pharmacokinetics of drugs in humans. Clin Pharmacokinet 2010;49(2):71–87.

[5] Brill MJE, et al. Impact of obesity on drug metabolism and elimination in adults and children. Clin Pharmacokinet 2012;51(5):277–304.

[6] Kiortsis DN, et al. Management of obesity-induced kidney disease: a critical review of the literature. Obes Facts 2012;5:821–32.

[7] Padwal R, et al. A systematic review of drug absorption following bariatric surgery and its theoretical implications. Obes Rev 2010;11:41–50.

[8] Yska JP, et al. Influence of bariatric surgery on the use and pharmacokinetics of some major drug classes. Obes Surg 2013;23:819–25.

[9] Esser N, et al. Inflammation as a link between obesity, metabolic syndrome and type 2 diabetes. Diabetes Res Clin Pract 2014;105:141–50.

[10] Apostolopoulos V, et al. The complex immunological and inflammatory network of adipose tissue in obesity. Mol Nutr Food Res 2016;60:43–57.

[11] Ruster C, et al. Adipokines promote chronic kidney disease. Nephrol Dial Transplant 2013;28(Suppl. 4):iv8–iv14.

[12] Bauer LA, et al. Vancomycin dosing in morbidly obese patients. Eur J Clin Pharmacol. 1998;54(80):621–5.

[13] Hall JE, et al. Obesity-induced hypertension: role of sympathetic nervous system, leptin, and melanocortins. J Biol Chem 2010;285:17271–6.

[14] Lyman GH, et al. Chemotherapy dosing in overweight and obese patients with cancer. Nat Rev 2013;10:451–9.

[15] Griggs JJ, et al. Appropriate chemotherapy dosing for obese adult patients with cancer: American Society of Clinical Oncology clinical practice guideline. J Clin Oncol 2012;30:1553–61.

[16] Lopes-Serrao MD, et al. Evaluation of chemotherapy-induced severe myelosuppression incidence in obese patients with capped dosing. J Oncol Pract 2011;7:13–17.

[17] Griggs JJ, et al. Undertreatment of obese women receiving breast cancer chemotherapy. Arch Intern Med 2005;165:1267–73.

[18] Baillargeon J, et al. L-Asparaginase as a marker of chemotherapy dose modification in children with acute lymphoblastic leukemia. Cancer 2005;104:2858–61.

[19] Tolbert J, et al. The challenge of obesity in paediatric leukaemia treatment: it is not just size that matters. Arch Dis Child 2015;100:101–5.

[20] Rosner GL, et al. Relationship between toxicity and obesity in women receiving adjuvant chemotherapy for breast cancer: results from cancer and leukemia Group B Study 8541. J Clin Oncol 1996;1:3000–8.

[21] Pai MP, et al. Treatment of bacterial infections in obese adult patients: how to appropriately manage antimicrobial dosage. Curr Opin Pharmacol 2015;24:12–17.

[22] Payne KD, et al. Dosing of antibacterial agents in obese adults: does one size fit all? Expert Rev Anti Infect Ther 2014;7:829–54.

[23] Schwartz SN, et al. A controlled investigation of the pharmacokinetics of gentamicin and tobramycin in obese subjects. J Infect Dis 1978;138:499–505.

[24] Bauer LA, et al. Influence of weight on aminoglycoside pharmacokinetics in normal weight and morbidly obese patients. Eur J Clin Pharmacol 1983;24:643–7.

[25] Traynor AM, et al. Aminoglycoside dosing weight correction factors for patients of various body sizes. Antimicrob Agents Chemother 1995;545–8.

[26] Vance-Bryan K, et al. Effect of obesity on vancomycin pharmacokinetic parameters as determined by using Bayesian forecasting technique. Antimicrob Agents Chemother 1993;436–40.

[27] Hamad GG, et al. The effect of gastric bypass on the pharmacokinetics of serotonin reuptake inhibitors. Am J Psychiatry 2012;169(3):256–63.

[28] Marzinke MA, et al. Decreased esitalopram concentrations post-Roux-en-Y gastric bypass surgery. Ther Drug Monit 2015;37:408–12.

CHAPTER 11

Special Issues in Therapeutic Drug Monitoring in Patients With Uremia, Liver Disease, and in Critically Ill Patients

Kamisha L. Johnson-Davis[1,2] and Amitava Dasgupta[3]
[1]Department of Pathology, University of Utah Health Sciences Center, Salt Lake City, UT, United States
[2]ARUP Institute for Clinical and Experimental Pathology, Salt Lake City, UT, United States
[3]Department of Pathology and Laboratory Medicine, University of Texas Health Science Center at Houston, Houston, TX, United States

CONTENTS

11.1 Introduction 245
11.2 Monitoring Free Drug Concentrations in Patients With Uremia, Liver Disease, and in Critically Ill Patients ..246
11.3 Altered Drug Disposition in Patients With Kidney Disease ..247
11.3.1 Altered Clearance of Renally Cleared Drugs in Patients With Kidney Disease 248
11.3.2 Altered Clearance of Nonrenally Cleared Drugs in Patients With Kidney Disease 249
11.4 Altered Drug Disposition in Hepatic Disease 250
11.5 Altered Drug Disposition in Critically Ill Patients 252
11.6 Application of TDM in Critically Ill Patients..........255
11.6.1 TDM of Antimicrobial Drugs in Critically Ill Patients.......255
11.6.2 TDM of Digoxin in Critically Ill Patients.......256

11.1 INTRODUCTION

Chronic kidney disease affects more than 20 million people in the United States, and approximately 500,000 patients require hemodialysis. An average dialysis patient may require more than 12 medications. In 2009, the U.S. Food and Drug Administration (FDA) published a survey on new drug applications approved between January 2003 and July 2003 that assessed the effects of renal impairment on disposition of 37 orally administered drugs as a part of FDA submission. Of 23 new drugs that are not eliminated by the kidneys, 13 of these drugs showed an average 1.5-fold increase in area under the concentration curve (AUC) in kidney-impaired patients compared to healthy subjects. Moreover, five drugs that are metabolized by the liver showed significant increases in AUC that required labeling recommendations for dosage adjustment for patients with renal impairment. Therefore, renal impairment also affects clearance of some drugs that have a nonrenal mechanism of clearance [1].

Liver disease can be either acute or chronic in nature. In the United States, the burden of liver disease is a serious public health issue. Liver disease is the fourth leading cause of death among young people (age group, 45–54 years) and one of the top 12 causes of death among all adults. Nonalcoholic fatty liver disease is the major hepatic disease among Americans, with an estimated 32.5 million people affected by it; of this population, 28.5 million adults have hepatic steatosis [2]. According to data available from the Centers for Disease Control and Prevention, an estimated 1.2 million people

W. Clarke & A. Dasgupta (Eds): Clinical Challenges in Therapeutic Drug Monitoring. DOI: http://dx.doi.org/10.1016/B978-0-12-802025-8.00011-8

11.6.3 TDM of Sedatives and Analgesics in Critically Ill Patients......257

11.7 Conclusions......257

References......257

in the United States are infected with hepatitis B, approximately 2.7 million people are infected with hepatitis C, and approximately 31,000 people die every year due to complications from cirrhosis of the liver. Altered drug disposition is frequently encountered in patients with liver disease; as a result, dosage of a drug must be carefully adjusted in these patient populations. If any of these patients are taking any medication that is routinely monitored, then therapeutic drug monitoring (TDM) is useful in dosage adjustments in these patients. Patients with liver disease usually have low serum albumin, the major drug-binding protein in serum, and as a result, protein binding is significantly reduced. Because free drug is pharmacologically active, dosage adjustment is needed to avoid toxicity. Moreover, liver disease also alters metabolism of drugs by the liver, the major detoxifying organ in the body.

Critically ill patients, with conditions such as sepsis, malnutrition, recent surgical procedure, shock, burns, organ failure, trauma, and exposure to toxins, may require life-support systems such as feeding tubes, fluid resuscitation, mechanical ventilation, and renal replacement therapy. These medical interventions will impact drug pharmacokinetics and pharmacodynamics. These patients often show signs of impaired liver and/or renal function, along with variable fluid status, which could lead to volume expansion. Edema can also occur due to capillary leakage, when fluid from the intravascular compartment enters the interstitial spaces. In addition, these patients may also have hypoalbuminemia. As a result, critically ill patients are at higher risk for adverse drug reactions that may result in increased length of stay in the hospital and higher costs. Failure to recognize altered pharmacokinetics and pharmacodynamics of certain drugs may complicate management of a critically ill patient. Leape et al. [3] commented that inclusion of a pharmacist as a member of a multidisciplinary team to manage intensive care unit (ICU) patients may reduce adverse drug reactions up to 66%.

11.2 MONITORING FREE DRUG CONCENTRATIONS IN PATIENTS WITH UREMIA, LIVER DISEASE, AND IN CRITICALLY ILL PATIENTS

Drug–protein binding may vary from 0% (eg, lithium) to 99% (eg, ketorolac), and it is only the unbound drug (free drug) that is pharmacologically active. Albumin is the major drug-binding protein in the serum, although other proteins, such as α_1 acid glycoprotein, lipoproteins, and globulins, are also capable of binding drugs. As a general rule, drugs that are minimally protein bound penetrate tissues better than those that are highly protein bound, but clearance of such drugs is also higher. However, for drugs that are less than 80–85% protein bound, differences may not be clinically significant. On the other hand, drugs that are strongly protein bound (≥90%) may differ markedly from minimally protein-bound drugs in terms of tissue penetration and half-life. The concentration of albumin and other protein may be altered under stress, surgery, pregnancy, and liver or kidney diseases; in such circumstances, monitoring free drug is more useful for strongly protein-bound drugs. For these patients, free drug levels correlate better with clinical picture than traditionally monitored total drug level [4].

In clinical laboratories, free concentrations of phenytoin, valproic acid, and carbamazepine are often monitored. Some laboratories also have the capability of monitoring free mycophenolic acid. All these drugs are bound mainly to serum albumin, and significantly elevated free concentrations of these drugs are frequently encountered in patients with uremia and liver disease, as well as critically ill patients. As a result, free concentrations of phenytoin, valproic acid, carbamazepine, and possibly mycophenolic acid should be monitored in these patients instead of the traditionally monitored total drug concentration. In addition, carbamazepine epoxide concentrations may be significantly elevated in patients with uremia, which may increase the risk of toxicity. This topic is discussed in detail in Chapter 4 "Monitoring Free Drug Concentration: Clinical Usefulness and Analytical Challenges."

11.3 ALTERED DRUG DISPOSITION IN PATIENTS WITH KIDNEY DISEASE

Chronic kidney disease is usually defined as an estimated glomerular filtration rate (eGFR) of less than 60 mL/min/1.73 m^2. Although kidney diseases may be encountered in any age group, older adults have a higher incidence of kidney diseases because GFR decreases by approximately 8 mL/min with each decade of life after 40 years of age due to loss of age-related renal mass and reduction in the number and size of nephrons. Therefore, additional care must be taken in prescribing drugs for the elderly. Many drugs are cleared by the kidneys, and as a result, dosage adjustment is needed for these drugs in patients with chronic kidney diseases. Although it may appear that a drug metabolized by the liver may not require dosage adjustment in uremic patients, both animal and human studies indicate otherwise because uremia may significantly alter activities of liver enzymes responsible for metabolism of many drugs. As a result, both renally cleared and nonrenally cleared drugs may require dosage adjustments in renally compromised patients. As mentioned previously, free fraction of a strongly albumin-bound drug may be significantly increased in these patients due to hypoalbuminemia as well as the presence of certain uremic toxins capable of displacing a strongly protein-bound drug from the binding site (please see Chapter 4, "Monitoring Free Drug Concentration: Clinical Usefulness and Analytical Challenges").

11.3.1 Altered Clearance of Renally Cleared Drugs in Patients With Kidney Disease

Creatinine clearance is widely used by clinicians to evaluate renal function of patients using formulas such as the Cockroft–Gault formula. Correlations have been established between creatinine clearance and clearance of digoxin, lithium, procainamide, aminoglycoside, and several other drugs, but creatinine clearance does not always predict renal excretion of all drug. However, serum creatinine may remain normal until GFR has declined by at least 50%. Nearly half of older patients have normal serum creatinine but reduced renal function, and 26% of older patients (aged 70 years or older) may have chronic renal insufficiency. Dosage adjustment based on renal function is recommended for many medications in patients with chronic kidney diseases as well as elderly patients, even for medications that exhibit large therapeutic windows [5]. The most common recommendation is often to reduce the dose or expand the dosing interval or use both approaches simultaneously [6].

Hanlon et al. studied clearance of renally cleared medications in 1304 elderly patients aged 65 years or older and observed that median eGFR based on the Cockroft–Gault formula was 67 mL/min, and 26.2% of patients had chronic kidney diseases. The authors observed that 1 in 10 older patients had potentially inappropriate prescription of primarily renally cleared medications when evidence-based consensus-derived criteria were used. The authors listed 21 drugs that were associated with dosage errors, and they also recommended that eight of these drugs should be avoided if GFR is very low. It is generally recognized that ranitidine and glyburide are the most common drugs with prescription problems in patients with kidney diseases [7]. Glyburide is associated with a higher risk of hypoglycemia in renally compromised patients [8]. Examples of drugs that should not be prescribed in renally compromised patients with GFR below certain thresholds are given in Table 11.1.

Other investigators have also reported inappropriate dosing of renally cleared drugs in patients with renal insufficiency. Papaioannou et al. reported that overall 42.3% of residents of long-term care facilities were at least once prescribed inappropriate dosage of a drug based on creatinine clearance. The authors concluded that renal function is often overlooked when prescribing a renally cleared drug in older long-term care residents [9]. Nielsen et al. commented that based on various published studies, inappropriate dosing of a renally cleared medication in patients with renal insufficiency varied from 19% to 67%. The authors implemented an electronic clinical support system (Ranbase) for appropriate dosing of drugs in renally compromised patients and observed that despite implementation of electronic prescribing

Table 11.1 Following Drugs Should be Avoided if Creatinine Clearance is Below the Stated Level

Drug	Creatinine Clearance (mL/min)
Chlorpropamide	<50
Colchicine	<10
Duloxetine	<30
Glyburide	<50
Meperidine	<50
Nitrofurantoin	<60
Probenecid	<50
Propoxyphene	<10
Sulfamethoxazole/trimethoprim	<15
Spironolactone	<30
Triamterene	<30

Source: From Ref. [7].

and automated reporting of eGFR, patients with renal insufficiency may still be exposed to inappropriate drug use [10].

Patients with chronic kidney disease or kidney failure may undergo hemodialysis or hemofiltration to filter waste products from the blood. For patients on hemodialysis, drugs that are not bound to plasma proteins can pass through the membrane filter and be removed or cleared from the blood. The rate of blood filtration is impacted by the duration of dialysis and flow rate of the dialysate. Hemoperfusion can also be used as treatment in overdose situations for low-molecular-weight drugs with a low volume of distribution [11,12]. Drugs that are highly protein bound to plasma proteins, such as phenytoin, valproic acid, carbamazepine, and digoxin, cannot be removed by hemodialysis. Thus, hemodialysis will not increase the removal of drugs that are highly protein bound.

11.3.2 Altered Clearance of Nonrenally Cleared Drugs in Patients With Kidney Disease

Chronic kidney disease may also affect clearance of drugs metabolized by the liver due to reduced activities of hepatic drug-metabolizing enzymes responsible for both phase I and phase II metabolism. Moreover, the activities of drug transporter proteins such as P-glycoprotein and organic anion transporting polypeptide are also altered in patients with chronic kidney disease. More than 75 commonly used drugs demonstrated altered nonrenal clearance in patients with chronic kidney disease. Most of these drugs are metabolized by phase I enzymes [involving the cytochrome P450 (CYP) mixed function oxidase family of enzymes], and relatively few drugs are subjected to phase II metabolism undergoing glucuronidation (diacerein, morphine, oxprenolol, and zidovudine). Procainamide and isoniazid undergo N-acetylation. Most of these drugs showed reduced nonrenal clearance, and certain drugs showed increased bioavailability (drugs undergoing first-pass metabolism in the intestinal mucosa or liver due to impaired activity of these enzymes causing reduced first-pass metabolism) [1]. Certain drugs, such as phenytoin, fosinopril, cefpiramide, nifedipine, bumetanide, and sulfadimidine demonstrated increased nonrenal clearance in patients with chronic kidney disease [13,14]. At least for phenytoin, the increased nonrenal clearance may be due to reduced protein binding of phenytoin with albumin causing a disproportionate increase in free fraction because only free drug is subjected to metabolism and clearance [15].

Animal studies have demonstrated significant downregulation (40–85%) of hepatic and intestinal metabolism mediated by cytochrome P450 enzymes in chronic renal failure [16]. Several mechanisms based on animal studies and a few human studies have been proposed for reduced enzymatic activities of hepatic drug-metabolizing enzymes in patients with chronic kidney disease. Animal studies have indicated that chronic kidney disease is associated with reduced expression of CYP genes and gene products (reduced messenger RNA and protein or reduced protein but no change in messenger RNA level). As a result, activities of CYP enzymes may be significantly reduced in patients with chronic kidney disease. High amounts of circulating uremic toxins, cytokines, and parathyroid hormones are also probably responsible for reduced activities of various CYP isoforms. Therefore, hepatic clearance of many drugs may be significantly reduced in patients with chronic kidney disease, thus increasing the possibility of adverse drug reactions in these patients [17]. Nolin et al. demonstrated that in patients with uremia, nonrenal clearance of fexofenadine was decreased up to 63% compared to that of controls. Moreover, there was a 2.8-fold increase in area under the plasma concentration curve of fexofenadine in uremic patients compared to healthy subjects used as the control group. However, clearance of midazolam was not affected, indicating that the function of CYP3A4 was not altered in patients with uremia. The authors further observed that the changes in hepatocytes and enterocyte protein expression were consistent with reduced clearance of fexofenadine, which is cleared through the activity of drug transport protein [18].

11.4 ALTERED DRUG DISPOSITION IN HEPATIC DISEASE

Liver dysfunction not only reduces clearance of a drug metabolized through hepatic enzymes or biliary mechanism but also affects plasma protein binding due to reduced synthesis of albumin and other drug-binding proteins.

Even mild to moderate hepatic disease may cause an unpredictable effect on drug metabolism. Portal-systemic shunting present in patients with advanced liver cirrhosis may cause significant reduction in first-pass metabolism of high-extraction drugs, thus increasing bioavailability as well as the risk of drug overdose and toxicity [19]. In addition, activities of several isoenzymes of CYP enzymes (CYP1A1, CYP2C19, and CYP3A4/5) are reduced due to liver dysfunction, whereas activities of other isoenzymes, such as CYP2D6, CYP2C9, and CYP2E1, may not be affected significantly. Therefore, drugs that are metabolized by CYP1A1, CYP3A4/5, and CYP2C19 may show increased blood levels in patients with hepatic dysfunction requiring dosage adjustment in order to avoid toxicity [20].

Whereas the phase I reaction involving CYP enzymes may be impaired in liver disease, the phase II reaction (glucuronidation) seems to be affected to a lesser extent, although both phase I and phase II reactions in drug metabolism are substantially impaired in patients with advanced cirrhosis. Currently, there is no universally accepted endogenous marker to access hepatic impairment, and the semiquantitative Child–Pugh score is frequently used to determine the severity of hepatic dysfunction and thus dosage adjustments, although there are limitations to this approach [19]. Nonalcoholic fatty liver disease is the most common chronic liver disease. This type of liver disease also affects activities of drug-metabolizing enzymes in the liver, with the potential to produce adverse drug reactions from standard dosage [21].

Mild to moderate hepatitis infection may also alter clearance of drugs. Trotter et al. [22] reported that total mean tacrolimus dose in the first year after transplant was lower by 39% in patients with hepatitis C compared to patients with no hepatitis C infection. Zimmermann et al. [23] reported that oral dose clearance of sirolimus was significantly decreased in subjects with mild to moderate hepatic impairment compared to controls, and the authors stressed the need for careful monitoring of trough whole blood sirolimus concentrations in renal transplant recipients exhibiting mild to moderate hepatic impairment. Wyles and Gerber reviewed the effect of hepatitis with hepatic dysfunction on antiretroviral therapy, especially highly active antiretroviral therapy (HAART), in patients with AIDS and commented that dosage of protease inhibitors indinavir, lopinavir, ritonavir, amprenavir, and atazanavir may require reduction in patients with liver disease, although hepatic dysfunction does not affect pharmacokinetics of nucleoside reverse transcriptase inhibitors because these drugs are not metabolized by liver enzymes [24]. Ho et al. reported two novel genetic variations in the promoter sequence of the CYP2D6 gene, 1822A→G and 1740C→T, in patients infected with hepatitis C that are associated with a lower activity of CYP2D6 enzyme that is responsible for metabolism of many drugs. Therefore, patients

Table 11.2 Six Questions to Manage Medications in Patients with Hepatic Disease

Question	Comment
1. Is the patient experiencing acute or chronic liver failure?	Alteration of pharmacokinetic and pharmacodynamic factors may be different in acute versus chronic liver failure. Drug dosing based on Child–Turcotte–Pugh score is often determined based on clinical studies in patients with chronic liver disease and cirrhosis rather than acute liver failure.
2. Does the drug have high hepatic first-pass metabolism?	If the extraction ratio of a drug is high, the rate-limiting step for first-pass metabolism is blood flow, but for drugs with low extraction ratio, the rate-limiting step is activities of liver enzymes. Care must be taken for dosage of drugs with high-extraction ratio.
3. Is the medication highly protein bound?	High free level of a strongly protein-bound drug may cause toxicity.
4. Is there a change in the volume of distribution for the medication?	Volume of distribution may be altered in liver disease and as a result dosage adjustment is needed.
5. Is the clearance of the medication significantly altered?	Even dosage of a medicine cleared by the kidney should be adjusted based on creatinine clearance, even in patients with liver disease.
6. Is there a pharmacodynamic interaction with the medication?	The physiological changes due to liver disease can lead to pharmacodynamic interactions that do not correlate with serum drug level. If a significant pharmacodynamic interaction is noticed, alternative therapy may be needed.

infected with hepatitis C virus may be at risk of drug-induced toxicity and adverse drug reaction from drugs that are metabolized via CYP2D6 [25].

Nguyen et al. described a protocol for managing medications in patients with liver disease based on six questions. Patients with hepatic failure may often experience renal insufficiency, which may further complicate dosing in such patients. If a drug undergoes hepatic first-pass metabolism, its bioavailability may be affected significantly in patients with hepatic diseases. Drugs such as morphine, nitroglycerine, doxepin, venlafaxine, fluorouracil, idarubicin, mercaptopurine, quetiapine, perphenazine, propranolol, nicardipine, verapamil, tacrolimus, and sumatriptan have a high-extraction ratio ($>70\%$) and may undergo hepatic first-pass metabolism. Other factors may also cause altered drug disposition in patients with liver disease [26]. The six questions suggested by the authors for managing medications in patients with hepatic disease are summarized in Table 11.2.

11.5 ALTERED DRUG DISPOSITION IN CRITICALLY ILL PATIENTS

Critically ill patients may have altered absorption, distribution, metabolism, and elimination of drugs. Moreover, many critically ill patients may have underlying renal insufficiency or hepatic dysfunction or both, further complicating disposition of a drug. Antimicrobial agents are among the most important and most commonly prescribed drugs in the management of critically ill patients. Despite available guidelines, failure of antimicrobial treatment may occur in critically ill patients due to altered pharmacokinetic and pharmacodynamic properties of these drugs. In general, hydrophilic antimicrobial agents (β-lactams, aminoglycosides, and glycopeptides) have a higher risk of interindividual pharmacokinetic variations than lipophilic antimicrobials (macrolide antibiotics, fluoroquinolones, tetracyclines, chloramphenicol, and rifampin) in critically ill patients. Hypoperfusion, which may decrease drug distribution to the tissues and bodily organs, is a common symptom in patients who are critically ill. As a result, hydrophilic antibiotics demonstrate higher fluctuations in serum concentrations requiring frequent dosing adjustment in critically ill patients compared to lipophilic antimicrobial agents. In addition, underexposure of drug may occur due to increased volume of distribution as a result of edema in patients with sepsis and trauma, as well as patients receiving fluid therapy or having indwelling post-surgical drainage issues. On the other hand, drug-induced toxicity due to overexposure may occur due to renal insufficiency [27].

Vancomycin is widely used in critically ill patients to treat life-threatening bacterial infection, including methicillin-resistant *Staphylococcus aureus*. The recommended trough vancomycin concentration to treat such severe infection is 15–20 μg/mL. However, in critically ill patients with severe sepsis or septic shock, reaching target vancomycin serum level may be challenging due to increased volume of distribution of vancomycin and augmented renal clearance. As a result, treatment failure may occur. In contrast, critically ill patients with acute kidney injury may experience toxicity from vancomycin therapy due to decreased renal clearance of vancomycin. Based on their study of 24 critically ill patients who received vancomycin, Blot et al. noted that there was a large interindividual variability in vancomycin pharmacokinetics and pharmacodynamics in these patients. The authors recommended reevaluation of current vancomycin dosing guidelines for critically ill patients in order to achieve better drug exposure and avoid vancomycin toxicity [28].

Many factors may affect pharmacokinetic parameters of a drug in a critically ill patient. Delayed gastric emptying, vasopressor use, and use of a feeding tube may significantly alter absorption of a drug. Fluid resuscitation, alteration in plasma protein binding, and tissue perfusion also may affect pharmacokinetic and pharmacodynamic characteristics of a drug. Moreover, drug metabolism and elimination may be significantly altered in critically ill patients [29]. Factors that affect drug disposition in critically ill patients are summarized in Table 11.3. Alterations in normal pharmacokinetics may increase drug bioavailability, resulting in high serum concentrations of the

Table 11.3 Causes of Altered Drug Disposition in Critically Ill Patients

Parameter	Comments
Oral absorption	Reduced gastrointestinal motility is common in critically ill patients and as a result oral bioavailability of a drug may be affected.
Subcutaneous absorption	Subcutaneous absorption of a drug may be reduced if a critically ill patient is receiving vasopressor.
Volume of distribution	Increased volume of distribution and reduced peak concentrations of hydrophilic drugs due to fluid resuscitation. Reduced tissue perfusion due to shock may reduce free drug concentration in peripheral tissue.
Drug–protein binding	Increase in free drug concentration and volume of distribution of drugs bound to albumin due to hypoalbuminemia, which is common among critically ill patients. However, free fraction is reduced for drugs that are bound to α_1 acid glycoprotein (AAG) because AAG is an acute phase reactant and its concentration is increased in critically ill patients.
Metabolism	Acute reduction in hepatic blood flow may reduce hepatic clearance of drugs with high-extraction ratio. Critical illness may inhibit activity of liver enzymes, thus causing reduced hepatic clearance of drugs with low extraction ratio.
Elimination	Acute kidney insufficiency may reduce renal clearance of drugs. Alternatively, augmented renal clearance (eg, in patients with sepsis) may increase renal clearance of a drug. Altered active transport of medications may occur in critically ill patients.

drug, which may cause normal drug doses to have toxic effects. In addition, bioavailability may be decreased, which will reduce serum drug concentrations and drug efficacy [30]. Pharmacokinetic changes in critically ill patients are not predictable, and there are no general rules that are applicable for modifying drug dosage in all critically ill patients.

Enterohepatic circulation (EHC) may also be impacted in critically ill patients. EHC occurs when drugs and/or drug metabolites are circulated from the liver and are excreted into the bile. From the bile, the compounds enter the gallbladder and then undergo reabsorption in the small intestines. The drug metabolite can be reabsorbed or converted back to the parent compound for reabsorption then transported back to the liver for systemic circulation [30]. Several physiological conditions will affect the rate of drug absorption in the gastrointestinal (GI) tract, including gastric emptying, pH, surface area, blood flow, GI disease, food consumption, and intestinal microflora [31]. EHC can increase the bioavailability of the drug, and the recycling process will also prolong the half-life of the drug [30]. Several drugs undergo EHC, including mycophenolic acid, morphine, warfarin, imipramine, amiodarone, indomethacin, cimetidine, lorazepam, cyclosporine, colchicine, rifampicin, azithromycin, chloramphenicol, and methotrexate [31].

11.6 APPLICATION OF TDM IN CRITICALLY ILL PATIENTS

Drug classes that are frequently administered to hospitalized patients include sedatives, analgesics, anticoagulants, hypoglycemics, vasopressors, immunosuppressants, anticonvulsants, antiarrhythmics, antidepressants, antineoplastic agents, central nervous system stimulants or depressants, antimicrobials, neuromuscular blocking agents, diuretics, proton pump inhibitors, and drug antidotes [32,33]. The pharmacokinetics of these drug classes can be unpredictable in critically ill patients; thus, pharmacokinetic assessments will assist clinicians in calculating the appropriate loading and maintenance dosage for optimal drug therapy. Patients with reduced drug clearance may require lower doses, whereas patients with increased drug clearance may require higher doses for treatment. TDM is not well established in patients with acute kidney injury caused by conditions such as inflammation, uremia, or drug-induced injury (ie, calcineurin inhibitors and aminoglycoside drugs). TDM may be utilized to reduce mortality rates from acute kidney injury, which range from 43% to 74% [34,35].

11.6.1 TDM of Antimicrobial Drugs in Critically Ill Patients

Antimicrobial therapies, such as antifungals, antiretrovirals, β-lactams, penicillins, cephalosporins, carbapenems, and quinolone drugs, are often administered in patients with severe infections. TDM of antimicrobials is employed to assess whether patients have optimal exposure to therapy to treat the infection and reduce the risk of microbial drug resistance and toxicity [36]. Toxic effects may include nephrotoxicity and ototoxicity for aminoglycoside drugs. The volume of distribution of antimicrobial therapy may increase during fluid resuscitation treatment, and the clearance of these drugs will decrease under physiological conditions that cause hypoperfusion to the kidneys and liver and thus elevate antimicrobial serum/plasma concentrations [36]. Studies have also demonstrated that antimicrobial drug elimination may also increase and lead to a decrease in serum concentrations, which may not be sufficient to fight infection [37,38]. Patients with complex infections may require multidrug therapy, and TDM can be implemented to identify drug-induced toxicities and to optimize treatment. Dose-adjustment strategies usually consist of assessing drug exposure using serial collections to establish the AUC or, in some cases, peak and/or predose (trough) serum/plasma collections [36].

Case Report: A 66-year-old male was admitted to the ICU to treat acute ischemia in his lower leg. The patient had preexisting comorbidities that included ischemic cardiomyopathy and peripheral arterial occlusive disease, hypertension, and type II diabetes mellitus. It was not possible to perform revascularization, and the patient had to undergo leg amputation. One week (day 7) after the amputation procedure, the patient developed fever, inflammation, and acute cholecystitis, which led to a cholecystectomy. On day 9, the patient's urine culture was positive for *Candida albicans*. On day 12, the patient's condition worsened, and he developed respiratory insufficiency and positive blood culture for *C. albicans* and *Enterococcus faecium*. Multidrug therapy with vancomycin, intravenous levofloxacin, and anidulafungin was administered. On day 19, the patient was stabilized, blood cultures were negative, and the patient was removed from antimicrobial therapy [39].

11.6.2 TDM of Digoxin in Critically Ill Patients

Digoxin is widely utilized in elderly patients to treat congestive heart failure and is eliminated from the body via the kidneys. Patients with renal impairment may experience prolonged intoxication of digoxin therapy due to an increase in the half-life of the drug [40]. Patients with diseases that cause plasma volume expansion, such as congestive heart failure, liver disease, uremia, and hypertension, also have increased formation of digoxin-like immunoreactive substances (DLIS) in comparison to healthy individuals [41–45]. Several published studies have demonstrated that digoxin quantification by immunoassay can be impacted by DLIS due to cross-reactivity of digoxin antibodies [46–60]. Consequently, a falsely elevated digoxin immunoassay result above the therapeutic range, caused by DLIS, may cause a clinician to make a dose reduction, which may lead to a decrease in drug efficacy. On the other hand, a falsely low digoxin immunoassay result may cause a clinician to increase the digoxin dosage, which may put the patient at risk for toxicity if the serum concentration exceeds the therapeutic range [52].

Case Report: An 87-year-old male with a history of congestive heart failure presented to the emergency department with a 3-day history of nausea and poor appetite. The patient was prescribed furosemide and digoxin. Upon physical exam, his blood pressure was 130/85 mm Hg; heart rate 52 beats/min, and electrocardiogram displayed irregular heartbeats. Laboratory tests were ordered, and the results were as follows: Blood urea nitrogen was 51 mg/dL (reference range, 7–18 mg/dL), and serum digoxin was 4.7 mg/L (therapeutic range, 0.8–2.0 mg/L). The patient was treated with digoxin-specific antibody fragments (DigiFab), and 1 h after treatment, heart rate was 68 beats/min. However, serial measurements of serum digoxin did not decrease despite antidote therapy. It was discovered that the digoxin assay antibodies cross-reacted with DigiFab, and centrifugation was necessary prior to digoxin analysis for accurate quantification. Three days after treatment, the patient's predose serum digoxin concentration was 1.6 mg/L, and antidote therapy was discontinued.

11.6.3 TDM of Sedatives and Analgesics in Critically Ill Patients

Sedatives and analgesics are frequently administered to patients in the ICU to provide pain relief for pre- and postsurgical procedures, promote relaxation, relieve anxiety and physical stress, induce sleep, and reduce mobility and agitation. Sedatives that are commonly utilized include benzodiazepines, opioids, and barbiturates [61]. These drugs may also be utilized for analgesia, in addition to nonsteroidal antiinflammatory antibodies. TDM can be utilized to reduce the risk of adverse effects, such as constipation and fatality caused by respiratory depression, hepatotoxicity, nephrotoxicity, portal hypertension, and GI bleeding. However, therapeutic guidelines are not well established [62]. The response to therapy of drugs that are extensively metabolized by the liver can also be impacted by liver disease. For example, the bioavailability of opioid therapy can vary in patients with chronic liver disease or combined hepatorenal diseases and can induce encephalopathy in critically ill patients [62,63].

11.7 CONCLUSIONS

TDM in critically ill populations is not well established, and standard drug therapy in these patients may increase the risk of drug-induced toxicity or therapeutic failure. The pharmacokinetics and pharmacodynamics of drug therapy are difficult to predict, and TDM is utilized to guide dosing to help reduce mortality, optimize drug efficacy, and minimize toxicity.

References

[1] Yeung CK, Shen DD, Thummel KE, Himelfarb J. Effects of chronic kidney disease and uremia on hepatic drug metabolism and transport. Kidney Int 2014;85:522–8.

[2] Lazo M, Hernaez R, Eberhardt MS, Bonekamp S, et al. Prevalence of nonalcoholic fatty liver disease in the United States: the Third National Health and Nutrition Examination Survey 1988–1994. Am J Epidemol 2013;178:38–45.

[3] Leape LL, Cullen DJ, Clapp MD, et al. Pharmacist participation on physician round and Adverse drug events in the intensive care unit. JAMA 1999;282:267–70.

[4] Scheife RT. Protein binding: what does it mean? DICP 1989;23(7–8 Suppl):S27–31.

[5] Terrell KM, Heard K, Miller DK. Prescribing to older ED patients. Am J Emerg Med 2006;24:468–78.

[6] Gabardi S, Abramson S. Drug dosing in chronic kidney disease. Med Clin North Am 2005;89:649–87.

[7] Hanlon JT, Wang X, Handler SM, Weisbord S, et al. Potentially inappropriate prescribing of primarily renally cleared medications for older Veterans Affairs Nursing Home patients. J Am Med Dir Assoc 2011;12:377–83.

[8] Krepinsky J, Ingram AJ, Class CM. Prolonged sulfonylurea-induced hypoglycemia in diabetic patients with end stage renal disease. Am J Kidney Dis 2000;35:500–5.

[9] Papaioannou A, Clarke JA, Campbell G, Bedard M. Assessment of adherence to renal dosing guidelines in long term care facilities. J Am Geriatr Soc 2000;48:1470–3.

[10] Nielsen AL, Henriksen DP, Marinakis C, Hellebek A, et al. Drug dosing in patients with renal insufficiency in a hospital setting using electronic prescribing and automated reporting of estimated glomerular filtration rate. Basic Clin Pharmacol Toxicol 2014;114:407–13.

[11] Blye E, Lorch J, Cortell S. Extracorporeal therapy in the treatment of intoxication. Am J Kidney Dis 1984;3:321–8.

[12] Vernon DD, Gleich MC. Poisoning and drug overdose. Crit Care Clin 1997;13(3):647–67.

[13] Dreisbach AW, Lertora JJ. The effect of chronic renal failure on hepatic drug metabolism and drug disposition. Semin Dial 2003;16:45–50.

[14] Elston AC, Bayliss MK, Park GR. Effect of renal failure on hepatic drug metabolism by the liver. Br J Anaesth 1993;71:282–90.

[15] Dasgupta A. Usefulness of monitoring free (unbound) concentrations of therapeutic drugs in patient management. Clin Chim Acta 2007;377:1–13.

[16] Dreisbach AW, Lertora JJ. The effect of chronic renal failure on drug metabolism and transport. Expert Opin Drug Metab Toxicol 2008;4:1065–74.

[17] Nolin TD, Naud J, Leblond FA, et al. Emerging evidence of the impact of kidney disease on drug metabolism and transport. Clin Pharmacol Ther 2008;83:898–903.

[18] Nolin TD, Frye RF, Le P, Sadr H, et al. ESRD impairs nonrenal clearance of fexofenadine but not midazolam. J Am Soc Nephrol 2009;20:2269–76.

[19] Verbeeck RK. Pharmacokinetics and dosage adjustment in patients with hepatic dysfunction. Eur J Clin Pharmacol 2008;64:1147–61.

[20] Villeneuve JP, Pichette V. Cytochrome P450 and liver disease. Curr Drug Metab 2004;5:273–82.

[21] Merrell MD, Cherrington NJ. Drug metabolism alterations in nonalcoholic fatty liver disease. Drug Metab Rev 2011;43:317–34.

[22] Trotter JF, Osborne JC, Heller N, Christians U. Effect of hepatitis C infection on tacrolimus dose and blood levels in liver transplant recipients. Aliment Pharmacol Ther 2005;22:37–44.

[23] Zimmermann JJ, Lasseter KC, Lim HK, Harper D, et al. Pharmacokinetics of sirolimus (rapamycin) in subjects with mild to moderate hepatic impairment. J Clin Pharmacol 2005;45:1368–72.

[24] Wyles DL, Gerber JG. Antiretroviral drug pharmacokinetics in hepatitis with hepatic dysfunction. Clin Infect Dis 2005;40:174–81.

[25] Ho MT, Kelly EJ, Bodor M, Bui T, et al. Novel cytochrome P450-2D6 promoter sequence variations in hepatitis C positive and negative subjects. Ann Hepatol 2011;10:327–32.

[26] Nguyen HM, Cutie AJ, Pham DQ. How to manage medications in setting of liver disease with the application of six questions. Int J Clin Pract 2010;65:858–67.

[27] Pea F, Viale P, Furlanut M. Antimicrobial therapy in critically ill patients: a review of pathophysiological conditions responsible for altered disposition and pharmacokinetic variability. Clin Pharmacokinet 2005;44:1009–34.

[28] Blot S, Koulenti D, Akova M, Bassetti M, et al. Does contemporary vancomycin dosing achieve therapeutic targets in a heterogeneous clinical cohort of critically ill patients? Data from multinational DALI study. Crit Care 2014;18:R99.

[29] Smith BS, Yogaratnam D, Levasseur-Franklin K, Formi A, et al. Introduction to drug pharmacokinetics in the critically ill patients. Chest 2012;141:1327–36.

[30] Roberts MS, Magnusson BM, Burczynski FJ, Weiss M. Enterohepatic circulation: physiological, pharmacokinetic and clinical implications. Clin Pharmacokinet. 2002;41(10):751–90.

[31] Gao Y, Shao J, Jiang Z, Chen J, Gu S, Yu S, et al. Drug enterohepatic circulation and disposition: constituents of systems pharmacokinetics. Drug Discov Today 2014;19(3):326–40.

[32] Blot SI, Pea F, Lipman J. The effect of pathophysiology on pharmacokinetics in the critically ill patient—concepts appraised by the example of antimicrobial agents. Adv Drug Deliv Rev 2014;77:3–11.

[33] John LJ, Devi P, John J, Arifulla M, Guido S. Utilization patterns of central nervous system drugs: a cross-sectional study among the critically ill patients. J Neurosci Rural Pract 2011;2 (2):119–23.

[34] Kolhe NV, Stevens PE, Crowe AV, Lipkin GW, Harrison DA. Case mix, outcome and activity for patients with severe acute kidney injury during the first 24 hours after admission to an adult, general critical care unit: application of predictive models from a secondary analysis of the ICNARC Case Mix Programme database. Crit Care 2008;12(Suppl. 1):S2.

[35] Abosaif NY, Tolba YA, Heap M, Russell J, El Nahas AM. The outcome of acute renal failure in the intensive care unit according to RIFLE: model application, sensitivity, and predictability. Am J Kidney Dis 2005;46(6):1038–48.

[36] Wong G, Sime FB, Lipman J, Roberts JA. How do we use therapeutic drug monitoring to improve outcomes from severe infections in critically ill patients. BMC Infect Dis 2014;14:228.

[37] Udy AA, Roberts JA, De Waele JJ, Paterson DL, Lipman J. What's behind the failure of emerging antibiotics in the critically ill? Understanding the impact of altered pharmacokinetics and augmented renal clearance. Int J Antimicrob Agents 2012;39(6):455–7.

[38] Udy AA, Varghese JM, Altukroni M, Briscoe S, McWhinney B, Ungerer J, et al. Subtherapeutic initial β-lactam concentrations in select critically ill patients. Chest 2012;142:30–9.

[39] Lichtenstern C, Wolff M, Arens C, Klie F, Majeed RW, Henrich M, et al. Cardiac effects of echinocandin preparations—three case reports. J Clin Pharm Ther 2013;38(5):429–31.

[40] Hazara AM. Recurrence of digoxin toxicity following treatment with digoxin immune fab in a patient with renal impairment. QJM 2014;107(2):143–4.

[41] Graves SW, Brown B, Valdes Jr R. An endogenous digoxin-like substance in patients with renal impairment. Ann Intern Med 1983;99:604–8.

[42] Gault MH, Vasdev SC, Longerich LL. Endogenous digoxin-like substance(s) and combined hepatic and renal failure. Ann Internal Med 1984;101:567–8.

[43] Cloix JF. Endogenous digitalis-like compounds. A tentative update of chemical and biological studies. Hypertension 1987;10:I67–70.

[44] Shilo L, Adawi A, Solomon G, Shenkman L. Endogenous digoxin-like immunoreactivity in congestive heart failure. Br Med J 1987;295:415–16 (Clin Res Ed.).

[45] Lackner TE, Lau BW, Parvin C, Valdes Jr R. Endogenous digoxin-like immunoreactivity in elderly patients with normal serum creatinine concentrations. Clin Pharm 1988;7:449–53.

[46] Greenway DC, Nanji AA. Falsely increased results for digoxin in sera from patients with liver disease: ten immunoassay kits compared. Clin Chem 1985;31:1078–9.

[47] Nanji AA, Greenway DC. Falsely raised plasma digoxin concentrations in liver disease. Br Med J (Clin Res Ed.) 1985;290:432–43.

[48] Soldin SJ. Digoxin—issues and controversies. Clin Chem 1986;32:5–12.

[49] Soldin SJ, Stephey C, Giesbrecht E, Harding R. Further problems with digoxin measurement. Clin Chem 1986;32:1591.

[50] Dasgupta A, Nakamura A, Doria L, Dennen D. Comparison of free digoxin and total digoxin: extent of interference from digoxin-like immunoreactive substances (DLIS) in a fluorescence polarization assay. Clin Chem 1989;35:323–4.

[51] Luke M, Dasgupta A. Digitoxinlike and digoxinlike immunoreactivities in sera of patients with uremia and liver disease as measured by fluorescence polarization immunoassays: poor correlation between digitoxinlike and digoxinlike immunoreactivities. Ther Drug Moni 1997;19:230–5.

[52] Wu SL, Li W, Wells A, Dasgupta A. Digoxin-like and digitoxin-like immunoreactive substances in elderly people. Impact on therapeutic drug monitoring of digoxin and digitoxin concentrations. Am J Clin Pathol 2001;115:600–4.

[53] Crossey MJ, Dasgupta A. Effects of digoxinlike immunoreactive substances and digoxin FAB antibodies on the new digoxin microparticle enzyme immunoassay. Ther Drug Moni 1997;19:185–90.

[54] Dasgupta A. Endogenous and exogenous digoxin-like immunoreactive substances: impact on therapeutic drug monitoring of digoxin. Am J Clin Pathol 2002;118:132–40.

[55] Dasgupta A, Schammel DP, Limmany AC, Datta P. Estimating concentrations of total digoxin and digoxin-like immunoreactive substances in volume-expanded patients being treated with digoxin. Ther Drug Monit 1996;18:34–9.

[56] Dasgupta A, Saldana S, Heimann P. Monitoring free digoxin instead of total digoxin in patients with congestive heart failure and high concentrations of digoxin-like immunoreactive substances. Clin Chem 1990;36:2121–3.

[57] Dasgupta A, Trejo O. Suppression of total digoxin concentrations by digoxin-like immunoreactive substances in the MEIA digoxin assay. Elimination of negative interference by monitoring free digoxin concentrations. Am J Clin Pathol 1999;111:406–10.

[58] Dasgupta A. Therapeutic drug monitoring of digoxin: impact of endogenous and exogenous digoxin-like immunoreactive substances. Toxicological Rev 2006;25:273–81.

[59] McMillin GA, Owen WE, Lambert TL, De BK, Frank EL, Bach PR, et al. Comparable effects of DIGIBIND and DigiFab in thirteen digoxin immunoassays. Clin Chem 2002;48:1580–4.

[60] Yamamoto T, Takano K, Sanaka M, Nomura Y, Koike Y, Mineshita S. Digoxin-like immunoreactive substance in the elderly patient and its impact on the TDx digoxin assay. Ther Drug Monit 1998;20:417–21.

[61] Payen JF, Chanques G, Mantz J, Hercule C, Auriant I, Legulliou JL, et al. Current practices in sedation and analgesia for mechanically ventilated critically ill patients. Anesthesiology 2007;106:687–95.

[62] Imani F, Motavaf M, Safari S, Alavian SM. The therapeutic use of analgesics in patients with liver cirrhosis: a literature review and evidence-based recommendations. Hepat Mon 2014;14(10):e23539.

[63] Okawa H, Ono T, Hashiba E, Tsubo T, Ishihara H, Hirota K. Use of bispectral index monitoring for a patient with hepatic encephalopathy requiring living donor liver transplantation: a case report. J Anesth 2011;25(1):117–19.

CHAPTER 12

Issues of Pharmacogenomics in Monitoring Warfarin Therapy

Jennifer Martin[1] and Andrew Somogyi[2]
[1]School of Medicine and Public Health, University of Newcastle, Newcastle, NSW, Australia
[2]Discipline of Pharmacology, School of Medicine, University of Adelaide, Adelaide, Australia

CONTENTS

12.1 Introduction 261
12.2 Potential for Pharmacogenomics of Warfarin......262
12.3 Pharmacology...263
12.3.1 CYP2C9 Status....263
12.3.2 VKORC Status.....264
12.3.3 Other Genetic Mutations......265
12.3.4 Nongenetic Factors Affecting Warfarin Dosing......266
12.3.5 Vitamin K 268
12.3.6 Diet......268
12.3.7 Age 268
12.3.8 Gender......269
12.3.9 Anthropometric Variables......269
12.4 Clinical Relevance 270
12.5 Cost-Effectiveness of Pharmacogenomics Testing in Warfarin Therapy......272
12.6 New Medications to Replace Warfarin......273
12.7 Conclusions 274
References 275

12.1 INTRODUCTION

Warfarin remains the most commonly prescribed anticoagulant drug for the prophylaxis and treatment of venous and arterial thromboembolic disorders, with more than 20 million scripts written each year in the United States alone. This volume of use persists despite newer oral anticoagulation medications (NOACs) now entering the market, predominantly because the population able to be treated for atrial fibrillation is increasing, particularly the elderly, and because there has been some disappointment with the benefit of the newer drugs over warfarin in clinical practice. Many of the warfarin comparators have no reversal agent, those that do have are very expensive, and many countries cannot afford either the drug or the reversal agent. Furthermore, because warfarin already has a wealth of efficacy data with regard to its use (ie, 50 years), is relatively cheap, and is routinely used by many patients with atrial fibrillation, there is a continued need to undertake research to improve its safety outcomes.

However, warfarin is a "difficult to use" drug due to a narrow therapeutic index. In addition, warfarin also suffers from multiple clinically important drug–drug interactions and an erratic safety profile—conditions that would impede its market approval if developed today. In reality, however, its use has been well managed by clinicians by using the international normalized ratio (INR) measurements and adjustment of dose to reduce the risk of bleeding and ensure efficacy. It is notable that the NOACs registration indication of one dose fits all also appears to not hold for at least dabigatran, for which physicians caring for elderly or comorbid patients are now recommended to have at least a single concentration measurement [1]. For warfarin, time in INR therapeutic range has been shown to directly correlate with

W. Clarke & A. Dasgupta (Eds): Clinical Challenges in Therapeutic Drug Monitoring. DOI: http://dx.doi.org/10.1016/B978-0-12-802025-8.00012-X

efficacy and bleeding [2]. However, with current management, patients remain on average within their target INR range for only 70% of the time [2], and bleeding events do occur. The 10-fold variability around dose, target INR, and side effects has therefore led to interest in whether testing for genetic variations in warfarin metabolism and target site could be useful for predicting the optimum dose, reducing bleeding risk, and reducing the time to achieve a therapeutic prothrombin time, expressed as the INR.

12.2 POTENTIAL FOR PHARMACOGENOMICS OF WARFARIN

Warfarin is a potential candidate for pharmacogenetic testing because it is a commonly used medication, has a narrow therapeutic window, and displays highly variable pharmacokinetics and responses between individuals. As a result, achieving and maintaining INRs within the therapeutic range that reduces adverse drug events can be difficult and time-consuming. Furthermore, warfarin has a mechanism of action and elimination pathways that involve enzymes that are polymorphic—variations that account for 30–50% of the variability in dosing. Thus, there has been much interest in whether testing for these enzyme variants improves warfarin safety and efficacy. Recent interest has been noted in measuring the comparative incremental benefit in safety and efficacy of pharmacogenetic testing either instead of or in addition to the benefit of the well-known pharmacodynamic endpoint of efficacy and safety—INR.

During approximately the past 15 years, a number of predominantly retrospective studies have examined the relationship between some or all pharmacogenetic variables, surrogate outcomes such as time in therapeutic range (TTR), and actual clinical outcomes; however, the effect sizes have been variable, and clinical outcomes were often not measured. Thus, there has been discussion as to the actual clinical relevance of these genetic tests. Although there remains some uncertainty, it is agreed that both the analytical validity of these tests has been met and there is strong evidence to support association between these genetic variants and required dose of warfarin [3]. In 2010, the U.S. Food and Drug Administration (FDA) recommended that prescribers refer to a table showing stable maintenance doses and dosing ranges observed in multiple patients having different combinations of genetic variants [4] derived from a study of a data set of more than 4000 patients [5]. This study showed that dosage recommendations that were based on the combined pharmacogenetic–clinical algorithm were significantly better than those that were based on an algorithm that used only clinical variables or those that were based on a fixed-dose strategy. The combined algorithm in this study included variables of age, height, weight, amiodarone use, race, presence of

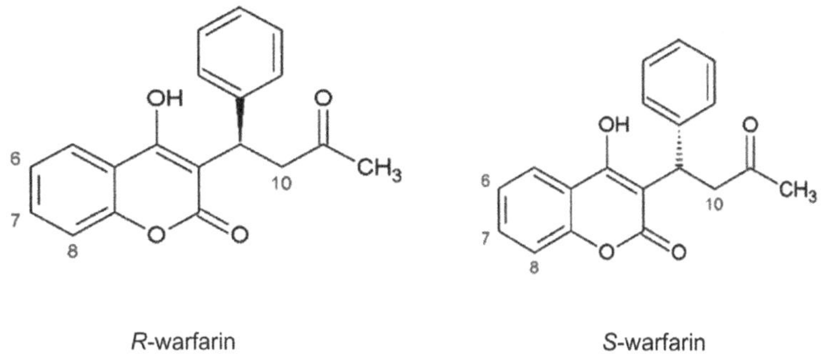

R-warfarin *S*-warfarin

FIGURE 12.1
Chemical structure of *R*- and *S*-warfarin.

enzyme inducer, as well as the genetic polymorphisms in the C1 subunit of the vitamin K 2,3-epoxide reductase complex (VKORC1) and CYP2C9 (an isoenzyme in the cytochrome P450 family of enzymes responsible for metabolism of the majority of drugs). It is available at http://www.warfarindosing.org. However, actual high-quality prospective randomized clinical trials powered by clinical outcome endpoints have not borne out the retrospective predictions that had predominantly used surrogate markers of clinical outcomes [6].

12.3 PHARMACOLOGY

Warfarin is an equal mixture of the enantiomers *S*-warfarin and *R*-warfarin, with *S*-warfarin being approximately three to five times more potent than *R*-warfarin (Fig. 12.1). It is generally considered that *S*-warfarin is the pharmacologically active enantiomer and that *R*-warfarin plays a small role only. Metabolism of *S*-warfarin occurs through the cytochrome P450 2C9 enzyme to the 6- and 7-hydroxy metabolites, whereas metabolism of the less potent *R*-warfarin occurs through CYP2C19 (8-hydroxy), CYP1A2 (6-hydroxy), CYP3A4 (10-hydroxy), and carbonyl reductase of the side chain ketone to form the alcohols 1 and 2 [7].

12.3.1 CYP2C9 Status

People who metabolize warfarin "normally" are homozygous for the usual (wild-type) allele *CYP2C9*1*. Two clinically relevant single nucleotide polymorphisms have been identified in CYP2C9 (*2[C430T] and *3[A1075C]). These result in reduced enzymatic activity (*2 has ~12% *S*-warfarin metabolic activity and *3 has <5% activity compared to the wild type) and therefore reduced *S*-warfarin metabolism. These single nucleotide polymorphisms are relatively common in Caucasians. Approximately 1% of the population is homozygous for *CYP2C9*2* and 20% are heterozygous carriers of this allele.

Table 12.1 Frequencies of the Two Major *CYP2C9* Variant Alleles in Various Populations

Population	*2 (%)	*3 (%)
Caucasian	15	7
Asian[a]	0	3
Indian	0	18
African	0	1
African American	0	1
Hispanic	7	6

[a]Asian: Han Chinese, Korean, and Japanese.
Source: Adopted from Martin J, Somogyi A. Pharmacogenomics and warfarin therapy. In: Dasgupta A, editor. Therapeutic drug monitoring: newer drugs and biomarkers. San Diego, CA: Academic Press; 2012 (Reprinted with permission).

The corresponding figures for *CYP2C9*3* are 0.4% and 8%. Another 1.4% of people are compound heterozygotes (*CYP2C9*2/*3*). However, there are wide interethnic differences in the frequencies of the *2 allele (practically absent in Asian and Indian populations and very low in Africans and African Americans) and the *3 allele (high incidence in Indians and low in Africans and African Americans) (Table 12.1).

Patients requiring a low dose of warfarin (1.5 mg daily or less) have a high likelihood of having a CYP2C9 variant allele (*2 or *3) and an increased risk of major bleeding complications [8]. A number of studies have shown that knowing the patient's genotype helps in achieving the target INR more quickly [9–11]. However, if this does not result in a reduction of hospital bed stay, reduction of bleeding, or another clinical endpoint, the relevance of reducing time to target INR by a small amount may not provide any meaningful benefit. Specifically, using this knowledge to predict dose may not necessarily even reduce bleeding events because CYP2C9 genotype per se predicts only 10–15% of dose variability [11]. Even after adjusting the warfarin dose for the variability in CYP2C9 status, there is still a considerable amount of dosing variability in patients who have similar CYP2C9 alleles. This variability appears to be partly attributable to genetic polymorphisms in the VKORC1 complex. The VKORC1 complex is the rate-limiting step in the vitamin K-dependent γ-carboxylation system that activates clotting factors. Warfarin exerts its anticoagulant effect by inhibiting VKORC1.

12.3.2 VKORC Status

A number of common polymorphisms in noncoding sequences have been identified in VKORC1, seven of which were found to be significantly associated with warfarin maintenance dosage, and five were in strong linkage

Table 12.2 Frequencies of the Two Major *VKORC1* Haplotypes in Various Populations

Population	A (Low Dose; %)[a]	B (High Dose; %)[b]
Caucasian	40	45
Asian[c]	90	10
Indian	45	45
African	15	50
African American	10	50
Hispanic	44	40

[a]Includes −1639G > A (rs9923231), 1173C > T(rs9934438); 2.7 ± 0.2 mg/day [11].
[b]3730G > A (rs7294); 6.2 ± 0.3 mg/day [11].
[c]Asian: Han Chinese, Korean, and Japanese.
Source: Adopted from Martin J, Somogyi A. Pharmacogenomics and warfarin therapy. In: Dasgupta A, editor. Therapeutic drug monitoring: newer drugs and biomarkers. San Diego, CA: Academic Press; 2012 (Reprinted with permission).

disequilibrium [12]. As such, low-dose (A) (2.7 ± 0.2 mg/day) and high-dose (B) (6.2 ± 0.3 mg/day) haplotypes were identified, and again these haplotypes showed substantial interethnic differences, with Caucasians and Indians having 40–45% A and 45–55% B, Chinese almost 90% A, and Africans 15% A and 50% B and a sizable missing haplotype (Table 12.2). Polymorphisms of this receptor are associated with a need for lower doses of warfarin [12]. The *VKORC1* genotype alone may explain up to 35% of the variability in warfarin dosage [13].

12.3.3 Other Genetic Mutations

It is likely that point mutations in the genes other than those for CYP2C9 or VKORC1 could explain some of the remaining variability in warfarin requirements. Evidence for this derives from at least two models that have demonstrated that the CYP2C9 and VKORC1 genotypes, together with known factors such as age and body size, only explain half to two-thirds of the interindividual variability in warfarin requirements [11,14]. Although this is an improvement on current nonpharmacogenetic algorithms, at least one-third of the variability is still unaccounted for. There are at least 30 other genes involved in the pharmacodynamics of warfarin that may explain this variability, including polymorphisms in apolipoprotein E, multidrug resistance 1, genes encoding vitamin K-dependent clotting factors, and possibly genes encoding additional components of the vitamin K epoxide reductase complex. Evidence for these is lacking. However, recently, polymorphisms in CYP4F2, an enzyme involved in vitamin K oxidation, have been associated with altered warfarin dosage requirements, although the contribution is minor at less than 3% in Caucasians but up to 10% in some Asian populations [15].

The first genome-wide association study, conducted in more than 1000 Swedish subjects, investigated approximately 326,000 markers. This study confirmed the major role of VKORC1 followed by CYP2C9 and the minor role of CYP4F2 [16]. However, as noted previously, having a mutant allele, although potentially predicting for altered warfarin metabolism, does not (1) actually predict warfarin metabolism (due to the myriad other genetic and environmental factors), (2) necessarily predict clinical outcome, or (3) imply that measuring it and altering dose based on this information improves clinical outcomes.

However, variant allele may well predict the likelihood of bleeding—that is, a pharmacodynamic rather than a pharmacokinetically important mutation. This is possible because although there is a good relationship between concentration and INR, and between INR and bleeding, there is still a large percentage (~50%) of variability noted between dose and INR that is currently unexplained.

12.3.4 Nongenetic Factors Affecting Warfarin Dosing

One of the difficulties with focusing solely on the effect of polymorphisms in the metabolizing pathways of *S*-warfarin and vitamin K epoxide reductase is that there are a number of important nongenetic factors that affect the INR and warfarin dosing requirements. Nongenetic factors associated with variation in warfarin requirements are summarized in Table 12.3. Factors associated with lower warfarin requirements are summarized in Table 12.4, and factors associated with higher warfarin requirements are given in Table 12.5. Age, racial group, and sex are well known, but increasingly recognized yet understudied is the effect of dietary and gut-derived vitamin K.

Table 12.3 Nongenetic Factors Associated With Variation in Warfarin Requirements

Factor	Effect
Dietary vitamin K (average plus daily intake)	Altering vitamin K intake alters warfarin requirements.
Variation in concomitant medications or CYP-interacting foods	Includes alteration in dose of drugs such as amiodarone and thyroxine, plus short courses of antibiotics, and intermittent ingestion of foods such as grapefruit juice or St John's wort.
Diarrhea	Reduce vitamin K recycling in gut wall.

Source: Adopted from Martin J, Somogyi A. Pharmacogenomics and warfarin therapy. In: Dasgupta A, editor. Therapeutic drug monitoring: newer drugs and biomarkers. San Diego, CA: Academic Press; 2012 (Reprinted with permission).

Table 12.4 Factors Associated With Lower Warfarin Requirements

Factor	Effect
VKORC1; −1639 AA	This genotype affects warfarin requirement less than GA or GG genotypes.
CYP2C9 *2 or *3 CYP2C9 *2 and *3	Both heterozygotes of *2 or *3 or homozygotes of *2 and *3 result in a reduced warfarin requirement.
Factor X insertion genotype	Mildly lower reduction.
Factor VII deletion genotype	Mildly lower reduction.
Reduced vitamin K intake	For example, if starving or in institutional care.
Some racial groups	May be independent or secondary to known racially divergent CYP2C9 or VKORC1 mutations, different diet, or additional factors.
Gender	Gender did not make any significant contribution to the regression models, but it is likely that the differences in warfarin requirements noted clinically are attributable to body size, with females in general being smaller than males.
Age	Reduced requirements for age may be secondary to altered distribution, altered receptor sensitivity, or reduced clearance.
Advanced malignancy	Reduced requirements may be due to liver metastases, lower body weight, and drug interactions.
Malabsorption syndromes	Affects vitamin K production and absorption in gut.
Liver disease	Affects synthetic functions of liver, including production of clotting factors and warfarin metabolism.
Heart disease	This causes hepatic congestion, resulting in abnormal liver function and reduced clotting factor synthesis.
Pyrexia	This increases warfarin sensitivity by enhancing the rate of degradation of vitamin K-dependent clotting factors.

Source: Adopted from Martin J, Somogyi A. Pharmacogenomics and warfarin therapy. In: Dasgupta A, editor. Therapeutic drug monitoring: newer drugs and biomarkers. San Diego, CA: Academic Press; 2012 (Reprinted with permission).

Table 12.5 Factors Associated With Higher Warfarin Requirements

Factor	Effect
Increased body weight	Higher total and lean body weight increase warfarin requirements, possibly through their effect on increasing body surface area.
Smoking	Increased metabolism, particularly of the *R*-enantiomer.
CYP2C9 inducers	Induces metabolism of the more potent *S*-enantiomer.

Source: Adopted from Martin J, Somogyi A. Pharmacogenomics and warfarin therapy. In: Dasgupta A, editor. Therapeutic drug monitoring: newer drugs and biomarkers. San Diego, CA: Academic Press; 2012 (Reprinted with permission).

12.3.5 Vitamin K

Vitamin K is an essential cofactor for the normal production of clotting factors II, VII, IX, and X. By inhibiting VKORC1, warfarin reduces the regeneration of vitamin K and thereby inhibits the activation of vitamin K-dependent clotting factors. It is known that a patient's vitamin K status when starting warfarin affects the time to reach a therapeutic INR. In addition, a daily dietary intake of more than 250 μg reduces warfarin sensitivity. Interesting from a therapeutic perspective is the finding that giving patients with an unstable INR daily doses of 150 μg of vitamin K decreases the variability of INR and increases the time in the target range [17].

12.3.6 Diet

Until recently, it had been assumed that all dietary effects on warfarin were due to vitamin K content. However, the role of other chemicals and vitamins in food may also have an effect on warfarin carboxylation or activity, perhaps by changing gut flora (and gut-derived vitamin K), pharmacodynamic effects on other blood coagulation pathways [eg, as seen with dong quai, garlic, papaya, and St John's wort (serotonin effect)], or competing reductase systems [18]. However, this effect has not been as rigorously studied, and instead most of the studied dietary effects are due to intake of vitamin K in foods such as green tea, turnips, avocados, and green leafy vegetables (eg, Brussels sprouts, broccoli, lettuce, and cabbage) [19]. Similarly, certain beverages can increase the effect of warfarin on bleeding outcomes, such as cranberry juice and alcohol intake. The latter may also contribute to a higher risk of bleeding due to increased likelihood and severity of falls and gastritis. The microbiome is another factor that has been well known to have an effect on warfarin bioavailability; the microbiome is affected by diet and other factors, such as antibiotic usage.

12.3.7 Age

The use of warfarin is expanding among the population older than 70 years of age in part because of the increasing prevalence of atrial fibrillation but also because of increased longevity and quality of life and increasing clinical experience with this drug in these populations. However, there is still limited actual evidence available in this population, although dosing regimens in most medical centers opt for a 5-mg per day or less dosage of warfarin as the starting dose in people older than age 70 years. The lower requirement for induction and maintenance dose is likely to be due to a number of factors, including drug interactions (due to higher likelihood of comorbidity and polypharmacy in this age group), reduced dietary vitamin K intake (especially in residential/nursing home subjects or in recipients of "meals on wheels"), and reduced absorption of vitamin K due to changes in bowel flora (which are also likely to be related to dietary changes). Reduced volume of distribution may be an issue if lower body fat or if low serum albumin occurs during the warfarinization period, affecting warfarin concentrations, but it is assumed that clearance is constant; the literature is mixed on the effect of age on clearance. For example, some studies include a population pharmacokinetic model showing that warfarin clearance is inversely related to age [20,21], especially the clearance of *R*-warfarin. However, many other studies do not show any significant difference in clearance with age, albeit these include small numbers of elderly subjects [22,23]. Measuring free as opposed to total warfarin clearance may be relevant to this debate because it has recently been demonstrated that the unbound clearance of *R*- and *S*-warfarin decreases by approximately 0.5% per year [24]. There is also an up to 30% decline in hepatic drug metabolism and cytochrome P450 content with age; the clinical effect of this on INR has not been clearly reported. However, there is a large amount of literature on the effect of weight and body size on clearance, and thus it is possible that the effect of age on warfarin is a surrogate of the previously discussed factors, such as diet and reduction in body weight or change in body composition with age.

A prospective study of 4616 patients, including 2359 patients older than age 80 years, showed that not only is the warfarin dose inversely related to age but also it is strongly associated with gender. The weekly warfarin dose declined by 0.4 mg/year [95% confidence interval (CI) = 0.37–0.44; $p < 0.001$] [25]. Among patients who were older than age 70 years, an initiation dose of 5 mg daily would have been excessive for 82% of women and 65% of men.

12.3.8 Gender

It has been well established that women in general need less maintenance dose of warfarin than men. For example, in the previously mentioned large study, females required 4.5 mg less warfarin per week than males (95% CI, 3.8–5.3; $p < 0.001$). The effects of age were clear (and additive) for gender; however, the effects of differences in lean versus fat body mass on this gender difference were not reported.

12.3.9 Anthropometric Variables

There is a large literature on the effect of body size variables on warfarin dosage requirements. By themselves, body surface area and body weight have significant correlations with warfarin dose; however, they do not appear to make a significant contribution to the regression model for dose once genotype and other factors are added [26]. In a later model, age, height, and CYP2C9 genotype were shown to significantly contribute to the more potent *S*-warfarin and total warfarin clearance, whereas only age and body size significantly contributed to *R*-warfarin clearance. The multivariate regression model including the variables of age, CYP2C9 and VKORC1 genotype, and height produced the best model for estimating warfarin dose ($R^2 = 55\%$) [14].

12.4 CLINICAL RELEVANCE

During approximately the past 5 years, the interest in routine pharmacogenetic testing for warfarin therapy introduced into clinical practice has not waned, predominantly because, as previously discussed, it does not predict all the variability in a patient's response to warfarin. Further well-conducted trials to document a clinical effect from better "numbers" have similarly not shown clinical benefit [27,28].

This is despite reasonable data from well-powered studies such as the parallel-design study CoumaGen-II [29], which included a comparative effectiveness research study of genotype-guided dosing of patients compared with a parallel standard dosing cohort of patients treated with warfarin in the same hospitals, in the same time frame, and managed by the same clinicians or anticoagulation service teams. In this study, in nearly all the endpoints tested, significant benefits with the pharmacogenetic-guided dosing, including out-of-range INRs, percentage of time in the therapeutic range (PTTR), and serious adverse events, were seen. Specifically, the pharmacogenetic cohort had a 10.3% absolute reduction in the out-of-range INRs at 1 month, with similar differences at 3 months. This finding was primarily attributable to significantly fewer INR values less than 1.5, which coincided with a significant 66% lower rate of deep vein thrombosis. The reduced number of out-of-range INRs led to a greater TTR in the pharmacogenetic cohort, with 69% and 71% PTTR at 1 and 3 months versus 58% and 59% TTR, respectively, in the parallel control group. Thus, the absolute improvements in PTTR were 10.5% and 12.6% at 1 and 3 months, respectively. Importantly, these benefits in the pharmacogenetic cohort accrued in a setting in which warfarin-treated patients were typically managed by standard protocol by an anticoagulation service/clinic. In addition, differences in serious adverse events were noted. At 90 days, these were 4.5% in the pharmacogenetic cohort and 9.4% in the parallel controls, primarily attributable to significantly lower rates of death and deep vein thrombosis and borderline significant lower rates of moderate/serious hemorrhage. These are relatively impressive differences in clinically important endpoints that suggested the potential value of pharmacogenetic-guided dosing of warfarin, although it had been noted that the parallel-group study design may have led to bias—specifically that the pharmacogenetic cohort was more aggressively managed because the treating doctors and clinicians were aware that these patients were in the pharmacogenetic arm.

Subsequently, several randomized controlled trials set out to overcome these potential flaws with observational or similar clinical trial designs and to robustly test whether time outside the therapeutic range was clinically relevant in a tightly controlled clinical trial setting. In particular, race was noted as a concern because it is known that warfarin pharmacogenetic dosing algorithms are most predictive in individuals of northern European ancestry [30,31].

Despite this, it is still likely that for some individual patients, pharmacogenetic testing has potential benefits, although it is known at a population level that routinely gene typing does not improve clinical outcomes. This includes patients who have more than one "high risk" mutation in the genes known to affect warfarin clearance (eg, *CYP2C9* or *VKORC*) and where knowledge of this would change the starting dose. It is especially relevant when starting treatment (ie, before the first dose) because this is when the risk of bleeding due to over-anticoagulation is high. Another group that may benefit is older people. Here, the induction regimens in current use are only moderately successful in achieving the target INR [32]. It also takes up to 14 days to reach a therapeutic INR for some people. In addition, once treatment is started using induction doses tailored for elderly patients, the contribution of *VKORC1* and *CYP2C9* genotypes in dose refinement appears to be negligible compared with two INR values measured during the first week of treatment [33]. Nevertheless, patients with the VKORC1 A/A haplotype had a significantly reduced time to first INR being in the therapeutic range and to the first INR of more than 4, as did those with the CYP2C9 *2 and *3 genotype; however, this was not a significant predictor of the time to the first INR within the therapeutic range. Both genotypes had a significant influence on the required warfarin dose after the first 2 weeks of therapy [34].

Interestingly, initiation regimen and long-term rules that have specifically been developed and included in a computerized dosage program improve quality of anticoagulation in elderly inpatients [35]. In nonelderly, even knowing the patient's *CYP2C9* and *VKORC1* status predicts less than half of the variation in the response to warfarin. Better predictions are achieved by incorporating pharmacogenetics into a dosing algorithm such as that based on the regression model of Sconce [14]. In this model, the variables age, height, and the *CYP2C9* and *VKORC1* genotypes were the best predictors for estimating the starting dose of warfarin. This algorithm also confirmed that the mean warfarin daily dose requirement is significantly lower with some genotypes.

This model is a marked improvement on current algorithms, but it still explains only approximately half of the variability in dose requirements. However, it has been shown that despite the shortcomings, a pharmacogenetics algorithm is clinically helpful to predict appropriate initial doses of warfarin in high-risk patients. In a study of approximately 900 patients, providing *CYP2C9* and *VKORC1* genotype information to the prescribers resulted in a 28% reduction in hospitalizations due to hemorrhage over 6 months [14].

There are other algorithms for warfarin dosing that take into account genetic and nongenetic factors, some of which are web based (http://www.warfarindosing.org). In addition, the FDA has now made specific dosing recommendation for warfarin based on genotype and nongenetic factors [36] and has made strong suggestions that genotype be considered when warfarin is to be initiated. Also, an analysis of 13 warfarin algorithms concluded that most performed well in the intermediate-dose range (21–49 mg/week) compared to the low (<21 mg/week) and high (>49 mg/week) dose ranges and that admixed population algorithms were in general better performers than race-specific algorithms [37].

12.5 COST-EFFECTIVENESS OF PHARMACOGENOMICS TESTING IN WARFARIN THERAPY

As with all new technologies, it is important to evaluate the incremental cost-effectiveness of pharmacogenetic testing versus standard clinical practice. Pharmacogenetic testing for warfarin is relatively cheap compared with other anticlotting medications, such as new clotting factor inhibitors, or new health technologies, such as nuclear medicine/magnetic resonance imaging fusion techniques. The extra costs of this warfarin service would have to include both the diagnostics of genotyping and clinical interpretative support. Analyzing *CYP2C9* and *VKORC* genes with the costs of clinical interpretation is estimated at AUD$100 per person, with a clinically realistic turnaround time (within 1 day). Because multiple platforms are now available, the costs are rapidly declining, and the use of point-of-care testing may prove superior in terms of cost and speed.

For the cost-effectiveness analysis, the efficacy of pharmacogenetic tests is measured as the reduction in the number of expensive adverse effects, time in hospital, and improvement in quality of life due to less frequent INR monitoring. None of this has been accurately quantified in a prospective study, but it is clear that even a reduction in hospital stay by 1 day would provide a sizeable cost saving. However, although testing seems to be relatively good value for the money, there are additional issues to consider—for example, the cost of screening all potential warfarin users and the cost of pharmacist or clinical pharmacologist time. It would also be an additional cost to current therapy, with INR testing still required albeit possibly less often. In addition, although the prevalence of heterozygotes is relatively high

(~30% for CYP2C9 depending on the ethnic group studied), patients with a null genotype (those likely to incur life-threatening and expensive adverse effects) are rare (~1%). The detection rate for a genotype associated with serious adverse events is therefore low. Lastly, clinical outcomes such as bleeding are rare in patients followed in anticoagulation clinics because warfarin therapy is closely monitored and individualized. Thus, the benefit is likely to be much smaller in tertiary centers and much larger for the small number of patients who live in rural or remote settings away from easily available medical care. Furthermore, the INR is a well-validated and inexpensive surrogate marker for warfarin effects that is already used widely in clinical practice. However, it is not helpful for predicting which dose of warfarin to use for starting anticoagulation, which is where pharmacogenetics testing could make a difference, especially in high-risk groups.

Comparative effectiveness research is a developing field. Its relationship with individualized medicine to combine pharmacogenetic and clinical markers of warfarin use and response is of interest for its ability to accurately estimate and quantify comparative efficacy and toxicity with other medicines [38]. Decision tree analyses have also been developed to evaluate the potential clinical and economic outcomes of using genotype data to guide the management of warfarin therapy. In a recent study, a decision tree was designed to simulate the clinical and economic outcomes of patients newly started on warfarin with either no genotyping or CYP2C9 genotyping prior to initiation of warfarin therapy. The total number of events and the direct medical cost per 100 patient-years in the genotyped and nongenotyped groups were 9.58 and $155,700 and 10.48 and $150,500, respectively. The marginal cost per additional major bleeding averted in the genotyped group was $5,778. The model was sensitive to the variation of the cost and reduction of bleeding rate in the intensified anticoagulation service and concluded that incorporating pharmacogenetic management into warfarin therapy is potentially more effective in preventing bleeding with a marginal cost, depending on the relative local costs and local effectiveness, which would vary depending on patient-specific factors [39]. Additional local clinical and epidemiological studies, including studies on the benefits in different racial groups, are needed to assess the association between genotype and the absolute risk of adverse effects before a cost-effectiveness analysis can be completed for warfarin pharmacogenetic testing [40].

12.6 NEW MEDICATIONS TO REPLACE WARFARIN

In addition to warfarin, other traditional anticoagulant drugs, including unfractionated heparin and low-molecular-weight heparins, have therapeutic issues such as slow onset of action, narrow therapeutic index, requirement for monitoring, and multiple food and drug interactions that make use expensive, slightly unpredictable, and time-consuming.

NOACs have been developed that target a single coagulation factor and have more predictable dose–response relationships than warfarin. These include direct thrombin inhibitors and factor Xa inhibitors. However, despite being more pharmacologically "pure," they need to have all or most of the following characteristics in order to replace warfarin use: equivalent or superior efficacy, an easily available antidote, predictable toxicities with no unexpected toxicities, and a reasonable cost. So far, data are emerging for some of these newer anticoagulants on both efficacy and safety grounds. Dabigatran is a DTI and is a new medication used in anticoagulation therapy. Oldgren et al. compared risk for stroke, bleeding, and death in patients with atrial fibrillation receiving either warfarin or dabigatran and observed that rates of stroke or systematic embolism in patients with dabigatran 150 mg twice daily and/or intracranial bleeding with dabigatran 150 or 11 mg twice daily were lower than those for patients receiving warfarin [41].

Selective inhibitors of specific coagulation factors represent a potentially exciting development in antithrombosis therapy and may in time replace warfarin use. Compared to conventional drugs such as warfarin, they have the potential to be as effective, safer, and easier to use. However, clinical evidence so far has not yet shown superiority to older anticoagulants in all spheres, although data are promising. In particular, studies examining the potential for warfarin replacement in people with atrial fibrillation are very exciting. However, the reality, in the prescribing world outside the clinical trial setting, is that some of these drugs, in the elderly especially but also those at the extremes of body weight, require therapeutic drug monitoring [1] and must have dosage reductions in renal impairment, which includes many inpatients. Thus, the benefit over warfarin for many of these groups is still not clear.

12.7 CONCLUSIONS

The variability in warfarin dosage requirements is multifactorial, although genetic polymorphisms play a role. Current warfarin dosing algorithms fail to take into account genetics and other individual patient and environmental factors. Theoretically, including these factors could help in predicting an individual's loading and maintenance doses for safer anticoagulation. However, linear regression analysis, taking into account genetic polymorphisms of CYP2C9 and VKORC1 (additive effect), body weight, body surface area, and height, has so far been able to capture only approximately half of the large inter- and intrapatient variation in dose requirements. Vitamin K status and alcohol intake, together with additional genetic factors, are likely to account for some of the remaining difference in warfarin requirements, but they still need to be studied in a regression analysis. For now, incorporation of age, body surface area, and *CYP2C9* and *VKORC1* genotype allows the best estimate of warfarin induction and maintenance dose. In addition, *CYP2C9* and *VKORC1* genotype explains more of the dose variability than the other clinical variables, and several large prospective clinical trials are underway [35]. Pharmacogenomics per se, apart from perhaps a few examples such as homozygosity for the *CYP2C9*3* alleles, although attractive, seems to be unable to explain enough of the variation to ensure it is a helpful tool to predict dosing for all-comers currently. It is clear, however, despite the well-conducted randomized trials showing no benefit of clinical endpoints from pharmacogenetic testing, that in certain individual situations, *CYP2C9* and *VKORC1* testing may be useful and warranted in determining the cause of unusual therapeutic responses to warfarin therapy or high-risk patients. Based on the clinical data, and although used in some research-based centers, routine incorporation of pharmacogenetics for warfarin use cannot be recommended.

References

[1] Shigeru O, Osamu M, Takaaki G, Takuhiro U, Shoyama YS, Hiroko K. Generation of an anti-Dabigatran monoclonal antibody and its use in a highly sensitive and specific enzyme-linked immunosorbent assay for serum dabigatran. Ther Drug Monit 2015;37:594–9.

[2] Wieloch M, Sjalander A, Frykman V, Rosenqvist M, Eriksson N, Svensson PJ. Anticoagulation control in Sweden: reports of time in therapeutic range, major bleeding, and thrombo-embolic complications from the national quality registry AuriculA. Eur Heart J 2011;32 (18):2282–9.

[3] Flockhart DA, O'Kane D, Williams MS, Watson MS, Flockhart DA, Gage B, et al. Pharmacogenetic testing of CYP2C9 and VKORC1 alleles for warfarin. Genet Med 2008;10:139–50.

[4] Charles B, Norris R, Xiaonian X, Hague W. Population pharmacokinetics of metformin in late pregnancy. Ther Drug Monit 2006;28:67–72.

[5] Klein TE, Altman RB, Eriksson RN, Gage BF, Kimmel SE, Lee MT. International Warfarin Pharmacogenetics Consortium. Estimation of the warfarin dose with clinical and pharmacogenetic data. N Engl J Med 2009;360:753–64.

[6] Pirmohamed M, Burnside G, Eriksson N, Jorgensen AL, Toh CH, Nicholson T, et al. A randomized trial of genotype-guided dosing of warfarin. N Engl J Med 2013;369:2294–303.

[7] Kaminsky LS, Zhang ZY. Human P450 metabolism of warfarin. Pharmacol Ther 1997;73:67–74.

[8] Aithal GP, Day CP, Kesteven PJ, Daly AK. Association of polymorphisms in the cytochrome P450 CYP2C9 with warfarin dose requirement and risk of bleeding complications. Lancet 1999;353:717–19.

[9] Wilke RA, Berg RL, Vidaillet HJ, Caldwell MD, Burmester JK, Hillman MA. Impact of age, CYP2C9 genotype and concomitant medication on the rate of rise for prothrombin time during the first 30 days of warfarin therapy. Clin Med Res 2005;3:207–13.

[10] Taube J, Halsall D, Baglin T. Influence of cytochrome P-450 CYP2C9 polymorphisms on warfarin sensitivity and risk of over-anticoagulation in patients on long-term treatment. Blood 2000;96:1816–19.

[11] Caldwell MD, Berg RL, Zhang KQ, Glurich I, Schmelzer JR, Yale SH. Evaluation of genetic factors for warfarin dose prediction. Clin Med Res 2007;5:8016.

[12] Rieder MJ, Reiner AP, Gage BF, Nickerson DA, Eby CS, McLeod HL. Effect of VKORC1 haplotypes on transcriptional regulation and warfarin dose. N Engl J Med 2005;352:2285–93.

[13] Bodin L, Verstuyft C, Tregouet DA, Robert A, Dubert L, Funck-Brentano C. Cytochrome P450 2C9 (CYP2C9) and vitamin K epoxide reductase (VKORC1) genotypes as determinants of acenocoumarol sensitivity. Blood 2005;106:135–40.

[14] Sconce EA, Khan TI, Wynne HA, Avery P, Monkhouse L, King BP, et al. The impact of CYP2C9 and VKORC1 genetic polymorphism and patient characteristics upon warfarin dose requirements: proposal for a new dosing regimen. Blood 2005;106:2329–33.

[15] Caldwell MD, Awad T, Johnson JA, Gage BF, Falkowski M, Gardina P, et al. CYP4F2 genetic variant alters required warfarin dose. Blood 2008;111:4106–12.

[16] Takeuchi F, McGinnis R, Bourgeois S, Barnes C, Eriksson N, Soranzo N, et al. A genome-wide association study confirms VKORC1, CYP2C9, and CYP4F2 as principal genetic determinants of warfarin dose. PLoS Genet 2009;5:e1000433.

[17] Sconce E, Avery P, Wynne H, Kamali F. Vitamin K supplementation can improve stability of anticoagulation for patients with unexplained variability in response to warfarin. Blood 2007;109:2419–23.

[18] Campbell P, Roberts G, Eaton V, Coghlan D, Gallus A. Managing warfarin therapy in the community. Aust Prescr 2001;24:86–9.

[19] Food and Drug Administration. Milestones. Available at: <http://www.fda.gov/AboutFDA/WhatWeDo/History/Milestones/ucm128305.htm>; Access date December 19, 2015.

[20] Mungall DR, Ludden TM, Marshall J, Hawkins DW, Talbert RL, Crawford MH. Population pharmacokinetics of racemic warfarin in adult patients. J Pharmacokinet Biopharm 1985;13:213–27.

[21] James AH, Britt RP, Raskino CL, Thompson SG. Factors affecting the maintenance dose of warfarin. J Clin Pathol 1992;45:704–6.

[22] Chan E, McLachlan AJ, Pegg M, Mackay AD, Cole RB, Rowland M. Disposition of warfarin enantiomers and metabolites in patients during multiple dosing with rac-warfarin. Br J Clin Pharmacol 1994;37:563–9.

[23] Routledge PA, Chapman PH, Davies DM, Rawlins MD. Factors affecting warfarin requirements. A prospective population study. Eur J Clin Pharmacol 1979;15:319–22.

[24] Jensen B, Chin P, Roberts RR, Maddison J, Begg E. Total and free clearance of R- and S-warfarin in elderly people. Basic Clin Pharmacol Toxicol 2010;107:354–5.

[25] Garcia D, Regan S, Crowther M, Hughes RA, Hylek EM. Warfarin maintenance dosing patterns in clinical practice: implications for safer anticoagulation in the elderly population. Chest 2005;127:2049–56.

[26] Kamali F, Khan TI, King BP, Frearson R, Kesteven P, Wood P, et al. Contribution of age, body size, and CYP2C9 genotype to anticoagulant response to warfarin. Clin Pharmacol Ther 2004;75:204–12.

[27] Pirmohamed M, Burnside G, Eriksson N, et al. A randomized trial of genotype-guided dosing of warfarin. N Engl J Med 2013;369:2294–303.

[28] Burmester JK, Berg RL, Yale SH, Rottscheit CM, Glurich IE, Schmelzer JR, et al. A randomized controlled trial of genotype-based Coumadin initiation. Genet Med 2011;13:509–18.

[29] Anderson JL, Horne BD, Stevens SM, Woller SC, Samuelson KM, Mansfield JW, et al. A randomized and clinical effectiveness trial comparing two pharmacogenetic algorithms and standard care for individualizing warfarin dosing (CoumaGen-II). Circulation 2012;125:1997–2005.

[30] Klein TE, Altman RB, Eriksson N, Gage BF, Kimmel SE, Lee MT, et al. Estimation of the warfarin dose with clinical and pharmacogenetic data. N Engl J Med 2009;360:753–64.

[31] Limdi NA, Wadelius M, Cavallari L, Eriksson N, Crawford DC, Lee MT, et al. Warfarin pharmacogenetics: a single VKORC1 polymorphism is predictive of dose across 3 racial groups. Blood 2010;115:3827–34.

[32] Oates AJ, Jackson PR, Austin CA, Channer KS. A new regimen for starting warfarin therapy in out-patients. Br J Clin Pharmacol 1998;46:157–61.

[33] Moreau C, Pautas E, Gouin-Thibault I, Golmard JL, Mahe I, Mulot C, et al. Predicting the warfarin maintenance dose in elderly inpatients at treatment initiation: accuracy of dosing algorithms incorporating or not VKORC1/CYP2C9 genotypes. J Thromb Haemost 2011;9:711–18.

[34] Schwarz UI, Ritchie MD, Bradford Y, Li C, Dudek SM, Frye-Anderson A, et al. Genetic determinants of response to warfarin during initial anticoagulation. N Engl J Med 2008;358:999–1008.

[35] Gouin-Thibault I, Levy C, Pautas E, Cambus JP, Drouet L, Mahe I, et al. Improving anticoagulation control in hospitalized elderly patients on warfarin. J Am Geriatr Soc 2010;58:242–7.

[36] Australian Government Treasury. Available at: <http://www.treasury.gov.au/documents/1239/HTML/docshell.asp?URL=06_Appendix_A.htm>; Access date December 19, 2015.

[37] Shin J, Cao D. Comparison of warfarin pharmacogenetics dosing algorithms in a racially diverse large cohort. Pharmacogenomics 2011;12:125–34.

[38] Epstein RS, Teagarden JR. Comparative effectiveness research and personalized medicine: catalyzing or colliding? Pharmacoeconomics 2010;28:905–13.

[39] You JH, Chan FW, Wong RS, Cheng G. The potential clinical and economic outcomes of pharmacogenetics-oriented management of warfarin therapy—a decision analysis. Thromb Haemost 2004;92:590–7.

[40] Garber AM, Phelps CE. Economic foundations of cost-effectiveness analysis. J Health Econ 1997;16:1–31.

[41] Connolly SJ, Ezekowitz MD, Yusuf S, Eikelboom J, Oldgren J, Parekh A, et al. Dabigatran versus warfarin in patients with atrial fibrillation. N Engl J Med 2009;361:1139–51.

Alternative Sampling Strategies for Therapeutic Drug Monitoring

Sara Capiau[1], Jan-Willem Alffenaar[2] and Christophe P. Stove[1]
[1]Laboratory of Toxicology, Department of Bioanalysis, Faculty of Pharmaceutical Sciences, Ghent University, Ghent, Belgium
[2]Department of Clinical Pharmacy and Pharmacology, University Medical Center Groningen, University of Groningen, Groningen, The Netherlands

CONTENTS

13.1 Introduction 279

13.2 The Ideal Alternative Matrix for Therapeutic Drug Monitoring 280

13.3 The Correlation Between Alternative Matrix and Systemic Levels..........................282

13.4 Alternative Specimen: Dried Blood Spots 284

13.4.1 Applications of Dried Blood Spots in Therapeutic Drug Monitoring 290

13.4.1.1 Pediatric Patients 290

13.4.1.2 The Elderly.......... 292

13.4.1.3 Pregnant Women 293

13.4.1.4 Anemic Patients.. 293

13.4.1.5 Psychiatric Patients 293

13.4.1.6 Remote and Resource-Limited Setting.. 294

13.4.1.7 Home Monitoring 294

13.4.1.8 Cystic Fibrosis Patients 296

13.5 Alternatives to Classical Dried Blood Spots 296

13.1 INTRODUCTION

Alternative sampling strategies have proven to be of added value for ample medical applications, such as disease diagnosis and monitoring [1–6], vaccination status evaluation [4], and even therapy optimization and individualization [7–10]. Indeed, minimally or noninvasive sampling methods may facilitate home-based therapeutic drug monitoring (TDM) [11], simplify TDM in specific populations such as a pediatric population [12], and promote TDM programs in remote settings and developing countries [13–15]. Furthermore, alternative sampling strategies facilitate pharmacokinetic (PK) studies in special patient populations [12], yielding important information that is currently still lacking for many drugs. A better understanding of PK can in turn help to reduce the need for TDM because dosage regimens can be optimized for these special patient populations. Moreover, various easily obtainable matrices have been employed for cytochrome P450 genotyping [16] and phenotyping [17], further contributing to therapy optimization and individualization. Therefore, alternative matrices have the potential to contribute to better patient management and follow-up and thus to help improve patient health care in general.

These so-called alternative sampling strategies refer to the patient-friendly sampling of nonconventional matrices as well as to the unconventional way of sampling traditional matrices such as whole blood, serum, and plasma. Matrices that have been evaluated and/or used for TDM purposes include dried blood spots (DBS), volumetric absorptive microsamples (VAMS), dried plasma spots (DPS), other dried matrix spots (DMS), oral fluid (OF), interstitial fluid (ISF), hair, tears, exhaled breath, sweat, and nasal mucus. In this chapter, the benefits of the previously mentioned alternative

W. Clarke & A. Dasgupta (Eds): Clinical Challenges in Therapeutic Drug Monitoring. DOI: http://dx.doi.org/10.1016/B978-0-12-802025-8.00013-1

13.5.1 Volumetric Dried Blood Spots 296
13.5.2 Dried Plasma Spots 297
13.5.3 Other Dried Matrix Spots 297
13.6 Paper Spray Ionization for the Analysis of Dried Blood Spots 298
13.7 Alternative Specimen: Oral Fluid .299
13.7.1 Applications of Oral Fluid for Therapeutic Drug Monitoring 302
13.7.1.1 Pediatric Patients 303
13.7.1.2 The Elderly 303
13.7.1.3 Pregnant Women 304
13.7.1.4 Psychiatric Patients 304
13.7.1.5 Patients With Renal Failure 305
13.7.1.6 Remote and Resource-Limited Settings 305
13.7.1.7 Home Monitoring 306
13.7.1.8 Cystic Fibrosis Patients 307
13.8 Alternative Specimen: Tears 307
13.8.1 Application of Tears for Therapeutic Drug Monitoring 308
13.9 Alternative Specimen: Interstitial Fluid 310
13.9.1 Application of Interstitial Fluid for Therapeutic Drug Monitoring 311
13.10 Alternative Specimen: Hair 312
13.10.1 Application of Hair Specimens for Therapeutic Drug Monitoring 314
13.11 Alternative Specimen: Exhaled Breath 316

sampling strategies are discussed, with particular attention to special patient populations such as neonates, children, pregnant women, and the elderly. Furthermore, some important limitations, disadvantages, and challenges of these techniques are highlighted, as well as the future perspectives they offer.

Because it is not our goal to provide an exhaustive overview of all TDM-related applications that have been developed using alternative matrices, selected examples are discussed to highlight the (potential) added value of alternative matrix sampling. For more detailed information regarding existing applications and existing analytical methods, the reader is referred to some published reviews [7–10,13,18–20]. Furthermore, it is important to note that some of the methods that are discussed in this chapter were originally developed for PK purposes, for example, but are also applicable to TDM. Moreover, this chapter provides a brief overview of some proposed applications without addressing in-depth whether TDM is advisable for selected compounds. In addition, although the use of alternative matrices has been valuable for drug monitoring in drug replacement programs, this is not within the scope of this chapter [21].

13.2 THE IDEAL ALTERNATIVE MATRIX FOR THERAPEUTIC DRUG MONITORING

In TDM, systemic drug concentrations are measured to evaluate whether a suitable drug exposure has been achieved. In this way, the risk for toxicity, inadequate efficacy, or therapy resistance is minimized, whereas compliance problems as well as PK abnormalities or drug interactions can be detected. Unfortunately, traditionally used sampling strategies (ie, venipuncture) are invasive and necessitate relatively large blood volumes that cannot always be justified in neonates or anemic patients, for example [12,22]. Moreover, outpatients are obliged to frequent a clinic or a doctor's office for blood draws, and samples often require a cold chain during transportation to the laboratory. Furthermore, for highly protein-bound drugs, or in special situations such as pregnancy, hypoalbuminemia, renal failure, or liver insufficiency, it is advised to measure the free plasma concentration of a drug [23]. Indeed, in these circumstances, the latter is better suited to predict drug toxicity and efficacy. However, these measurements are laborious and technically challenging; hence, they are not easily incorporated in routine clinical laboratories.

Alternative matrices have gained much attention in TDM because they are obtained in a minimally or noninvasive manner and may provide a more convenient way of assessing total or even free [23] systemic drug concentrations, depending on the type of alternative matrix. Ideally, a matrix for TDM fulfills the criteria summarized in Table 13.1. Briefly, the sample collection should be noninvasive and should require only a small sample volume to be

Table 13.1 Overview of the Criteria of the Ideal Alternative Matrix for Therapeutic Drug Monitoring

Suitability Criteria for an Alternative Matrix
Sample collection
Noninvasive Requires little sample volume Straightforward Allows for self-sampling/home monitoring Applicable to everyone Robust Economic
Storage and shipment
Stable at ambient conditions No biohazard
Analysis
Limited preanalytical phase Automatable Economic Allows coanalysis of other relevant analytes
Results
Reproducible Correlate with systemic (free) drug concentrations Cave: Small intra- and interindividual variability in this correlation is required. Should not be time-dependent. Should not be concentration-dependent. Should not be influenced by additional variables such as pH and flow rate. Affect clinical decision making in the same way as traditional matrices

13.11.1 Application of Exhaled Breath for Therapeutic Drug Monitoring.......................317

13.12 Alternative Specimen: Sweat 318
13.12.1 Application of Sweat for Therapeutic Drug Monitoring.......................319

13.13 Alternative Specimen: Nasal Mucus 319

13.14 Toward Routine Implementation..........319

13.15 Conclusions 322

Acknowledgment 322

References 323

applicable to vulnerable patient populations such as neonates. The collection procedure should also be universal (ie, applicable to everyone) and straightforward enough to allow unsupervised patient self-sampling in a home setting. Preferably, the sample collection is robust so that variables arising during sample collection have no impact on the analysis result. Importantly, sample collection should be inexpensive to be considered for routine use. The target analyte(s) should be stable in the alternative matrix for prolonged periods of time, both at room temperature and under transport conditions. Moreover, the sample matrix should represent no biohazard to facilitate transport of samples via regular mail services. The analysis of the sample should be as straightforward as possible (ie, no or limited sample preparation), should not require expensive consumables, and should be automatable. In addition, whenever other analytes are generally requested by clinicians concurrently with the therapeutic drug of interest, these should also be measurable in the alternative matrix because otherwise there would still be a need to draw a venous blood sample, limiting the usefulness of the alternative sampling strategy. Obtained results in the context of TDM should be quantitative, reproducible, and should correlate with systemic levels—either free or total, depending on the type of alternative matrix. This correlation should show small intra- and interindividual variability and should be independent of the analyte concentration, time after drug administration, as well as any other variable (either introduced by the sampling method or inherent to the patient). Most important, results obtained via alternative matrix analysis should not influence clinical decision making any differently than traditionally obtained results would do so.

Unfortunately, the ideal alternative matrix does not exist because in reality every matrix has its own limitations, challenges, and inherent disadvantages. The goal is to find those applications in which the benefits outweigh the drawbacks and to develop approaches that eliminate or minimize the impact of these matrix-specific issues on the analytical result and on the corresponding clinical decision.

13.3 THE CORRELATION BETWEEN ALTERNATIVE MATRIX AND SYSTEMIC LEVELS

As mentioned previously, a prerequisite for the usefulness of alternative matrices for TDM is of course the existence of a correlation between the drug level of the alternative matrix and the (free) systemic plasma or whole blood concentration. As can be seen from Fig. 13.1, the free plasma concentration can diffuse into other tissues or biological fluids, indicating that the drug

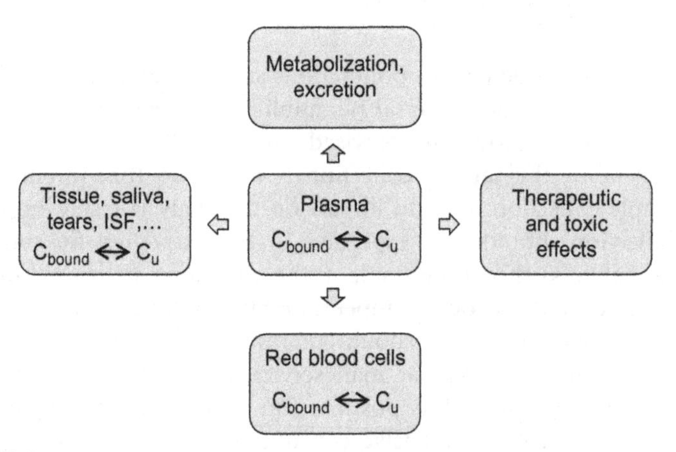

FIGURE 13.1

The equilibrium between the free plasma fraction and alternative matrices (with C_u, the unbound concentration, and C_{bound}, the bound concentration of a compound).

levels of the alternative matrix are—at least to some degree—determined by the corresponding free plasma concentration. However, the drug levels of the alternative matrix are generally also affected by multiple other variables. First, the protein content of these matrices codetermines their total drug concentration. Only when drug-binding protein content is low, such as in tears, can measured drug levels be similar to plasma free drug concentrations. Because drugs generally require passive diffusion to enter an alternative matrix, drug levels are also defined by the molecular weight and lipophilicity of the compound. Moreover, the pH of the matrix and the pKa of the compound are of key importance for the equilibrium between both compartments. In acidic matrices such as sweat and OF, basic compounds will become ionized and, hence, be subject to ion trapping. Another factor that is known to affect the measured concentration in, for example, OF, tear fluid, and sweat is the flow rate. Taking the previously mentioned factors into account, drug concentrations in alternative matrices are rarely equal to (free) systemic concentrations, but they can correlate with systemic concentrations. However, the latter has to be evaluated on a case-by-case basis. To be truly useful in the context of TDM, it is crucial that this correlation is constant both intra- and interindividually. When only the intraindividual variability is small enough, the alternative matrix can only be used for patient follow-up after having established reference levels for that patient. As discussed previously, the ratio between both matrices might be altered by differences in flow rate and pH, for example, and hence also by the sampling procedure. Therefore, sampling procedures often need to be standardized as much as possible. In addition, the observed correlation between alternative matrix and systemic levels may be time-dependent or concentration-dependent, complicating result interpretation and limiting the practicability of the sampling strategy.

Whether a sufficiently good correlation is present between the alternative matrix and the systemic concentrations needs to be evaluated for every compound. Importantly, this correlation has to be strong enough to allow prediction of corresponding systemic levels for individual patients because the latter will be employed to make therapeutic decisions (with the allowed deviation between true and back-calculated systemic concentrations depending on the specific target analyte). Therefore, a proper statistical evaluation of patient data obtained during a bridging study is essential. Although a high Pearson correlation coefficient can be a first indication of a good correlation between both matrices, as a stand-alone parameter, this is still insufficient proof. In addition, an artificially high Pearson correlation coefficient can be obtained by including a broad range of analyte concentrations, even when the alternative matrix is actually not a good surrogate for the assessment of systemic levels. In addition, often only a small amount of patient samples is included in bridging studies, limiting the validity of the obtained results.

The influence of different (matrix-specific) parameters on alternative matrix levels, and hence alternative matrix/systemic compartment correlation, needs to be evaluated. Whenever there is such an influence, this should be dealt with by standardizing the sample collection so that these variables are within acceptable limits (eg, the use of printed concentric circles on DBS cards to control the volume effect [24]) or in by alternative means (eg, hematocrit (Hct) correction to compensate for Hct bias in DBS analysis [25]).

13.4 ALTERNATIVE SPECIMEN: DRIED BLOOD SPOTS

An alternative sampling technique that has gained much attention for TDM purposes is DBS sampling [7,8]. Indeed, the collection of a drop of capillary blood from a finger or heel stick onto a filter paper offers many advantages. First, it is a very simple, inexpensive, and minimally invasive way of acquiring a representative sample, typically requiring only approximately 5–35 μL of blood. Furthermore, after proper drying, the analytes in DBS are generally stable at ambient conditions for prolonged periods of time, and the dried blood represents only limited biohazard, facilitating transport under ambient conditions.

A major opportunity provided by DBS sampling is the possibility of home-based patient self-sampling with subsequent transport of the dried samples to the laboratory via regular mail services [26]. For outpatients, this approach can reduce the frequency of clinic visits, which is undoubtedly more convenient, especially for active patients. Furthermore, this reduction may be beneficial to patients' health (eg, in cystic fibrosis (CF) patients) and may contribute to an improved quality of life for chronically ill patients [27]. Moreover, self-sampling allows DBS to be collected at any time of day. This is convenient, for example, when trough levels need to be measured before the next dose because this is often early in the morning or late in the evening (ie, when clinics are closed). Another advantage is ease of collecting multiple blood samples over a time period if needed. Home-based sample collection might also be preferable for children and patients with mental, psychological, or developmental disorders because obtaining samples in a familiar environment may be less stressful [28,29]. In addition, the convenience of home-based TDM offers the opportunity of more frequent patient follow-up, which in turn may lead to improved therapy adherence and hence reduced health care costs [30]. Indeed, it has been proven multiple times that the lack of therapy adherence results in many avoidable costs to the patient as well as to society due to (1) the unnecessary changing to stronger and more expensive therapies; (2) increased patient morbidity, invalidity, and mortality; and (3) the emergence of therapy-resistant pathogens (in the case of infection treatment).

Another setting in which the use of DBS may contribute significantly to patients' health is the setup of TDM programs in developing countries and other resource-limited areas [13–15], where limited accessibility is currently a major hurdle for the implementation of TDM. DBS samples can be collected in the field and sent to a central laboratory for analysis under ambient transport conditions. Consequently, the availability of a centrifuge, cooled storage, or even electricity may no longer be a necessity. In addition, DBS present a reduced biohazard compared to traditional samples, which results in safer sample handling, for example, in areas where HIV is highly prevalent.

Aside from the clear advantages for outpatient TDM, DBS sampling also possesses several benefits for hospital-based TDM, albeit only in specific circumstances. First, DBS sampling can be an easy solution for compounds with stability problems. However, stability during drying and storage needs to be assessed on a case-by-case basis because the use of DBS does not alleviate stability issues for all compounds. Furthermore, in particular patient populations, such as neonates and severely anemic patients, DBS sampling may be preferred over classic venous sampling because it is less invasive and requires a much smaller sample volume. Other inpatients for which DBS may be a worthy alternative are people with poor venous access (eg, the elderly or people with phlebitis) and patients with fear of needles [3]. The advantage of DBS sampling is in the latter cases obviously limited to those patients for whom no additional venous blood collection needs to be performed. Therefore, several analytical methods have been developed that can determine therapeutic drugs and/or other relevant parameters, such as creatinine, on the same DBS (extract) [3,31–33].

In addition, DBS preparation can also be employed as an easy form of sample collection and/or sample pretreatment, when preparing either capillary or venous DBS. An example is the on-card stabilization via derivatization of thiorphan, the active metabolite of the antidiarrheal drug racecadotril [34]. By directly collecting a drop of blood onto filter paper, pretreated with the derivatizing agent 2-bromo-3′-methoxyacetophenone, sample collection can be greatly simplified compared to the traditional sample collection, which requires complex handling procedures at the clinical site. In addition, when samples need to be sent to another laboratory for analysis, DBS can be prepared from classic venous samples because this facilitates more practical and economic sample transport.

DBS are mainly regarded as a suitable alternative for routine TDM checks of stable patients or for resolving less urgent issues. When TDM results need to be reported urgently, the DBS format is of course deemed less suitable because a certain drying period is required (typically at least 2 h). However, the drying process can also be accelerated using, for example, microwave-assisted drying, which takes only minutes [35].

Although DBS sampling can be valuable in the context of TDM, it also has some challenges. The most widely discussed issue in this regard is the so-called Hct effect, which consists of multiple aspects [36]. First, as the Hct affects the viscosity of blood, it also influences the spreading of the blood on/through the filter paper—with samples with a lower Hct spreading out further compared to samples with a higher Hct. Consequently, whenever a fixed-size DBS punch is made, this will contain more blood when the sample has a higher Hct compared to samples with a lower Hct. This phenomenon may lead to overestimation of drug levels in samples with a high Hct, whereas samples with a low Hct may be subject to underestimation of the drug levels. Second, the Hct may affect analyte recovery and potentially also matrix effects [36]. To cope with the Hct effect on sample spreading, multiple solutions have been postulated, which have been elaborately discussed by De Kesel et al. [36,37]. Briefly, the Hct effect on spreading can be avoided by preparing volumetric DBS and analyzing the entire sample. However, the deposition of a fixed volume of blood requires the use of capillaries or pipets, which might be feasible in a laboratory setting but is too complicated to allow for patient self-sampling. To resolve this issue, multiple formats have been developed to allow for the generation of volumetric dried blood samples starting from a nonvolumetric drop of blood [38–40]. Another option (which is also applicable to nonvolumetric DBS) is to estimate the Hct of a DBS, based on its potassium content, and to apply a potassium content-dependent (ie, a Hct-dependent) correction factor to the obtained DBS-based result to compensate for the expected Hct effect [24,25]. Recently, we developed a noncontact alternative to the potassium method that allows one to estimate the Hct of a DBS by merely scanning its surface (Capiau et al., unpublished data). A third solution that has been suggested is the use of specially designed substrates on which sample spreading is not (or at least to a lesser extent) influenced by the Hct [41]. Finally, it has been proposed to use in situ-generated DPS instead of DBS [42–44]. Although some of these approaches are promising, more experience is required to build up sufficient confidence regarding the obtained results.

To avoid the Hct effect on extraction recovery, it is advisable to optimize the sample preparation procedure over a wide range of Hct values. To attain sufficient and reproducible extraction efficiencies, some have used heated extraction or even specially pretreated filter paper substrates [45–47]. In addition, remaining variability in extraction efficiency can be accounted for by using an internal standard (IS) spray [48]. The latter can be applied to a DBS using an automated system. Unfortunately, only a few laboratories have such resources at their disposal. Therefore, typically, the IS is included in the extraction solvent, which entails that differences in extraction efficiency cannot be accounted for. Alternatively, the filter paper can also be impregnated with IS prior to DBS preparation [48,49]. However, the latter approach is not practical to perform and requires homogeneous distribution of the IS throughout the filter paper. In addition, this distribution should not be affected by the deposition of the blood sample.

Aside from the previously mentioned analytical aspects of the Hct effect, there is also a physiological aspect to this issue [36]. Specifically, the Hct may influence the distribution of a compound between red blood cells (RBCs) and plasma and hence the correlation between DBS and plasma levels [50]. This is of great importance in DBS-based data interpretation because generally obtained results need to be converted to their corresponding plasma values. Indeed, traditionally used therapeutic intervals are almost always established using plasma or serum. Theoretically, new DBS-based therapeutic intervals could be established, but this is expensive and time-consuming. Furthermore, physicians are familiar with existing therapeutic intervals. To convert DBS-based results into the corresponding plasma values, a conversion factor is often empirically determined. However, because this fixed correlation factor does not take the patient's Hct into account, this approach might be an oversimplification [51]. Alternatively, the average Hct of the target population (as well as the plasma protein binding of the compound) can be used to calculate the blood-to-plasma ratio [52]. This simplification is obviously only acceptable when the Hct range of that population is quite narrow. However, when a DBS method has to be applied on both healthy subjects and on special patient populations, such as anemic patients, neonates, children, pregnant women, and elderly, the Hct range will be much wider and the Hct of each DBS sample has to be determined and needs to be taken into account in the DBS–plasma conversion. In addition, the blood-to-plasma ratio may also be concentration-dependent, as was observed for the antiepileptic drug topiramate [53]. Obviously, establishing blood-to-plasma ratios should always be done using clinical samples because in vitro experiments may sometimes yield deviating results [53].

Another essential issue for the analysis of capillary DBS (ie, DBS collected directly from a heel or finger stick) is the possible difference between capillary and venous drug levels. The correlation between both levels must be assessed on a case-by-case basis. Therefore, to determine the usefulness of a developed DBS method, the latter should always be applied to true capillary patient samples. Several factors may influence the correlation between capillary and venous drug levels. On the one hand, it needs to be taken into account that capillary blood contains both venous and arterial blood and that arteriovenous concentration differences have been observed for certain compounds [54,55]. These differences depend on drug characteristics (molecular size, lipid solubility, and degree of protein binding) as well as on time after drug administration (after distribution, an equilibrium should be attained) [54]. A time-dependent distribution between both matrices has been observed, for example, for the antiviral drug oseltamivir [56]. Moreover, differences between capillary and venous samples may be due to the contamination of the capillary blood sample with ISF. To minimize the admixture of ISF with capillary blood, the first drop of blood is discarded during DBS sample collection and "milking" the finger too extensively is dissuaded. In addition, discrepancies between capillary and venous Hct have been observed previously and may contribute to differences in drug levels, especially when measured via DBS [57]. Lastly, Hct measurements performed via repeated capillary sampling are known to be more variable compared to Hct measurements obtained via repeated venous sampling [57]. Again, this emphasizes the importance of knowing a patient's Hct.

Other DBS-specific variables that need to be evaluated during method development and validation include potential sources of contamination, the volume effect, and sample homogeneity [36,58]. Contamination during sample preparation (eg, during the punching step) should be eliminated during method optimization. However, contamination issues occurring during sample collection can never be excluded, especially when sampling is performed by the patient. The likelihood of the latter, however, can be minimized by properly educating patients through demonstration videos and by instructing the patients or their caregivers to thoroughly wash and disinfect hands prior to the finger prick. To verify whether this type of contamination occurred, a second DBS or a blank DBS punch near the sampled DBS can be analyzed, whenever necessary [59,60]. The extent of the volume effect, which occurs due to differences in filter paper saturation, needs to be determined during analytical validation. Importantly, an acceptable volume range should be established at low, medium, and high Hct. To evaluate whether the volume effect is within acceptable limits in patient samples, filter paper with concentric circles can be employed for DBS sampling [24]. In this format, the diameter of the inner circle corresponds to the smallest volume that is still acceptable for the lowest Hct evaluated, whereas the diameter of the outer circle corresponds to the largest volume that is still acceptable for the highest Hct. As long as samples are within these concentric circles, the impact on target analyte quantitation should be acceptable. Also, the evaluation of DBS homogeneity experiments should be performed at different volumes and different Hct values. When a noticeable difference can be observed between quantitation on central and peripheral punches, only central punches can be employed for further analysis. Another option that can be evaluated in such a case is to use a larger punch size. The latter obviously does not influence DBS homogeneity per se, but it might make the analysis less susceptible to location-dependent differences.

Because both the volume effect and DBS homogeneity depend on the type of filter paper used, results obtained with a different type of filter paper than the one used during validation may be unreliable [61]. Furthermore, a different type of filter paper may also affect blood spreading, sample stability, and extraction efficiency [62]. In addition, extraction of each type of filter paper may yield its own specific interferences [63], thereby potentially affecting matrix effect and method sensitivity. Therefore, the choice of filter paper should be made carefully during method development and should definitely remain constant throughout validation and method application. Intra- and inter-lot differences of several types of filter paper, on the other hand, have proven to be acceptable even for quantitative applications [64].

Another challenge in DBS analysis is to obtain sufficient sensitivity when starting from such a small sample volume (a 3-mm punch corresponds to approximately 3 μL). Especially for compounds with low circulating concentrations, this might be challenging at trough values. Generally, sufficient sensitivity can be obtained using gas chromatography (GC) or liquid chromatography (LC) coupled to mass spectrometry, although LC coupled to ultraviolet (UV) or fluorescence detection, immunoassays, and many other techniques have also been employed, depending on the intended application [65]. Importantly, the lower limit of quantitation should be attainable at realistic volumes (ie, starting from a maximum of 25–30 μL). When this is not feasible, punch stacking might be a valuable option [66]. Also, on the other end of the dynamic range, DBS analysis may be challenging. Whenever a sample is above the upper limit of quantitation, it needs to be diluted into the dynamic range. However, for DBS analysis, this requires dilution with blank DBS extract, which is very cumbersome. To facilitate this step, IS-tracked dilution has been proposed [66,67]. In addition, alternatives that do not require the physical dilution of the DBS sample have been suggested and encompass mass spectrometer signal dilution (in which the signal that reaches the detector is artificially decreased by using suboptimal mass spectrometer parameters) or the use of the signal of naturally occurring, less abundant isotopes [68].

In addition, the evaluation of DBS sample stability might be challenging. In some cases, this can be best evaluated on true clinical samples (ie, via incurred sample stability). The stability of certain antiretroviral drugs, for example, which phosphorylate and accumulate in RBCs in vivo and are released from RBCs and dephosphorylated to generate additional parent compounds in vitro, cannot be mimicked in spiked samples [69]. Importantly, not just the stability under laboratory conditions should be evaluated. Whenever samples will have to endure more extreme temperatures (eg, in a mailbox during summer), these conditions also need to be assessed during method validation. Concerning the application of DBS methods in developing countries, not only the effect of elevated temperatures but also the effect of humidity and sunlight exposure should be evaluated. Although DBS generally show enhanced analyte stability, this is not always sufficient, and other approaches such as filter paper pretreatment or heat stabilization to stop enzymatic degradation need to be employed [70,71]. Even then, DBS may not alleviate all stability issues, and they remain unsuitable for the analysis of volatile or air-sensitive compounds.

13.4.1 Applications of Dried Blood Spots in Therapeutic Drug Monitoring

The first DBS-based method for TDM was published in 1978 by Albani et al., and evaluated the use of DBS for theophylline monitoring [72]. Since then, numerous other DBS-based TDM applications have been published, especially during the past decade [7,8]. Although DBS will never completely replace traditional matrices in TDM, they may provide valuable data that could not be (easily) obtained using traditional approaches, particularly in special patient populations or in certain challenging settings. To illustrate this, selected applications are discussed next. In addition, many different classes of therapeutic drugs have already been quantitated in DBS. These are summarized in Table 13.2, along with the most relevant advantages that DBS sampling offers for each of these classes.

13.4.1.1 Pediatric Patients

DBS sampling is minimally invasive and requires only a limited sample volume. Therefore, it is an interesting sampling technique for neonates and children [12]. Obviously, much experience exists with DBS sampling in newborn screening programs [5]. However, DBS can also be employed for subsequent therapy monitoring. This has been done, for example, for dose optimization and monitoring of nitisinone, the treatment for tyrosinemia type I [89]. In this case, other parameters, such as succinylacetone and tyrosine levels, can be measured simultaneously to evaluate therapy effectiveness. Another use of DBS in children is the monitoring of busulfan, used in the myeloablative conditioning regimen prior to hematopoietic stem cell transplantation [84]. Here, DBS could be prepared either from venous blood collected from a central venous line (after properly flushing it with saline to avoid contamination from infused busulfan) or from finger pricks, as an alternative to using peripheral cannulation to rule out potential sample contamination. Moreover, home-based sample collection by a child's parents has also been performed. The feasibility of such an approach was first demonstrated by Rattenbury et al. in 1988 for the analysis of theophylline [90]. Since then, multiple other home-based applications have been

Table 13.2 Overview of the Classes of Therapeutic Drugs Determined in DBS with the Associated Specific Advantages of DBS

Drug Class	Specific Advantages of DBS
Antibiotics [15,27,73]	Cystic fibrosis patients [27] Home monitoring of, for example, tobramycin Reduced number of hospital visits Increased quality of life Reduced risk of infection transmission to other CF patients Tuberculosis: transport under ambient conditions to centralized reference laboratory (also in resource-limited settings) [15]
Antidepressants [28]	Psychiatric facility: transport under ambient conditions to laboratory Frequent home-based adherence monitoring
Antidiabetics [33]	Frequent home-based adherence monitoring HbA1c and creatinine can be evaluated on DBS as well [3,33]
Antiepileptics [11,74,75]	Sampling immediately after seizure (in patients who are difficult to control) [11] Facilitates more frequent monitoring, for example, during pregnancy [75] Neonates and children: small sample volume [74]
Antihypertensive drugs [76,77]	Frequent home-based adherence monitoring
Antimalarials [14]	Resource-limited settings (transport/stability) Reduced biohazard Parasitic load monitoring [14] Chloroquine: blood-based therapeutic interval [14]
Antimycotics [73, 234]	Home-based TDM Children: small sample volume
Antipsychotics [28,78]	Frequent home-based adherence monitoring Psychiatric facility: Convenient sampling: patients might be less frightened [28] Transport under ambient conditions to laboratory
Antiretrovirals [13,69,71]	Reduced biohazard Resource-limited settings (transport/stability) Recent and long-term adherence can be evaluated for tenofovir, for example [69] Viral load testing possible [79] Genotypic drug resistance testing possible [79] Anonymity [26]
Antivirals [80]	Reduced biohazard Ribavirin (hepatitis C treatment): anonymity [26]
Bronchodilators [81]	Sampling after asthma attack (in patients who are difficult to control)
Immunosuppressants [31,32,45,82,83]	Abbreviated AUC (eg, tacrolimus): ease of timing, serial sampling [83] Cyclosporin A monitoring at C2: ease of timing [82] Blood-based therapeutic intervals for most immunosuppressants

Continued...

Table 13.2 Overview of the Classes of Therapeutic Drugs Determined in DBS with the Associated Specific Advantages of DBS *Continued*

Drug Class	Specific Advantages of DBS
Oncolytics [52,62,84,85]	Long-term home-based monitoring (eg, tamoxifen) [85] Anemic patients (common in cancer patients): low sample volume [22] Busulfan (myeloablation in children): low sample volume [84]
Pulmonary hypertension [86]	Treatment of pediatric patients: low sample volume [86]
Cholesterol control [77]	Frequent home-based adherence monitoring
Various other drugs (selected examples)	Caffeine (apnea of prematurity): limited sample volume [87] Deferasirox (anemic patients): limited sample volume [88] Nitisinone (tyrosinemia I treatment) [89] Neonates and children: limited sample volume, minimally invasive Simultaneous evaluation of tyrosine and succinylacetone Sodium oxybate: timing of sampling (late at night) [59] Thiorphan: more convenient sample collection [34]

CF, cystic fibrosis; DBS, dried blood spots; TDM, therapeutic drug monitoring.

evaluated, such as the adherence assessment of children to antiepileptic treatment [74] and the monitoring of immunosuppressant therapy in child solid organ transplant recipients [32].

In addition, the use of DBS sampling in neonates, infants, and children may provide an opportunity to obtain better age-specific dosing regimens, thereby optimizing therapy efficacy and safety and limiting the need for TDM [12]. Indeed, current dosing regimens are often derived from data obtained in an adult population because obtaining age-specific PK data via traditional approaches would require the collection of unethical amounts of blood. Moreover, the ease of sample handling, transport, and storage facilitates multicenter pediatric PK studies, which are often required given the limited sampling pools available. In addition, pediatric nurses are receptive to the use of a method with which they are familiar and that can be fairly easily performed in the smallest infants with minimal training. Overall, DBS sampling fits well in an infant-, parent-, and staff-friendly PK study design [87].

13.4.1.2 The Elderly

In the elderly population, venipuncture may be challenging due to age-related changes in the skin (eg, dermal thinning or reduced elasticity) and the blood vessels (eg, atherosclerotic narrowing or the emergence of fragile subsurface blood vessels). Moreover, osteoarthritis may hinder straightening of the arm for venipuncture. Therefore, DBS collection may be more convenient in this population. Furthermore, reducing the number of hospital visits for elderly patients—by having DBS collected by the patients or their caregivers—may contribute to the patients' quality of life and may also be logistically easier to organize for the families of the patients. In addition, many other parameters that are relevant to monitor in an elderly population can also be determined on DBS. For example, DBS can be employed for the evaluation of glycemic control, renal function, cardiovascular risk, vitamin B_{12} status, thyroid function, iron status, and vitamin D status, as well as for cancer screening [3].

13.4.1.3 *Pregnant Women*

Strong changes in the level of antiepileptic drugs such as lamotrigine are known to occur during pregnancy and the postpartum period, which can lead to more frequent seizures during pregnancy and potentially toxic levels postpartum. Because it is currently not possible to predict when these changes will occur exactly, and because proper seizure control is essential for the mother's health as well as for the well-being of the fetus, these patients need to be closely monitored to maintain their pre-pregnancy drug levels. Therefore, it has been suggested to collect DBS weekly at home to monitor lamotrigine levels in a more convenient way. Use of this approach in pregnant women demonstrated its applicability in a real-life setting [11,75].

13.4.1.4 *Anemic Patients*

A potentially valuable DBS-based application in anemic patients is the measurement of deferasirox, an iron chelating agent used in the treatment of transfusional iron overload in chronic anemia patients. Deferasirox levels need to be closely monitored in these patients to ensure therapeutic efficacy, and DBS sampling may be an ideal format to do so because it requires only very small volumes of blood, which will not aggravate the patients' anemia. Although the method developed by Nirogi et al. to measure deferasirox in DBS could be valuable in this context, it must be noted that the latter was developed to facilitate a PK study in rats and has not been evaluated in humans [88].

13.4.1.5 *Psychiatric Patients*

Because conventional sampling is often perceived as frightening by psychiatric patients, DBS sampling may provide a valuable alternative in this patient population [28]. Moreover, the use of DBS sampling may help improve patient cooperation during sample collection. In addition, because no syringes are required, this sampling procedure may contribute to safety in a psychiatric facility. In addition, for psychiatric facilities that do not have their own laboratory, samples need to be sent to a centralized laboratory for analysis; in this case, the stability and convenient transport provided by the DBS may be of added value. The use of DBS sampling also facilitates (frequent) home-based adherence monitoring, which can be very relevant in this population because adherence issues are highly prevalent. Importantly, because samples can be obtained in a familiar environment, this might be less stressful for patients. The feasibility of DBS sampling in a psychiatric population was demonstrated by Patteet et al. for the analysis of a diverse panel of antipsychotics [28].

13.4.1.6 Remote and Resource-Limited Setting

A major hurdle in the setup of TDM programs in remote settings is accessibility. Many patients live far away from a hospital and therefore samples should be obtainable "in the field". Other logistical issues regarding sample collection in resource-limited settings include the unavailability of centrifuges or even electricity and the feasibility and cost of refrigerated transport. Here as well, DBS can provide a solution because sampling can be performed conveniently at every location, no centrifuges are required, and often transport to a centralized laboratory can be performed under ambient conditions. This is the case, for example, for TDM of the antiretroviral telaprevir. Whereas patient samples obtained via venipuncture typically require immediate centrifugation, acidification of the plasma fraction, and freezing to stabilize the equilibrium between stereoisomers of telaprevir, DBS can be easily collected on a filter paper card preimpregnated with 20% citric acid [71].

In addition, the reduced biohazard of DBS is a strong added value in developing countries because HIV and malaria are often highly prevalent. Therefore, numerous DBS-based methods have been developed for the monitoring of antiretrovirals and antimalarials [13,14]. Also, viral load monitoring, genotypic resistance testing, and parasitic load determination can also be performed on DBS [14,79]. Another example for which DBS can be valuable is the setup of a TDM program for tuberculosis (TB) treatment. In this context, DBS have been suggested to facilitate the setup of a network of selected, quality-controlled reference laboratories that receive samples of difficult-to-treat patients from throughout the world. Such an approach could help improve a patient's prognosis, prevent further development of multidrug-resistant TB (caused by underexposure due to PK variability or nonadherence), and reduce overall TB treatment costs [15,91,92]. In addition, DBS may be an ideal format for specific PK studies that can be performed only in developing countries. Bartelink et al., for example, used DBS to facilitate a PK study in food insecure HIV-infected pregnant and breast-feeding women for multiple antiretroviral drugs [93].

13.4.1.7 Home Monitoring

Because DBS sampling is fairly straightforward, it can be performed at home by the patient or by his or her caregivers. Subsequently, samples can be sent to the laboratory for analysis, and the results can be sent to the patient's attending physician. In this way, results can already be available when the patient visits the doctor. Alternatively, when no visit is scheduled, the results can be reported to the patient via telephone, mail, or an online system. The latter can be especially valuable for the frequent monitoring of outpatients on a long-term treatment. In this context, DBS sampling has been suggested, for example, for the follow-up of patients on an adjuvant breast cancer treatment with tamoxifen because it has been shown that DBS are able to effectively identify underexposure of this drug [85]. Other possibilities for DBS sampling at home include more frequent home-based adherence monitoring, which may help to reduce health care costs [30].

Another important advantage of home-based sampling is the fact that samples can be obtained anytime during the day. This is of particular relevance when the ideal sampling time is outside of regular clinic operating hours—for example, in the case of trough-level monitoring of prolonged-release tacrolimus [94]. Also, when timing of the sampling needs to be accurate, DBS may be easier to collect than venous samples because collection of the latter in a hospital setting might be difficult to organize logistically. Therefore, the collection of DBS has been evaluated, for example, for the C2 (ie, 2-h post-dosing) monitoring of cyclosporine A [82]. When so-called abbreviated area under the curve (AUC) estimation using serial sampling is required, proper timing of venipuncture might be even more difficult [83,95]. In addition, patients have to wait in the medical center until the last sample is obtained. Therefore, Cheung et al. proposed the use of DBS for tacrolimus monitoring at C_0, C_2, and C_4 because the AUC_{0-12} estimated from DBS collected by patients correlated well with the AUC_{0-12} estimated from traditionally obtained venous samples ($r^2 = 0.96$) [83]. Home-based sample collection also provides the opportunity to collect samples when adverse effects occur or whenever the effect of the drug proves to be inadequate (eg, immediately after an epileptic seizure or an asthma attack). In this way, valuable information can be obtained that a physician would not be able to acquire in any other way. This strategy was employed by Edelbroek to identify the reason for intermittent diplopia in a 63-year-old woman who was taking multiple antiepileptic drugs at the time [11].

Obviously, home-based DBS sampling is only helpful if the patient does not require venous sampling to assess other parameters. Therefore, it is important that other relevant biomarkers can also be determined on DBS, ideally even on the same DBS extract. In addition, the quality of the DBS samples has to be adequate to allow reliable quantitation. Fortunately, it has been shown multiple times that the majority of DBS samples have sufficient quality and that samples collected at home by the patient yield comparable results to those of DBS-based monitoring performed at specialized centers [82].

Home-based DBS sampling has much potential, but it may not be suitable for all patients. The DBS approach works only when patients are properly instructed, motivated, and willing to collect DBS themselves (as has already been demonstrated for the routine DBS-based monitoring of antiepileptic drugs in Dutch outpatient clinics [96]). Moreover, it is possible that a method is not applicable in certain patient subpopulations for which the method was not validated. In addition, home-based TDM will not be applicable for all drugs or drug-related questions because turnaround time is higher when samples have to be sent via regular postal services.

13.4.1.8 Cystic Fibrosis Patients

A special population for which home monitoring may be particularly favorable is CF patients. Not only might the reduction of hospital visits contribute to the patients' quality of life but also it may reduce the risk of transmission of their infection to other CF patients [27]. In addition, CF patients often have a difficult venous access, making venipuncture challenging [97]. Home-based DBS monitoring has already been performed in routine practice for the determination of the aminoglycoside tobramycin in CF patients [27]. Importantly, this strategy should be performed only when tobramycin is intravenously administered because finger prick blood can be contaminated with tobramycin after nebulized administration, even after thorough handwashing [98]. Home-based monitoring has also been proposed for other types of antibiotics used in CF patients [73].

13.5 ALTERNATIVES TO CLASSICAL DRIED BLOOD SPOTS

There are several alternatives to classical dried blood spots. These are addressed in this section.

13.5.1 Volumetric Dried Blood Spots

The most convenient way to avoid the Hct effect on DBS spreading is to analyze an entire fixed-volume DBS instead of a partial punch of a nonvolumetric DBS. However, depositing a fixed volume of blood on a filter paper is not easily performed by patients because this requires the use of pipettes or handheld volumetric capillaries. To resolve this issue, several approaches have been suggested that facilitate the preparation of volumetric DBS starting from a nonvolumetric drop of blood. A first approach is the system developed by Leuthold et al., in which a standard DBS card can be positioned and that utilizes integrated capillaries to transport a fixed volume of blood to the filter paper when a blood droplet is offered to the inlet of the system [39]. Similarly, Lenk et al. used chip-based microfluidics to generate volumetric DBS on regular filter paper [38]. Alternatively, VAMS devices can be employed to prepare volumetric dried blood samples [40]. These devices consist of a plastic handle with a hydrophilic polymer at the tip and absorb a fixed volume of blood independent of patient Hct. However, experiments in our laboratory showed that although no Hct effect could be discerned on the volume of blood absorbed, the VAMS devices seemed to be more prone to an Hct effect on extraction recovery [99]. Although all these formats seem very promising, applications are absent or still limited, so further investigation is warranted.

13.5.2 Dried Plasma Spots

To avoid the Hct effect, some investigators have suggested using DPS instead of DBS. The former have the same advantages as DBS regarding biosafety, stability, and ease of storage. In addition, the obtained results are more easily compared to existing reference ranges and therapeutic intervals, which are generally established using plasma or serum. Unfortunately, the preparation of DPS generally requires a centrifugation step (either when samples are obtained in the traditional manner or with special precision capillaries), which is not feasible in the real-life home-based TDM context. Nonetheless, this approach may still be of added value in specific cases due to the improved analyte stability and reduced transport and storage costs (eg, when samples need to be transported to another lab for the required analysis). To circumvent the need for a centrifuge, several blood separation devices have been proposed for in situ DPS generation. The first one is a membrane-based device introduced by Li et al., which consists of a separation membrane that withholds the cellular material, letting the liquid part of the blood flow through so it can be collected onto a filter [42,43]. A second device, commercialized as Noviplex, is based on a similar principle but yields volumetric DPS [44]. In addition, a novel blood separation device, which is composed of a spiral-shaped filter paper, has been commercialized as HemaSpot-SE [100]. Although the previously mentioned approaches seem promising, published applications are still lacking. Importantly, it has also not yet been proven that the blood filtrate obtained is equivalent to plasma, and at least some of these devices have shown an Hct effect on the filtration process and on DPS-based quantitation [43].

13.5.3 Other Dried Matrix Spots

The collection of biological fluids on filter paper has not been limited to whole blood and plasma but, rather, has also been suggested for multiple other matrices. In the context of the analysis of therapeutic drugs, the use of dried tear spots has been proposed for the quantitation of tobramycin [101], whereas dried OF spots have been employed for the analysis of lidocaine [102]. Dried OF spots could facilitate cost-effective sample transport and storage, as well as provide increased sample stability. Because it can be difficult to identify the exact location of the spot on the filter paper for translucent matrices such as OF and tears, it has been suggested to impregnate the filter paper with a substance that changes color after contact with a biological sample to facilitate correct analysis [103]. However, the use of these DMS in clinical practice seems limited.

13.6 PAPER SPRAY IONIZATION FOR THE ANALYSIS OF DRIED BLOOD SPOTS

To facilitate direct analysis of dried matrices, multiple ambient ionization techniques such as direct analysis in real time [104], desorption electrospray ionization [105], and paper spray ionization (PSI) [106] have been evaluated. In PSI, a small, triangular filter paper containing a DBS is sprayed with a wetting solution, which flows through the DBS, resulting in real-time extraction. Subsequently, a high voltage is applied to the filter paper via a copper clamp, effecting the formation of gas phase ions at the sharp tip of the filter paper. These ions are immediately transported to the MS, and the resulting signal is recorded in function of time, yielding a so-called chronogram. The ratio of the integrated AUC of the target's and the IS's chronogram is used as a measure of compound concentration [106,107].

Although this approach facilitates high-throughput analysis and potentially can be adopted even for point-of-care (POC) testing [108], the lack of sample preparation or a separation technique may have a substantial effect on assay specificity and sensitivity. Despite this, the performance is still sufficient to cover the therapeutic ranges of most compounds, according to Manicke et al. [106]. Moreover, the efficiency of PSI can be improved in various ways, such as by optimizing the filter paper material [109], the wetting solution [109], as well as the sharpness of the filter paper tip [107]. In addition, to further enhance throughput, a cartridge has been developed which can hold up to 14 filter paper tips that can be analyzed successively [110].

For PSI to be implementable in a POC setting (eg, via combination with a portable MS such as the Mini 12 [111]), the samples need to be analyzed wet, after the addition of a coagulant (which might be prespotted on the filter paper), or following a fast drying procedure [108,112]. Moreover, because the IS cannot be added to the blood sample before spotting, the IS needs to be added to either the wetting solvent or the filter paper prior to the application of the blood droplet [106]. In this context, a three-layered

IS-containing cartridge has been developed by Siebenhaar et al. [113]. Although there is a growing literature on the use of PSI for TDM purposes, including the analysis of imatinib, citalopram, paclitaxel, and tacrolimus, these generally only include a proof of concept and typically lack an appropriate analytical validation (eg, the evaluation of volume and Hct effect) [106,108,112]. In addition, in most cases, the IS is added to the blood sample before spotting, and only volumetric DBS are analyzed. Therefore, although promising results have been published, more research is necessary to demonstrate that PSI is able to deliver quantitative results comparable to those of traditional samples.

13.7 ALTERNATIVE SPECIMEN: ORAL FLUID

OF consists of saliva (the aqueous secretion produced by the major salivary glands—the glandula parotis, submandibularis, and sublingualis), the secretions of the accessory glands, gingival fluid, enzymes, other proteins, electrolytes, bacteria, epithelial cells, oronasopharyngeal secretions, and food debris [114]. To enter the OF, drugs have to cross the capillary wall, the basement membrane, as well as the cell membrane of the epithelial cells. This process is believed to be predominantly governed by passive diffusion, but other transport mechanisms, such as active transport, ultrafiltration, and pinocytosis, may also play a role [114].

Because only the free fraction of a therapeutic drug can migrate into the OF, the latter has often been proposed as a convenient alternative to assess the free plasma concentration of a therapeutic drug. This might be particularly relevant for TDM of highly protein-bound drugs, as well as for the monitoring of drug levels in special patient populations such as pregnant women, hypoalbuminemic patients, or patients with liver disease [23]. In addition, OF is readily available and easy to collect without the need for qualified personnel, even in difficult populations such as children, people with poor venous access, or people with fear of needles.

OF can be collected via drooling, expectoration, suction, swabbing, or the use of commercial and noncommercial collectors [115–117]. In addition, samples can be collected without stimulation (via passive drooling), with mechanical stimulation (eg, via movement of the tongue or by chewing of parafilm or paraffin wax), or with chemical stimulation (eg, via the use of citric acid or a lemon juice drop). Importantly, although major differences exist in the performance between OF collection devices, no particular one seems to consistently outperform the others [116]. Nonetheless, as discussed later, the choice of the collection method may be pivotal to the success of an OF-based TDM method [117].

A prerequisite for the usefulness of OF-based TDM is of course the existence of a (robust) correlation between the drug levels in OF and the (free) plasma levels. However, the amount of a therapeutic drug present in OF, and hence the correlation between OF and plasma levels, generally depends on many variables, such as the compound molecular weight, size, lipophilicity, pKa, and ionization status, as well as salivary flow rate and metabolism, plasma and salivary pH, and protein binding [114,117]. Importantly, the used collection method will affect the ratio between the OF and plasma drug levels because stimulation (either mechanical or chemical) will change the flow rate and thereby the OF composition and pH [117]. The effect of the employed collection method on the observed ratio was demonstrated by Lomonaco et al. for the determination of warfarin in OF [118].

Furthermore, OF flow rate and/or composition are influenced by multiple factors, such as circadian rhythm, diet, age, emotions, taste, smell, hormonal changes, the intake of medication with a parasympathetic or sympathetic (side) effect, and the presence of certain disease states (eg, infections increase the permeability of capillaries, and kidney failure influences the salivary pH by increasing the amount of urea) [114,119]. In addition, the ratio between the OF and plasma drug levels may be concentration- or time-dependent. The latter was demonstrated by Wintergerst et al. for the quantitation of the antiretroviral indinavir, where a better correlation was observed at later time points, possibly due to a slow equilibrium between both matrices [120]. A similar time-dependent phenomenon was proposed by Berkovitch et al. to explain why a good correlation was found between OF and plasma levels of gentamicin after once-daily administration, whereas no clear correlation could be observed when gentamicin was administered twice or thrice daily [121].

To minimize the influence of the parameters described previously, standardization of the sample collection method and correct timing of the sampling are of utmost importance. However, standardization might be difficult in specific patient populations, such as neonates and children [122]. Another strategy that has been suggested to improve the correlation between the measured drug levels in both matrices is to correct for the salivary pH when the diffusion of the drug is pH-dependent [123].

Due to the variables described previously, often a poor correlation between both matrices is observed or the obtained correlation displays an inter- and/or intraindividual variability that is too large to reliably derive the corresponding (free) plasma drug levels from the drug levels measured in OF [10,18,115]. Although, generally, unionizable drugs (or at least drugs that do not ionize within the salivary pH range) are regarded as optimal candidates for OF-based TDM, it must be evaluated on a case-by-case basis during clinical validation whether the correlation between both matrices is adequate. However, even when no clear correlation can be observed between both matrices, it has been suggested that OF sampling can still be employed for short-term adherence monitoring, as long as the therapeutic drug is consistently present in OF in quantifiable amounts. In addition, the influence of the collection method (as well as other parameters) may explain the variability in reported OF:plasma ratios and correlation coefficients in different studies. Moreover, note that sometimes OF drug levels are compared to total plasma drug levels, whereas other studies compare OF drug levels with free plasma drug levels, which may obviously yield different results.

The collection method or collection device that is employed affects not only the obtained correlation but also other parameters, such as analyte stability and recovery [117,124,125]. Therefore, stability should be evaluated in neat OF and on the collector, both over a relevant pH range and time period. Furthermore, recovery from collection devices can be concentration-dependent and should be determined over the entire concentration range [115]. Also, the extent of analyte adsorption may be influenced by the used collector or stimulating aid (eg, parafilm [126]). To improve stability and to reduce adsorption, some collection devices contain buffers, preservatives, and surfactants. However, these may cause matrix effects, which should be taken into account during method development [20,115]. Also, other materials used during sample collection, such as paraffin, may yield interferences during analysis [115]. In addition, it has been postulated that the use of certain collection strategies may improve method reproducibility. For example, chewing of absorbent materials may help to standardize salivary flow rate and thereby reduce method variability [124].

As stated previously, in addition to an optimal collection method, correct timing is very important for OF sampling. Samples should be collected after steady state is reached and when an equilibrium between blood and OF is obtained. Because contamination from oral intake of medication may lead to erroneous results during the first several hours after ingestion, it is advised to collect samples either before the next dose or after a sufficiently long "wash-out" period [9,117]. Moreover, contamination issues may be aggravated if the patient has dental caries because these may lead to the formation of so-called drug pockets [9]. To reduce the risk of contamination, it has been suggested to ask donors to rinse their mouth before sample collection [9,127].

Although the collection of OF is easy to perform in most patients, sometimes it may be challenging to collect sufficient sample volume due to xerostomia or dry mouth [114,119,125]. This issue may be age-related or can be caused by certain pathologies, medications, drugs of abuse, or anxiety. In addition, when using an OF collector, the volume of OF is unknown, which hampers OF-based quantitation [115,125]. Moreover, it is incorrect to assume that a fixed sample volume is collected, which is consistently recovered from the collection device [117]. To derive the collected OF volume, it has been proposed to weigh the OF collectors before and after sampling, which is of course tedious to perform [115]. Another strategy is to use a collection device such as the Greiner Bio-One Saliva Collection System (SCS), which contains a dye in the collection solution that allows one to determine the collected OF volume photometrically [124].

Not only the sample volume but also the sample quality is of utmost importance. Samples can be contaminated with blood, which can drastically impact OF-based quantitation [9]. Especially for drugs with a very low OF: whole blood ratio, even a minor contamination with blood may yield extremely erroneous results. This can be a particularly important issue in people with gingivitis or gingival hyperplasia [9]. To minimize the risk of blood contamination, it is advised to collect samples before teeth brushing. Furthermore, OF samples may contain food debris or other contaminants [9,115]. Interestingly, it has been stated that unstimulated OF is more likely to contain debris [125], whereas OF samples collected after stimulation with citric acid yield cleaner samples and better chromatography [20]. Another practical issue is the fact that OF samples may be quite viscous, which leads to sample inhomogeneity, hampers pipetting, and causes the formation of bubbles when samples are agitated [20,115,128]. To resolve this issue, some have used sonication to break up the mucin [127], whereas others have employed buffers to dilute the sample and to reduce sample viscosity [128]. Again, the choice of the collection method may impact sample quality. Gröschl et al., for example, noticed that neat OF samples remained sticky even after centrifugation, whereas samples collected with certain OF collectors yielded a cleaner matrix, potentially by removing mucopolysaccharides from the OF matrix [124]. Consequently, to warrant sample quality, the collection method should be optimized during method development, and colored or cloudy patient samples should always be discarded [9].

Another technical limitation for OF analysis might be the required method sensitivity [115]. Because OF drug levels (at least theoretically) correspond to the plasma free fraction, quite low drug levels can be anticipated. This is particularly true when trough levels are measured and/or when a dilution buffer is used during collection. For certain compounds that are subject to ion trapping in OF, this issue might be less relevant.

13.7.1 Applications of Oral Fluid for Therapeutic Drug Monitoring

The noninvasive nature and the ease of OF collection may be favorable in certain patient populations and settings. This is illustrated next with the discussion of selected examples of OF-based TDM methods. Classes of therapeutic drugs that have been analyzed in OF include antibiotics [18,121,129], antidepressants [130], antiepileptics [9,131,132], antimalarials [133,134], antimycotics [135,136], antipsychotics [137–140], antiretrovirals [120,141,142], immunosuppressants [127], and oncolytics [95,143].

13.7.1.1 Pediatric Patients

Because OF collection is noninvasive and painless, it may represent a valuable alternative for TDM in neonates, infants, and children. Furthermore, it may also provide a stress-free alternative in children who are afraid of needles or who have poor venous access. In addition, OF collection is preferred over classic venous sampling not only by most pediatric patients but also by their parents [144]. Because caffeine levels need to be monitored weekly during treatment of apnea in premature infants, it was evaluated whether OF would provide a good alternative to traditional TDM by comparing OF levels in infants with their corresponding plasma levels [145]. To facilitate OF collection, a nonwoven gauze attached to a cotton swab was employed, onto which the infants could suckle. Although unstimulated collection and stimulated collection using a citric acid-impregnated gauze were more convenient to perform, the best correlation ($r = 0.89$) between OF and plasma levels was obtained when OF production was stimulated by depositing citric acid in the cheek pouch 5–10 min prior to OF collection. Another application of OF analysis in pediatric patients is the analysis of busulfan. Because conventional TDM requires serial blood sampling from a central (or peripheral) venous line, which is time-consuming and potentially distressing for both the children and their parents, OF collection could provide an interesting noninvasive alternative. The feasibility of such an approach was evaluated by Rauh et al. and a good correlation was observed between AUCs determined from OF and plasma busulfan levels. Interestingly, although a Salivette collection device was generally used to collect OF, in patients younger than age 2 years a modified medical pacifier was employed in which the teats had been perforated and a DBS filter paper had been inserted [95].

13.7.1.2 The Elderly

Because albumin levels tend to be lower in the elderly, it is indicated to measure the free plasma levels of highly protein-bound drugs in this population [23]. Therefore, OF analysis may be a good alternative whenever OF levels correspond to the free plasma levels. This was demonstrated for the analysis of phenytoin in elderly patients [146]. In addition, the elderly are more likely to have difficult venous access, which further lends support to the use of OF in this population [3]. On the other hand, the elderly more frequently present with xerostomia, which may hamper OF collection.

Indeed, the elderly typically take more chronic medications, including anticholinergic medication or medication with anticholinergic side effects, which is a major risk factor for dry mouth [119].

13.7.1.3 Pregnant Women

As discussed previously, the levels of antiepileptic drugs may change drastically during pregnancy and the postpartum period. In this context, Herkes and Eadie demonstrated that the frequent collection of OF samples at home may be a feasible approach to follow up phenytoin plasma levels in pregnant women before and after delivery [131]. Although samples were acquired more frequently during this study, the authors proposed that pregnant women could collect OF at home once a week to minimize the risk of inadequate drug levels. In theory, similar results could be obtained via frequent plasma monitoring; however, this is generally hindered by logistical issues. In addition, OF phenytoin levels correspond well with free plasma levels [9], which is relevant for TDM in pregnant women because pregnancy affects phenytoin protein binding [147].

13.7.1.4 Psychiatric Patients

The fear and anxiety that accompany blood sampling in schizophrenic patients could potentially be avoided by using OF collection instead. In addition, a collection technique that does not require needles may be regarded as safer in a psychiatric facility. In some schizophrenic patients, however, OF collection might be challenging because certain antipsychotics (or other comedications) can have anticholinergic side effects, including a reduction in salivary flow rate [148]. OF-based TDM of a diverse panel of antipsychotics was evaluated by comparing drug levels in OF and serum collected from psychiatric patients. Unfortunately, OF-to-serum concentration ratios showed a fairly large variability for all of the investigated antipsychotics, limiting the practical usefulness of the approach. Nonetheless, OF collection could potentially still be employed to screen for therapy nonadherence [137]. Another compound that has been evaluated in this context is lithium (Li). Unfortunately, here as well, the correlation between OF and serum levels was not strong enough to make OF-based TDM a worthy alternative for classic Li TDM [138]. However, it has been postulated that although interindividual variability in the OF:serum ratio is too large to allow OF-based TDM, intraindividual variability is much smaller, and once an individual ratio is determined, this could be safely used in clinically stable patients on prophylactic Li therapy [139]. In addition, El-Mallakh et al. found that when mucinous material was removed from the OF (via dialysis), the correlation with plasma levels increased dramatically [140].

13.7.1.5 Patients With Renal Failure

Renal failure can have a pronounced effect on plasma protein binding. This was observed by Reynolds et al., who determined total and free phenytoin levels in seven patients with chronic renal failure and found that these patients had low total plasma phenytoin levels, whereas the free plasma levels were disproportionally high. Interestingly, drug levels determined in OF correlated well with the plasma-free drug concentrations, indicating that OF could be an adequate matrix to monitor phenytoin in patients with renal failure [132].

13.7.1.6 Remote and Resource-Limited Settings

Because no specialized personnel or equipment (eg, centrifuges) are required to obtain OF samples, OF collection could potentially be of interest in resource-limited settings. In addition, OF samples represent less of a biohazard than blood, which is particularly important in areas where HIV is often endemic. However, to be applicable in resource-limited settings such as developing countries, the analyte obviously needs to be stable in OF under field conditions. Unfortunately, this parameter, although crucial, is typically not evaluated.

To facilitate TDM of antiretroviral drugs in developing countries, multiple methods have been developed to analyze these drugs in OF. An example is the semiquantitative method developed by George et al. to detect nevirapine underexposure [142]. To render this method more field-compatible, thin-layer chromatography (TLC) with UV detection was employed due to its simplicity and low cost. Although this method originally proved to reliably identify those patients with subtherapeutic drug levels, interestingly, when Lamorde et al. tried to employ the same method, they could not visualize any spots on the TLC plates [141]. Because the reason for this problem could not be identified, these authors employed LC-UV instead. However, the latter technique will probably be available only in specialist centers and therefore requires samples to be sent for analysis. Furthermore, also for the analysis of antimalarials, several OF-based methods have been developed. Although for some compounds, such as mefloquine, a good correlation ($r = 0.88$) was observed between OF and systemic levels [133], monitoring of other compounds, such as chloroquine and proguanil, has been deemed unreliable [134]. Another disease that is endemic in developing countries and for which OF-based analysis could help in setting up TDM programs is TB. The analysis of linezolid and clarithromycin in multidrug-resistant TB patients, for example, showed that the obtained AUC_{0-12} values or AUC_{0-24}:minimum inhibitory concentration ratios did not significantly differ from the ones calculated from simultaneously obtained serum samples [129]. Although this method was not specifically developed for use in developing countries, it might be valuable in this setting as well.

13.7.1.7 Home Monitoring

As discussed previously, home-based TDM has multiple advantages for the patient, the patient's caregiver, and the clinician. The latter was well demonstrated by the questionnaire by Baumann [235] that revealed that clinicians who treat epilepsy patients would find it very valuable if a "real-time" concentration could be obtained when a seizure or adverse effect occurs at home.

Mycophenolic acid and fluconazole are two compounds for which home-based TDM using OF samples has been proposed. Mendonza et al. found a good relationship between mycophenolic acid concentrations in OF and the corresponding total ($r = 0.909$) and unbound ($r = 0.910$) plasma concentrations [127]. However, timing of sample collection seemed to be crucial because brushing of the teeth was suggested to have contaminated the majority of the early collected samples with blood. Van der Elst et al. observed a good correlation ($r = 0.96$) between the OF and serum levels of fluconazole [135]. Moreover, fluconazole in OF was stable for at least 17 days when stored at room temperature, and recovery from the collection device was not affected by storage at 20°C for up to at least 6 days. Therefore, fluconazole monitoring in OF might be feasible in patients receiving prolonged courses of antifungal treatment at home. Koks et al. also found a good correlation ($r^2 = 0.80$) between the levels of fluconazole in OF and plasma [136]. However, they concluded that it was still insufficient to derive the exact plasma fluconazole levels from the corresponding OF levels and hence that OF fluconazole levels could only be employed for compliance monitoring or for the semiquantitative prediction of plasma fluconazole levels.

The feasibility of setting up a home-based TDM program using OF samples has been assessed by Tennison et al. for phenytoin, carbamazepine, and phenobarbital [149]. Not only were these compounds stable during transport but also patients were able to collect the OF samples themselves. Similar stability results were obtained for the newer antiepileptics as well [150]. In addition, it has been demonstrated that OF samples collected at home can be returned to the laboratory in a reasonable amount of time [150]. However, it was also noted that in certain situations (eg, to assess adverse effects), samples would have to be mailed via overnight carriers to allow sufficiently short turnaround time. To facilitate proper sample collection at home, the choice of the collection device might be crucial. The Quantisal, for example, has a volume indicator that informs the patient whether a sufficient sample volume has been collected, and it may thus help to minimize the amount of samples with inadequate sample volume. However, the use of other collection devices, such as the SCS, may be too complex and therefore less suitable for outpatient monitoring [124]. In addition, dried sample formats may facilitate home-based TDM due to increased sample stability and ease of transport.

13.7.1.8 Cystic Fibrosis Patients

Although home-based OF analysis of antibiotics would be convenient for CF patients, studies have suggested that this approach is not suitable for TDM of tobramycin and gentamycin [151,152]. In addition, CF patients tend to provide more viscous OF samples, which may complicate their analysis [114].

13.8 ALTERNATIVE SPECIMEN: TEARS

Lacrimal fluid or tear fluid, which is composed of proteins, mucins, lipids, electrolytes, epithelial cells, and ambient contaminants [153,154], has been proposed as a suitable alternative matrix for free plasma concentration monitoring [23,155]. This suggestion is based on the existing equilibrium between the free plasma fraction and the free fraction in tears, as well as on the low and constant protein content of lacrimal fluid [154,156]. Tear fluid offers the advantage over OF (which is also used for free fraction monitoring) that it is more homogeneous and has a more constant composition, particularly after stimulation [155]. Furthermore, contrary to OF, tears are subject to only relatively small changes in pH [154,156], implying a less pronounced, although still relevant, effect on transfer of ionizable drug into tear fluid [157].

Tear fluid can be obtained in multiple ways. First, direct tear collection can be performed by holding a capillary in the conjunctival sac [153]. This approach, however, requires a certain skill because the capillary needs to be held in place while avoiding injuries. Moreover, the eye has to be open for the entire collection procedure, which is unpleasant for the patient. Consequently, this technique cannot be utilized in noncooperative patients and children, unless tears can be sampled from the cheek [158]. Alternatively, tears can be collected in an indirect manner using Schirmer strips (ie, a filter paper) [153]. To that end, a small part of a filter paper strip is folded at 90 degrees and positioned in the lower conjunctival sac for several minutes, during which time the patient's eyes are closed. Afterwards, the tear fluid can be recovered from the strip by centrifugation. The latter sampling method is not only more feasible for non-ophthalmologists [153] but also preferred by patients and yields more reproducible results [159]. Other indirect tear sampling methods include the use of absorbing materials such as sponges, swabs, filter paper discs, and cotton thread [159,160].

Although "basal tears" (ie, unstimulated tears) can be sampled, it is also possible to stimulate tear production before sample collection [154]. The latter is often performed to obtain sufficient sample volume, especially when direct sampling techniques are employed, or to reduce collection times. These stimulated or "reflex" tears can be triggered by (1) noncontact methods such as looking into the sunlight, stimulation of the gag reflex, or emotions; (2) mechanical irritation caused by the use of the Schirmer strip, for example; or (3) chemical stimulation of the conjunctival or nasal mucosa with cigarette smoke, onion vapor, ammonia vapor, Vicks VapoRub, formaldehyde vapor, etc. [154,157,158,161,162]. Because stimulation activates different lacrimal glands compared to standard physiological circumstances, the composition of the tear sample will be altered [154]. Furthermore, differences in flow rate between reflex tears and basal tears may contribute to discrepancies in measured drug concentrations [157,160]. Hence, the influence of tear stimulation on measured drug levels should always be evaluated. Especially the reproducibility of this phenomenon is of key importance because dilution caused by stimulation may be time-dependent. To help improve sampling method reproducibility, it has been advised to standardize both collection time and the volume withdrawn [159].

Although lacrimal fluid may provide an alternative way of assessing the free fraction of a drug, it has certain drawbacks. First, tear sampling does not allow for patient self-sampling. Furthermore, the sampling itself requires a relatively long time, and the stimulation of tear production can be noxious. Moreover, the composition of tear fluid and/or the volume produced can be influenced by age, flow rate, disease state, intake of (therapeutic) drugs, the use of contact lenses, the employed collection technique, and the collection site [154,160]. Also, differences in sample collection methods or lacrimal pH may yield contradictory results [157,160]. In addition, because only a limited sample volume is available, tear analysis requires sensitive equipment.

13.8.1 Application of Tears for Therapeutic Drug Monitoring

Tear fluid analysis has been discussed mostly in the context of antiepileptic drug monitoring. Specifically, the comparison between tear and OF-based therapeutic drug analysis has gained attention. Both for carbamazepine and for phenobarbital, tear and plasma concentrations ($r = 0.75$ and 0.95, respectively) correlated better than OF and plasma concentrations ($r = 0.38$ and 0.60, respectively) [156]. Similar conclusions were obtained by Monaco et al., although the correlation coefficients found by the latter were less strong [161]. Furthermore, a better correlation was found between tear and cerebrospinal fluid (CSF) concentrations than between OF and CSF levels ($r = 0.86$ and 0.87 vs $r = 0.64$ and 0.44), favoring the use of tears instead of OF for TDM purposes. Similarly, a good correlation was found between tears and plasma levels as well as tears and CSF levels for both diphenylhydantoin and primidone. However, only for diphenylhydantoine did tear levels outperform OF levels [163]. Although for valproic acid (VPA) tear concentrations correlated better with plasma and CSF levels than did OF levels, the correlation coefficients were still relatively low ($r = 0.31$ and 0.61, respectively) [164]. Nakajima et al., on the other hand. found a correlation coefficient of $r = 0.864$ with total plasma concentrations and $r = 0.936$ with free plasma concentration [165]. Whereas the former group collected 100 μL of tears after exposure to cigarette smoke or sniffing formaldehyde, the latter group employed Schirmer strips for tear collection, requiring only a very small sample volume (3.37 ± 0.41 μL), which may have contributed to the observed contradictory results. Furthermore, Nakajima et al. utilized on-card derivatization (ie, direct derivatization to fluorobenzyl ester derivatives on the filter paper) followed by extraction and GC analysis, whereas the former employed an enzyme-multiplied immunoassay technique. Lastly, for ethosuximide the correlation coefficients were similar for OF and tears versus plasma levels ($r = 0.73$ and 0.74, respectively), whereas for both matrices no correlation could be found with CSF levels. The latter might have been due to the very limited sample number ($N = 6$) [166].

Furthermore, tears have been used for TDM of other drugs as well. Stoehr et al., for example, employed tear sampling for the analysis of theophylline [158]. Although a good correlation was obtained between tear levels and either total ($r = 0.94$) or free plasma levels ($r = 0.88$), the method did not allow adequate free plasma level determination based on tear levels. Using the tear:free plasma concentration ratio as a conversion factor, only 50% of the patient samples had a predicted unbound concentration within a 1-μg/mL deviation, limiting the clinical usefulness of this approach. Nakajima et al. also obtained similar results, with Pearson correlation coefficients of 0.8406 and 0.968 between tear and total, respectively free, plasma concentrations, respectively [167]. In addition, Steele et al. demonstrated that tear methotrexate concentrations correlated better with serum levels ($r = 0.714$) than did OF levels ($r = 0.557$) [168].

Although a correlation has been shown between tear and (free) plasma levels for multiple compounds, the clinical significance of tear sampling for TDM purposes is at best limited. Moreover, sampling is generally time-consuming, does not allow for self-sampling, and is not easily applicable in children, a population for which alternative, noninvasive sampling would be particularly beneficial.

Recently, alternative ways of measuring tear concentrations have been proposed. Although not developed in view of TDM, a quantitative chromatographic method has been set up for the determination of tobramycin of dried tear spots of 15 μL [101]. Furthermore, the use of contact lens sensors has been advocated for glucose monitoring in the lacrimal fluid of diabetics [169]. In the future, this promising technique is likely to be expanded to the detection of other disease biomarkers. Moreover, the analysis of exogenous compounds, such as therapeutic drugs, present in lacrimal fluid may be envisaged because TDM programs would greatly benefit from noninvasive, continuous drug monitoring.

13.9 ALTERNATIVE SPECIMEN: INTERSTITIAL FLUID

Water and small, nonprotein-bound solutes such as therapeutic drugs are continuously exchanged between whole blood and the ISF under the influence of hydrostatic and osmotic pressure. Therefore, ISF forms an interesting alternative matrix for TDM because its composition is closely related to plasma [170]. ISF can be sampled minimally invasively using reverse iontophoresis or microneedles [19,171]. Other methods of ISF sampling include clinical microdialysis [172,173], as well as the retrieval of fluid from induced skin blisters [155]. However, due to their less patient-friendly nature, the latter sampling methods are not within the scope of this chapter.

The feasibility of using ISF as a valid alternative for blood-based TDM was demonstrated by Kiang et al. [170]. To this end, a diverse group of therapeutic drugs were administered to rabbits, and the T_{max}, C_{max}, and AUC were assessed in both blood and ISF. Although comparable results were found for some compounds, ISF-based TDM did not seem feasible for others, indicating that the utility of ISF analysis needs to be evaluated on a case-by-case basis.

The most widely discussed technique regarding ISF-based TDM is reverse iontophoresis [19]. This sampling method extracts both charged and neutral drugs from the subdermal region in a noninvasive manner. To this end, a small electric current is sent through the skin after applying two electrodes, causing both electromigration and electroosmosis due to solvent drag. Eventually, positive and neutral molecules will accumulate at the cathode and negative components at the anode. As electroosmosis causes water migration from the anode toward the cathode, the extraction of positively charged molecules is facilitated, whereas the extraction of negatively charged molecules is hampered. In general, reverse iontophoresis is therefore most favorable for small hydrophilic or positively charged compounds [19].

In reverse iontophoresis, the amount of drug excreted across the skin is assumed to correlate with the subdermal levels, which in turn need to correlate with the plasma concentrations. However, it needs to be taken into account that the measured ISF drug level can be dependent on the site of sampling [174]. Moreover, certain drugs may form a skin reservoir, which has to be depleted before a concentration can be measured that is proportional to systemic levels [175]. After collection, the sample can be analyzed in the lab or biosensors can be incorporated in the reverse iontophoretic system allowing ambulatory monitoring [19].

Despite the noninvasive nature of this sampling technique, it may cause some minor adverse effects, such as erythema or edema. Furthermore, reverse iontophoresis is fairly costly and necessitates complicated technology. Moreover, it may require calibration with a whole blood sample, and results can be affected by contamination from sweat [19].

13.9.1 Application of Interstitial Fluid for Therapeutic Drug Monitoring

Because the noninvasive nature of this technique offers the possibility of frequent sampling, even in vulnerable populations, Sekkat et al. evaluated the feasibility of monitoring caffeine and theophylline in neonates [176]. In this context, they evaluated the effect of the functionality of the stratum corneum barrier on measured drug levels using tape-stripped porcine skin as a model. The authors concluded that at least for certain compounds, such as caffeine and theophylline, which are transferred via electroosmosis and electromigration, the maturity of the skin drastically influences the obtained concentration.

Most research regarding the use of reverse iontophoresis for TDM purposes has been performed in vitro or ex vivo. However, a rare in vivo application was published by Leboulanger et al. regarding Li monitoring in patients with bipolar and schizo-affective disorders ($N = 23$) [175]. Interestingly, the correlation between extracted Li and serum concentrations was poor after 30 min of extraction, whereas it improved dramatically after 60, 90, and 120 min. This observation indicated the presence of a drug reservoir in the skin and demonstrated the relevance of sufficient extraction time. Furthermore, the effect of using sodium (Na) as an IS was evaluated. Na was selected for this purpose because its extraction flux is constant. Using a training set of 10 patients, the Li_{ISF}/Li_{serum} and $(Li_{ISF}/Na_{ISF})/Li_{serum}$ ratios were established and subsequently used to predict serum concentrations for the 13 remaining patients. This yielded excellent results, with the Na-normalized values leading to a smaller error on the serum level estimation. However, for applicability in routine practice, miniaturization of the extraction system would be required, as well as a shorter extraction time.

A more convenient alternative for ISF sampling is the use of microneedles [171]. Previously, these devices have been employed for drug or vaccine delivery, but recently their potential for ISF or blood sampling has been recognized. Microneedles have micrometer-scaled protrusions ranging from 50 to 150 μm for ISF sampling and slightly larger needles (1000–1500 μm) for blood sampling. Biological fluid can be collected in hollow microneedles and either analyzed off-site or on-site using integrated detection systems. Alternatively, microneedle materials have been employed that undergo transition into a gel phase upon fluid collection [177]. Using the latter technique, the risk for blockage or breakage of the needles, and hence nonbiodegradable substances getting stuck in the skin, is avoided. Microneedles not only may be preferred over traditional blood sampling in a pediatric population [178] but also offer the potential for patient home sampling. However, when performing ISF sampling, attention must be paid to the fact that ISF levels may lag behind systemic levels. Furthermore, as discussed for reverse iontophoresis, skin reservoirs may exist for certain molecules, which must be depleted before microneedle sampling can be performed successfully.

13.10 ALTERNATIVE SPECIMEN: HAIR

Hair analysis has been proposed as a convenient tool for long-term compliance monitoring because it can provide retrospective information about a patient's drug exposure during the past weeks or months [179]. This could be particularly beneficial for the detection of so-called white coat compliance. Indeed, patients who only adhere to therapy soon before a hospital visit could be mistakenly considered as compliant based on their plasma levels, whereas their aberrant long-term adherence could be picked up using hair analysis [180]. However, contradictive opinions about the usefulness of hair analysis for TDM can be found in the literature [181].

Although the exact incorporation mechanisms are not yet fully elucidated, therapeutic drugs are believed to enter the hair root via passive diffusion from the surrounding capillaries during the anagen phase. As the hair grows, they then become sequestered in the hair shaft during keratinization. Other routes by which drugs can access the hair are via sweat and sebum excretion, as well as via external contamination. Drug incorporation is favored for lipophilic and alkaline compounds and is influenced by many variables, as will be discussed later [179,182].

Hair samples are preferably collected from the vertex posterior because the hairs in this region are more likely to be in the anagen phase, display a more stable growth rate, and are less affected by age- and sex-related differences [179,182]. Furthermore, samples have to be cut as close as possible to the scalp and have to be stored at room temperature protected from sunlight. Although hair is generally an easily accessible matrix, the collection of hair samples may be challenging in certain populations due to cultural traditions

[183], spiritual beliefs, and superstition [184]. However, by properly informing the patient about the goal of the hair analysis and the fate of the hair sample, this issue can be overcome. The feasibility of hair collection in developing countries was confirmed by Gandhi et al. [185]. Moreover, acceptability of hair analysis may depend on the amount of hair that needs to be collected. The required sample size may vary from a single hair to approximately 200 mg [182,186]. In addition, scalp hair is not always available. As an alternative, axial, pubic, or other hairs could be collected. However, drug levels can be affected by the collection site due to differences in growth rate, for example. In this respect, Liu et al. showed that pubic hair could not substitute for scalp hair in tenofovir monitoring [187].

Because drug concentrations in hair are low (typically picogram/milligram to nanogram/milligram), sensitive analytical techniques such as LC-MS/MS are required [179]. Moreover, hair analysis necessitates an elaborate preanalytical phase comprising multiple decontamination steps to remove external contamination, followed by hair homogenization, hair extraction, and additional sample cleanup procedures. These manual steps make hair analysis tedious and time-consuming [179].

Furthermore, the interpretation of quantitative and even qualitative hair analysis results is complicated by a multitude of factors. First, inadequate removal of external contamination leads to overestimation of hair levels or false-positive results. Moreover, it has been postulated that traditionally used washing steps may actually promote the incorporation of external contaminants in hair instead of resulting in their removal (Cuypers et al., unpublished data) [188]. In addition, hair treatments such as bleaching, coloring, perming, and hair flattening may damage the hair matrix, making it more accessible for external contaminants [179]. On the other hand, an increased accessibility may also cause leakage of incorporated drugs, yielding lower hair levels and even false-negative outcomes. In addition, high temperatures and noxious agents used during cosmetic hair treatment can have a direct deleterious effect on therapeutic drugs [179]. Other influences on the extent of drug incorporation are hair color (ie, melanin content), ethnicity, gender, age, and the amount of hairs in the anagen phase. To minimize the effect of hair color on the obtained quantitative results, it has been suggested to normalize drug levels to the hair melanin content [189].

As can be deduced from the previously mentioned issues, the correlation between systemic and hair drug levels is subject to ample variables that cause large interindividual differences. Therefore, it is impossible to evaluate strict therapy adherence based on absolute hair levels [181]. However, because intraindividual variation is expected to be less pronounced, relative changes in compliance could potentially be monitored using segmental hair analysis, in which the patient acts as his or her own control [180]. To that end, consecutive 1-cm segments of hair are analyzed and drug levels compared. Each segment corresponds to approximately 1 month, with the most proximal segment corresponding to the most recent drug exposure. Although a brief interruption of therapy would not be noticeable using this approach, large variations in drug-taking behavior (so-called drug holidays or white coat compliance) or complete noncompliance could be picked up.

Nonetheless, segmental hair analysis is accompanied by challenges and limitations. First, the evaluation of recent drug intake is not feasible [179] because hair requires 7–10 days to emerge from the scalp [190]. Furthermore, segmental hair analysis is subject to a washout phenomenon, meaning that hair levels progressively decrease along the hair shaft. Attempts have been made to compensate for this phenomenon based on a correction factor established in adherent, healthy volunteers. However, this factor is tedious to determine and susceptible to interindividual variability due to differences in, for example, hair structure and hair hygiene rituals. Moreover, although similarities have been observed in the temporal profile of hair concentrations and dose regimens, correlating drug levels in a certain segment to a specific moment in time remains challenging because hair growth rates differ both inter- and intraindividually [191]. In addition, it has been observed that a single dose intake can yield detectable drug levels in multiple consecutive or even all hair segments instead of only the segment encompassing the time of drug intake [192,193].

13.10.1 Application of Hair Specimens for Therapeutic Drug Monitoring

Williams et al. found that carbamazepine levels in the proximal 1-cm hair segment of adherent inpatients correlated with administered dose ($r = 0.581$) [180]. However, due to considerable interpatient differences, absolute compliance could not be assessed based on hair levels. On the other hand, a relatively stable individual carbamazepine incorporation rate (intrapatient coefficient of variation of $15 \pm 5.2\%$) was observed, indicating that segmental hair analysis could be utilized to detect compliance changes. Therefore, this approach was employed to reveal carbamazepine or lamotrigine therapy cessation during pregnancy, which is known to occur due to fear of teratogenic effects [194]. To that end, hair segments corresponding to periods before and during pregnancy were compared after correction for the washout phenomenon. In addition, segmental hair analysis in postmortem samples of epilepsy patients showed a larger intraindividual variability in antiepileptic hair concentrations in sudden unexplained death victims compared to the ones determined in epilepsy inpatients and outpatients [195]. These findings suggest that long-term compliance monitoring could help to prevent the occurrence of sudden unexplained death in epilepsy patients.

Furthermore, Müller et al. investigated the usefulness of hair analysis for cyclosporine A compliance monitoring [191]. Although a correlation ($r^2 = 0.57$) was observed between cyclosporine A hair levels and the average of blood trough levels measured within the corresponding time frame, hair levels did not allow to determine drug intake. However, the observed hair concentration profile aligned with the dosage regimen, suggesting segmental hair analysis can be helpful in the detection of major noncompliance or drastic changes in drug-taking behavior [191]. Moreover, hair analysis has received attention in psychiatric patients and patients with behavioral problems because compliance can be very challenging in these populations. In this context, Uematsu et al. found a significant correlation between haloperidol hair and plasma levels ($r = 0.772$) as well as haloperidol daily dose ($r = 0.555$) [196]. Also, relatively large interindividual variations were observed, limiting the usefulness for TDM purposes. On the other hand, the same group demonstrated that haloperidol hair levels showed noticeable changes following (strong) dosage adjustments or therapy discontinuation [197,198], again indicating that hair analysis could be valuable in detecting pronounced changes in therapy adherence. Furthermore, analytical methods have been developed to determine atomoxetine, methylphenidate, and their respective metabolites in hair of children and adolescents treated for attention deficit hyperactive disorder (ADHD) [199,200]. Although no correlation could be observed between daily dose and hair drug levels for either compound, the authors concluded that hair analysis could still be valuable to monitor long-term compliance.

Another domain in which hair analysis has been extensively investigated is TDM of (prophylactic) antiretroviral therapy. Because single plasma levels are known to show pronounced intraindividual variation, the analysis of multiple plasma samples over the course of several days is generally advised to assess average antiretroviral drug exposure. Because hair analysis is believed to provide an indication of average drug exposure, it has been proposed as a more convenient alternative [187].

To establish the usefulness of hair analysis for tenofovir compliance monitoring, two, four, or seven doses per week were administered to healthy volunteers to mimic different degrees of therapy adherence [187]. Although mean hair levels increased with the amount of doses received, the accompanying 95% concentration ranges clearly overlapped. Also, for 3 of 23 patients, hair levels did not always increase with dosage frequency. It should be noted that only individuals with dark hair were included in this study. Furthermore, Baxi et al. demonstrated that tenofovir and emtricitabine hair levels correlated significantly with medication event monitoring system caps openings ($r = 0.50$ and 0.58, respectively), plasma ($r = 0.41$ and 0.51, respectively), and peripheral blood mononuclear cell levels ($r = 0.43$ and 0.50, respectively). Hence, they concluded that hair levels could be utilized for compliance monitoring [201]. Interestingly, these correlations were noticeably more pronounced after 16 weeks compared to 8 weeks ($r = 0.61-0.86$).

In addition to the use of hair levels to monitor compliance to antiretroviral therapy, multiple studies have evaluated the correlation between these levels and therapeutic outcome. Bernard et al. demonstrated that mean indinavir hair levels were significantly higher in treatment responders (24.4 ± 16 μg/g) than in nonresponders (12.9 ± 8.6 μg/g), further confirming their previous results [202]. Comparable data were obtained by Duval et al., who showed that although a large overlap was observed between the hair levels of both groups, hair levels predicted viral response more adequately than a single plasma trough level [203]. Similarly, Huang et al. found higher mean lopinavir and efavirenz hair levels in responders than in nonresponders (1.6 vs 0.3 and 3.4 vs 0.68 ng/mg) [204]. Others investigators have taken it one step further and have set up clinical decision trees based on antiretroviral hair levels. For example, Van Zyl et al. developed an algorithm based on lopinavir plasma and hair levels to verify the cause of therapy failure in second-line regimens in a cost-effective manner [205]. Only when both recent and average drug exposure levels were adequate (ie, ≥ 1 μg/mL in plasma and ≥ 3.63 ng/mg in hair) did the authors advise genotyping to confirm resistance. Furthermore, Gandhi et al. established a decision tree to prolong the durability of the first-line antiretroviral treatment with atazanavir, especially in settings in which viral load monitoring is not available [206]. Specifically, the authors recommended analyzing hair samples 4–6 weeks after therapy initiation. If the measured drug level is 1.78 ng/mg or less, this may be indicative of impending treatment failure or resistance development, and either additional compliance motivation or an appropriate PK intervention is required. Also, in the case of disease progression or suspected PK changes, hair level assessment is advocated.

13.11 ALTERNATIVE SPECIMEN: EXHALED BREATH

Not only volatile but also nonvolatile exogenous compounds can be detected in exhaled breath. Indeed, after systemic uptake, drugs can enter the lungs via the bloodstream, after which they may become incorporated in aerosol particles, generated during exhalation from the fluid lining of the respiratory tract [207]. To this end, drugs need to pass the capillary wall, the interstitial space, and the alveolar epithelial cells [208]—a process that favors small, lipophilic compounds. Moreover, these lipophilic compounds are more likely to concentrate at the liquid surface and vaporize more easily, enhancing their concentration in aerosol particles [208].

Exhaled breath is an extremely accessible sample matrix that can be collected in a bag, via adsorption to a sorbent material, collection on a filter, or by breath condensation. Aside from these collection methods that are suitable for off-line analysis, direct breath analysis has also been performed. Concentrations of nonvolatile compounds in exhaled breath are generally in the picogram to nanogram per liter range and hence require very sensitive, mass spectrometry-based analytical systems. In this respect, a collection method that allows for preconcentration, as occurs with breath adsorption, may be advantageous [208].

13.11.1 Application of Exhaled Breath for Therapeutic Drug Monitoring

Beck et al. evaluated the usefulness of D-amphetamine and methylphenidate measurements in exhaled breath for compliance monitoring in patients receiving oral therapy for ADHD [207]. Patients were sampled within 24 h after last dose intake, and the microparticles in their exhaled breath were captured on a polymer filter. The latter was attached to a bag to somewhat standardize the amount of breath that was collected. Subsequently, the filters were extracted and the extracts analyzed for their drug content. For D-amphetamine, no correlation between breath level and dose could be observed ($n = 9$). On the other hand, a correlation coefficient of $r = 0.84$ was found between methylphenidate levels in exhaled breath and dose ($n = 8$).

Furthermore, direct breath analysis using extractive electrospray ionization (EESI) has been employed for the monitoring of VPA [209]. After VPA intake, extra peaks were observed at m/z 143 and 160, which the authors attributed to the presence of respectively 4-OH-VPA-γ-lactone (m/z 143), which was used as a biomarker of VPA exposure, and its noncovalent ammonium complex (m/z 160). When the EESI signal at m/z 143 was plotted against the free VPA concentration in blood, a good correlation coefficient ($r^2 = 0.89$) was observed ($n = 6$). On the other hand, a different group reported on the use of 3-heptanone as a biomarker of VPA exposure in exhaled breath [210]. However, despite 3-heptanone levels being elevated in all VPA-treated patients, these correlated with neither VPA serum concentrations nor VPA dose. Because 3-heptanone occurs naturally in exhaled breath, a possible explanation for this observation may be that other factors also affect the breath 3-heptanone level [209].

A completely different strategy to assess therapy adherence using a breath test was developed by Morey et al. [211]. Specifically, these authors proposed to coformulate the active compound of a drug with a taggant, which after systemic uptake is metabolized into a volatile adherence marker that can be measured using a portable, miniature gas chromatograph. In this context, it was demonstrated that either 2-butyl acetate or 2-pentyl acetate added to a vaginal gel leads to sufficient levels of the esters and corresponding alcohols and ketones in exhaled breath to assess vaginal gel adherence. Similarly, 2-butanol was used as a taggant to assess compliance to oral therapy [212]. The taggant 2-butanol was metabolized to 2-butanone and appeared in exhaled breath of all participants ($N = 8$) soon after capsule intake. The concentration measured after 10, 15, 20, or 30 min yielded an area under the curve of 1 (with a 95% confidence interval of 1.00-1.00) using a cutoff of 9.7, 5.4, 5.2, or 3.7 parts per billion, respectively.

13.12 ALTERNATIVE SPECIMEN: SWEAT

The first report of therapeutic drugs being detected in sweat dates back to the 19th century, when quinine excretion via perspiration was observed [213]. However, a patch for more convenient sweat collection was not developed until 1980 [214]. Because these patches were occlusive and led to skin irritation, local pH changes, and alterations in colonizing bacteria, they have more recently been replaced by nonocclusive patches, which allow the diffusion of water vapor, oxygen, and CO_2. These patches can be left on for more than 1 week, assessing cumulative drug exposure. Alternatively, swabs or wipes can be employed for sweat collection, providing information about more recent drug exposure (<24 h) [215].

Drugs are mainly incorporated into sweat via diffusion from the capillary system surrounding the sweat glands. Furthermore, transdermal migration of drugs and sebum excretion also contribute to the measured drug concentration in sweat. Because these processes favor more lipophilic components, mostly the parent compound is detected. In addition, alkaline drugs can accumulate in sweat due to its acidic pH (mean pH of 6.3) [215]. Because eccrine and apocrine sweat glands as well as sebum glands are unequally distributed throughout the body, sweat drug concentrations are dependent on sample location. Furthermore, these concentrations can be affected by sweat flow, which in turn depends on emotional status, ambient temperature, and physical activity. This entails that the collected sweat volume is unknown, impeding quantitative analysis. Moreover, problems with drug stability in the sweat patch and the risk of accidental patch contamination upon patch application or removal further complicate interpretation of sweat-based results [215]. Because drugs are also excreted via fingertips [216], proper precautions should be taken to avoid contamination during manual sweat patch manipulation. Due to the high inter- and intraindividual variability in sweat drug concentrations, sweat testing can only provide qualitative results [215]. This implies that only total noncompliance could be detected using this sampling method.

13.12.1 Application of Sweat for Therapeutic Drug Monitoring

The antiepileptic drugs phenytoin and phenobarbitone have been detected in sweat [217]. For phenytoin, a good correlation with the free plasma concentration was reported, whereas phenobarbitone concentrations seemed to depend on sweat flow. Furthermore, clozapine quantitation in sweat patches has been investigated for compliance monitoring in schizophrenic patients [218]. However, although a correlation coefficient of 0.589 was found between clozapine sweat concentrations and administered dose, large interindividual variations were noted, indicating that sweat is not suitable to assess patient compliance.

Sweat-based noninvasive compliance monitoring in ADHD patients has been evaluated by Marchei et al. Methylphenidate was found in 8-h post-dose sweat patches ($n = 3$), whereas ritalinic acid could not be determined. However, a fairly large interindividual variability in methylphenidate concentrations in patients receiving the same dose was reported [219]. Furthermore, it was estimated that only 0.08% of the administered methylphenidate dose was excreted via sweat during a 24-h period, indicating the necessity of a sensitive analytical technique [236]. In a second set of experiments, the determination of the norepinephrine reuptake inhibitor atomoxetine and its metabolites 4-OH-atomoxetine and *N*-desmethylatomoxetine in sweat was assessed [237]. Again, only the parent compound could be quantified in sweat, whereas both metabolites could not be detected [220]. This method was applied to six pediatric patients in a follow-up study, and peak atomoxetine concentrations appeared not to correlate with dose or with patient's body surface [221].

13.13 ALTERNATIVE SPECIMEN: NASAL MUCUS

A study in which nasal mucus was used to monitor oral drug therapy has been published in the scientific literature [222]. This alternative approach was suggested to be of value for the assessment of drug efficacy and toxicity during hyposmia treatment with theophylline. To this end, nasal discharge was directly and noninvasively collected in a plastic tube, and theophylline levels in nasal mucus and serum were compared. A Pearson correlation coefficient of 0.63 was found, and nasal theophylline concentrations varied between 53% and 97% of the corresponding serum concentrations.

13.14 TOWARD ROUTINE IMPLEMENTATION

Although the use of certain alternative sampling strategies may be appealing for TDM applications, routine implementation is still limited due to some practical and technical challenges (Velghe et al., submitted for publication). First, method development typically takes longer when using alternative matrices because additional variables need to be assessed [58]. Unfortunately, these variables are not included in standard validation guidelines. Therefore, white papers covering matrix-specific best practice guidelines have been published for certain matrices (although not yet for all) [58]. Another potential limitation for the setup of alternative TDM methods might be the availability of sufficient matrix [20]. However, if commutability is verified with spiked buffer solutions, the latter might be employed instead.

Furthermore, matrix-specific issues (eg, the Hct effect in DBS analysis and the influence of salivary pH in OF analysis) exist, which can lead to erroneous results and complicate data interpretation. However, as discussed previously, multiple solutions have been suggested to cope with these issues. Another practical hurdle is the fact that existing therapeutic intervals are generally established using serum or plasma. Thus, either new intervals need to be set up for the specific alternative matrix or bridging studies have to be conducted to correlate drug levels in both matrices.

Regrettably, thorough clinical validation is often lacking because generally only a limited number of samples or even only spiked samples have been included in these studies. Moreover, the correlation between drug levels in the alternative and systemic matrix is often only evaluated in healthy volunteers, whereas this may be significantly different in the target population. In addition, the effect (and potential benefit) of the use of an alternative matrix on the clinical decision process is only rarely evaluated [52]. Also, studies on patient outcome and economic impact are limited at best. In this context, however, Gorodischer et al. calculated that OF-based TDM could be less expensive than traditional TDM due to less nursing time, fewer visits to doctors' offices or hospitals, and the requirement of fewer sampling materials [144].

Because the analysis of alternative matrices often includes multiple manual steps, automation is required in a routine setting to reduce hands-on time and to increase turnaround time and throughput. In addition, automation excludes human errors and improves laboratory safety. In DBS analysis, for example, the tedious punching step can be replaced by (semi)automated punching devices, and sample preparation as well as the generation and spotting of calibrators and quality controls (QCs) can be automated using readily available liquid handling systems [223,224]. Furthermore, completely automated DBS/DPS analyzers have become commercially available [225–227]. These can either be used as automated sample preparation devices or be directly coupled to standard LC-MS/MS configurations. Also, for the analysis of other alternative matrices, such as OF, commercially available workstations can be employed to automate laborious sample pre-treatment steps as much as possible [228].

Moreover, setting up a QC program might be particularly challenging for alternative matrices. QC materials in native matrix (or in an adequate surrogate matrix) are often not commercially available. Although standard kits exist for TDM purposes (eg, for the determination of immunosuppressants via LC-MS/MS [229]), these are not necessarily suitable for methods using alternative matrices. For DBS methods, for example, the calibrator and QC materials should also have the same viscosity as true blood to have similar spreading properties. The development of more matrix-specific kits, encompassing calibrators and QCs in a suitable matrix as well as, for example, IS, extraction solvents, and mobile phases, would hence be favorable. Importantly, some alternative matrices also have special QC requirements, which may complicate quality assurance programs. For DBS, for example, it is advisable to include different Hct levels [58], whereas for hair analysis it is important to include different hair types [230]. In addition, external quality assurance schemes typically do not circulate specimens in alternative matrices. However, initiatives exist to set up such schemes. The Association for Quality Assessment in Therapeutic Drug Monitoring and Clinical Toxicology, for example, is setting up a pilot proficiency testing (PT) program for TDM of immunosuppressive drugs in DBS [231]. An additional complicating factor that needs to be taken into account in PT programs for alternative matrices is the different types of substrates that are used to collect samples (eg, different types of filter paper cards in DBS sampling and different collection devices in OF sampling). Therefore, to facilitate these types of programs, harmonization/standardization will be important in the future.

Not only the analytical method but also the patient samples have to be of sufficient quality. A first step toward achieving this is guaranteeing the quality of the substrate on which the sample is collected. For DBS, the quality criteria have been recorded in the Clinical and Laboratory Standards Institute guideline NBS01-A6 [232]. The latter has greatly contributed to the reduction of batch-to-batch variability in filter paper and hence method imprecision. Furthermore, the quality of a DBS is also evaluated before analysis, either by experienced laboratory personnel or by optical scanning instruments [233]. Such instruments objectively evaluate spot size, symmetry, and uniformity to distinguish acceptable, marginally acceptable, and unacceptable DBS. Furthermore, it needs to be taken into account that in the case of home sampling there is no control of (timing of) sample collection and storage conditions. Although the collection and proper storage of these samples is not difficult, quality issues with patient samples may pose problems. Therefore, independent of the type of matrix that is collected, sample quality in the case of home sampling can be maximized only via proper patient education using, for example, demonstration folders and movies.

13.15 CONCLUSIONS

The use of alternative sampling strategies clearly provides valuable opportunities for TDM. Particularly in special patient populations such as neonates, children, cystic fibrosis patients, and patients with poor venous access, alternative sampling strategies can be of great added value. Moreover, alternative sampling strategies may play a pivotal role in the setup of TDM programs in remote and resource-limited settings, where traditional sampling approaches are typically not applicable due to logistical and economic issues. In addition, the ease of collection and transport of certain alternative matrices facilitates home-based TDM, in which samples are collected by the patient or his or her caregiver and transported to the laboratory via regular mail services. The latter is not only more convenient for patients and their caregivers but also allows samples to be collected when, for example, epileptic seizures or adverse effects occur, yielding information that could not be obtained by the clinician in any other way. On the other hand, each matrix also has specific limitations and challenges that have to be taken into account. However, recent (and future) developments may provide answers to some of these challenges.

Although alternative matrices will of course never replace classic blood samples in TDM, these alternative specimens may be extremely valuable whenever they can offer (additional) information that could not be (easily) obtained using traditional approaches. Furthermore, certain alternative matrices can also be used for cytochrome P450 genotyping and phenotyping, as well as for the follow-up of disease-specific biomarkers, thereby further contributing to patient treatment optimization. Importantly, more studies are needed in which a proper clinical validation of alternative sampling strategies is performed to allow a more widespread implementation of these strategies in routine TDM programs. Moreover, the effect on clinical decision making, patient outcome, and medical costs needs to be thoroughly evaluated. Only when this results in a positive balance can a given alternative sampling strategy become routinely implemented. In addition, more proficiency programs will need to be set up to ascertain that the used methods are of adequate quality.

Acknowledgment

S. Capiau acknowledges the FWO Research Foundation—Flanders for granting her a Ph.D. fellowship.

References

[1] Nunes LA, Mussavira S, Bindhu OS. Clinical and diagnostic utility of saliva as a non-invasive diagnostic fluid: a systematic review. Biochem Med 2015;25(2):177–92.

[2] Wester VL, van Rossum EF. Clinical applications of cortisol measurements in hair. Eur J Endocrinol 2015;173(4):1–10.

[3] Lakshmy R, Tarik M, Abraham RA. Role of dried blood spots in health and disease diagnosis in older adults. Bioanalysis 2014;6(23):3121–31.

[4] Lehmann S, Delaby C, Vialaret J, Ducos J, Hirtz C. Current and future use of "dried blood spot" analyses in clinical chemistry. Clin Chem Lab Med 2013;51(10):1897–909.

[5] la Marca G. Mass spectrometry in clinical chemistry: the case of newborn screening. J Pharm Biomed Anal 2014;101:174–82.

[6] Sieg A, Wascotte V. Diagnostic and therapeutic applications of iontophoresis. J Drug Target 2009;17(9):690–700.

[7] Edelbroek PM, van der Heijden J, Stolk LM. Dried blood spot methods in therapeutic drug monitoring: methods, assays, and pitfalls. Ther Drug Monit 2009;31(3):327–36.

[8] Wilhelm AJ, den Burger JC, Swart EL. Therapeutic drug monitoring by dried blood spot: progress to date and future directions. Clin Pharmacokinet 2014;53(11):961–73.

[9] Patsalos PN, Berry DJ. Therapeutic drug monitoring of antiepileptic drugs by use of saliva. Ther Drug Monit 2013;35(1):4–29.

[10] Langman LJ. The use of oral fluid for therapeutic drug management-clinical and forensic toxicology. Oral Based Diagn 2007;1098:145–66.

[11] Edelbroek PM. Therapeutic drug monitoring van anti-epileptica. Pharm Weekbl 2002;137 (14):519–24.

[12] Pandya HC, Spooner N, Mulla H. Dried blood spots, pharmacokinetic studies and better medicines for children. Bioanalysis 2011;3(7):779–86.

[13] ter Heine R, Beijnen JH, Huitema AD. Bioanalytical issues in patient-friendly sampling methods for therapeutic drug monitoring: focus on antiretroviral drugs. Bioanalysis 2009;1 (7):1329–38.

[14] Taneja I, Erukala M, Raju KS, Singh SP, Wahajuddin M. Dried blood spots in bioanalysis of antimalarials: relevance and challenges in quantitative assessment of antimalarial drugs. Bioanalysis 2013;5(17):2171–86.

[15] Sotgiu G, Alffenaar JW, Centis R, D'Ambrosio L, Spanevello A, Piana A, et al. Therapeutic drug monitoring: how to improve drug dosage and patient safety in tuberculosis treatment. Int J Infect Dis 2015;32:101–4.

[16] Wijnen PA, Op den Buijsch RA, Cheung SC, van der Heijden J, Hoogtanders K, Stolk LM, et al. Genotyping with a dried blood spot method: a useful technique for application in pharmacogenetics. Clin Chim Acta 2008;388(1–2):189–91.

[17] De Kesel PM, Lambert WE, Stove CP. Alternative sampling strategies for cytochrome P450 phenotyping. Clin Pharmacokinet 2015; Epub ahead of print.

[18] Kiang TK, Ensom MH. A qualitative review on the pharmacokinetics of antibiotics in saliva: implications on clinical pharmacokinetic monitoring in humans. Clin Pharmacokinet 2015; Epub ahead of print.

[19] Nair AB, Goel A, Prakash S, Kumar A. Therapeutic drug monitoring by reverse iontophoresis. J Basic Clin Pharm 2011;3(1):207–13.

[20] Mullangi R, Agrawal S, Srinivas NR. Measurement of xenobiotics in saliva: is saliva an attractive alternative matrix? Case studies and analytical perspectives. Biomed Chromatogr 2009;23(1):3–25.

[21] Cone EJ. Legal, workplace, and treatment drug testing with alternate biological matrices on a global scale. Forensic Sci Int 2001;121(1–2):7–15.

[22] Thavendiranathan P, Bagai A, Ebidia A, Detsky AS, Choudhry NK. Do blood tests cause anemia in hospitalized patients? J Gen Intern Med 2005;20(6):520–4.

[23] Dasgupta A. Clinical utility of free drug monitoring. Clin Chem Lab Med 2002;40(10):986–93.

[24] Capiau S, Stove VV, Lambert WE, Stove CP. Prediction of the hematocrit of dried blood spots via potassium measurement on a routine clinical chemistry analyzer. Anal Chem 2013;35(5) 659–659.

[25] De Kesel PM, Capiau S, Stove VV, Lambert WE, Stove CP. Potassium-based algorithm allows correction for the hematocrit bias in quantitative analysis of caffeine and its major metabolite in dried blood spots. Anal Bioanal Chem 2014;406(26):6749–55.

[26] Tanna S, Lawson G. Self-sampling and quantitative analysis of DBS: can it shift the balance in over-burdened healthcare systems? Bioanalysis 2015;7(16):1963–6.

[27] Apotheek Haagse Ziekenhuizen en HagaZiekenhuis ontwikkelen patiëntvriendelijke methode voor bloedcontrole (consulted via <https://www.hagaziekenhuis.nl/over-hagaziekenhuis/actueel/nieuws/2009/apotheek-haagse-ziekenhuizen-en-hagaziekenhuis-ontwikkelen-pati%C3%ABntvriendelijke-methode-voor-bloedcontrole.aspx>). (27.11.15).

[28] Patteet L, Maudens KE, Stove CP, Lambert WE, Morrens M, Sabbe B, et al. Are capillary DBS applicable for therapeutic drug monitoring of common antipsychotics? A proof of concept. Bioanalysis 2015;7(16):2119–30.

[29] Tron C.J.M., Wijma R.A., van der Nagel B.C.H., Dierckx B., Verhulst F.C., Dieleman G.C. et al. Application of dried blood spots combined with ultra-high performance liquid chromatography-mass spectrometry for the identification and quantification of the antipsychotics risperidone, aripiprazole, pipamperone and their major metabolites. Presented at the 14th IATDMCT meeting, Rotterdam, The Netherlands, 2015.

[30] Tanna S, Lawson G. Dried blood spot analysis to assess medication adherence and to inform personalization of treatment. Bioanalysis 2014;6(21):2825–38.

[31] Koster RA, Greijdanus B, Alffenaar JW, Touw DJ. Dried blood spot analysis of creatinine with LC-MS/MS in addition to immunosuppressants analysis. Anal Bioanal Chem 2015;407(6):1585–94.

[32] Koop DR, Bleyle LA, Munar M, Cherala G, Al-Uzri A. Analysis of tacrolimus and creatinine from a single dried blood spot using liquid chromatography tandem mass spectrometry. J Chromatogr B 2013;926:54–61.

[33] Scherf-Clavel M, Högger P. Analysis of metformin, sitagliptin and creatinine in human dried blood spots. J Chromatogr B 2015;997:218–28.

[34] Mess JN, Taillon MP, Côté C, Garofolo F. Dried blood spot on-card derivatization: an alternative form of sample handling to overcome the instability of thiorphan in biological matrix. Biomed Chromatogr 2012;26(12):1617–24.

[35] Mercolini L, Mandrioli R, Protti M, Conca A, Albers LJ, Raggi MA. Dried blood spot testing: a novel approach for the therapeutic drug monitoring of ziprasidone-treated patients. Bioanalysis 2014;6(11):1487–95.

[36] De Kesel PM, Sadones N, Capiau S, Lambert WE, Stove CP. Hemato-critical issues in quantitative analysis of dried blood spots: challenges and solutions. Bioanalysis 2013;5(16):2023–41.

[37] De Kesel PM, Capiau S, Lambert WE, Stove CP. Current strategies for coping with the hematocrit problem in dried blood spot analysis. Bioanalysis 2014;6(14):1871–4.

[38] Lenk G, Sandkvist S, Pohanka A, Stemme G, Beck O, Roxhed N. A disposable sampling device to collect volume-measured DBS directly from a fingerprick onto DBS paper. Bioanalysis 2015;7(16):2085–94.

[39] Leuthold LA, Heudi O, Déglon J, Raccuglia M, Augsburger M, Picard F, et al. New microfluidic-based sampling procedure for overcoming the hematocrit problem associated with dried blood spot analysis. Anal Chem 2015;87(4):2068–71.

[40] Denniff P, Spooner N. Volumetric absorptive microsampling: a dried sample collection technique for quantitative bioanalysis. Anal Chem 2014;86(16):8489–95.

[41] Mengerink Y, Mommers J, Qiu J, Mengerink J, Steijger O, Honing M. A new DBS card with spot sizes independent of the hematocrit value of blood. Bioanalysis 2015;7 (16):2095–104.

[42] Li Y, Henion J, Abbott R, Wang P. The use of a membrane filtration device to form dried plasma spots for the quantitative determination of guanfacine in whole blood. Rapid Commun Mass Spectrom 2012;26(10):1208–12.

[43] Sturm R, Henion J, Abbott R, Wang P. Novel membrane devices and their potential utility in blood sample collection prior to analysis of dried plasma spots. Bioanalysis 2015;7 (16):1987–2002.

[44] Kim JH, Woenker T, Adamec J, Regnier FE. Simple, miniaturized blood plasma extraction method. Anal Chem 2013;85(23):11501–8.

[45] Hempen CM, Maarten Koster EH, Ooms JA. Hematocrit-independent recovery of immunosuppressants from DBS using heated flow-through desorption. Bioanalysis 2015;7 (16):2018–29.

[46] van der Heijden J, de Beer Y, Hoogtanders K, Christiaans M, de Jong GJ, Neef C, et al. Therapeutic drug monitoring of everolimus using the dried blood spot method in combination with liquid chromatography-mass spectrometry. J Pharm Biomed Anal 2009;50(4):664–70.

[47] Blessborn D, Römsing S, Annerberg A, Sundquist D, Björkman A, Lindegardh N, et al. Development and validation of an automated solid-phase extraction and liquid chromatographic method for determination of lumefantrine in capillary blood on sampling paper. J Pharm Biomed Anal 2007;45(2):282–7.

[48] Abu-Rabie P, Denniff P, Spooner N, Chowdhry BZ, Pullen FS. Investigation of different approaches to incorporating internal standard in dbs quantitative bioanalytical workflows and their effect on nullifying hematocrit-based assay bias. Anal Chem 2015; 87(9):4996–5003.

[49] Mommers J, Mengerink Y, Ritzen E, Weusten J, van der Heijden J, van der Wal S. Quantitative analysis of morphine in dried blood spots by using morphine-d3 pre-impregnated dried blood spot cards. Anal Chim Acta 2013;774:26–32.

[50] Rowland M, Emmons GT. Use of dried blood spots in drug development: pharmacokinetic considerations. AAPS J 2010;12(3):290–3.

[51] Rhoden L, Antunes MV, Hidalgo P, Álvares da Silva C, Linden R. Simple procedure for determination of valproic acid in dried blood spots by gas chromatography-mass spectrometry. J Pharm Biomed Anal 2014;96:207–12.

[52] Jager NG, Rosing H, Schellens JH, Beijnen JH, Linn SC. Use of dried blood spots for the determination of serum concentrations of tamoxifen and endoxifen. Breast Cancer Res Treat 2014;146(1):137–44.

[53] Shank RP, Doose DR, Streeter AJ, Bialer M. Plasma and whole blood pharmacokinetics of topiramate: the role of carbonic anhydrase. Epilepsy Res 2005;63(2–3):103–12.

[54] Mohammed BS, Cameron GA, Cameron L, Hawksworth GH, Helms PJ, McLay JS. Can finger-prick sampling replace venous sampling to determine the pharmacokinetic profile of oral paracetamol? Br J Clin Pharmacol 2010;70(1):52–6.

[55] Murphy JE, Peltier T, Anderson D, Ward ES. A comparison of venous versus capillary measurement of drug concentration. Ther Drug Monit 1990;12(3):264–7.

[56] Instiaty I, Lindegardh N, Jittmala P, Hanpithakpong W, Blessborn D, Pukrittayakamee S, et al. Comparison of oseltamivir and oseltamivir carboxylate concentrations in venous plasma, venous blood, and capillary blood in healthy volunteers. Antimicrob Agents Chemother 2013;57(6):2858–62.

[57] Ashley EA, Stepniewska K, Lindegardh N, Annerberg A, Tarning J, McGready R, et al. Comparison of plasma, venous and capillary blood levels of piperaquine in patients with uncomplicated falciparum malaria. Eur J Clin Pharmacol 2010;66(7):705–12.

[58] Timmerman P, White S, Globig S, Lüdtke S, Brunet L, Smeraglia J. EBF recommendation on the validation of bioanalytical methods for dried blood spots. Bioanalysis 2011;3 (14):1567–75.

[59] Ingels AS, Hertegonne KB, Lambert WE, Stove CP. Feasibility of following up gamma-hydroxybutyric acid concentrations in sodium oxybate (xyrem (r))-treated narcoleptic patients using dried blood spot sampling at home an exploratory Study. CNS Drugs 2013;27(3):233–7.

[60] Stove CP, Ingels AS, De Kesel PM, Lambert WE. Dried blood spots in toxicology: from the cradle to the grave? Crit Rev Toxicol 2012;42(3):230–43.

[61] Ren X, Paehler T, Zimmer M, Guo Z, Zane P, Emmons GT. Impact of various factors on radioactivity distribution in different DBS papers. Bioanalysis 2010;2(8):1469–75.

[62] Jager NG, Rosing H, Schellens JH, Beijnen JH. Determination of tamoxifen and endoxifen in dried blood spots using LC-MS/MS and the effect of coated DBS cards on recovery and matrix effects. Bioanalysis 2014;6(22):2999–3009.

[63] Chen X, Zhao H, Hatsis P, Amin J. Investigation of dried blood spot card-induced interferences in liquid chromatography/mass spectrometry. J Pharm Biomed Anal 2012;61:30–7.

[64] Luckwell J, Denniff P, Capper S, Michael P, Spooner N, Mallender P, et al. Assessment of the within- and between-lot variability of Whatman (TM) FTA (R) DMPK and 903 (R) DBS papers and their suitability for the quantitative bioanalysis of small molecules. Bioanalysis 2013;5(21):2613–30.

[65] Tanna S, Lawson G. Analytical methods used in conjunction with dried blood spots. Anal Methods 2011;3(8):1709–18.

[66] Shi Y, Jiang H. Assay dynamic range for DBS: battles on two fronts. Bioanalysis 2011;3 (20):2259–62.

[67] Liu G, Snapp HM, Ji QC. Internal standard tracked dilution to overcome challenges in dried blood spots and robotic sample preparation for liquid chromatography/tandem mass spectrometry assays. Rapid Commun Mass Spectrom 2011;25(9):1250–6.

[68] Mannu RS, Turpin PE, Goodwin L. Alternative strategies for mass spectrometer-based sample dilution of bioanalytical samples, with particular reference to DBS and plasma analysis. Bioanalysis 2014;6(6):773–84.

[69] Castillo-Mancilla JR, Zheng JH, Rower JE, Meditz A, Gardner EM, Predhomme J, et al. Tenofovir, emtricitabine, and tenofovir diphosphate in dried blood spots for determining recent and cumulative drug exposure. AIDS Res Hum Retroviruses 2013;29(2):384–90.

[70] Blessborn D, Sköld K, Zeeberg D, Kaewkhao K, Sköld O, Ahnoff M. Heat stabilization of blood spot samples for determination of metabolically unstable drug compounds. Bioanalysis 2013;5(1):31–9.

[71] Verweij-van Wissen CP, de Graaff-Teulen MJ, de Kanter CT, Aarnoutse RE, Burger DM. Determination of the HCV protease inhibitor telaprevir in plasma and dried blood spot by liquid chromatography-tandem mass spectrometry. Ther Drug Monit 2015;37(5):626–33.

[72] Albani M, Toseland PA. Simple rapid gas-chromatography method for determination of theophylline in dried whole-blood on filter-paper cards. Neuropadiatrie 1978;9(1):97–9.

[73] Hofman S, Bolhuis MS, Koster RA, Akkerman OW, van Assen S, Stove C, et al. Role of therapeutic drug monitoring in pulmonary infections: use and potential for expanded use of dried blood spot samples. Bioanalysis 2015;7(4):481–95.

[74] Shah NM, Hawwa AF, Millership JS, Collier PS, Ho P, Tan ML, et al. Adherence to antiepileptic medicines in children: a multiple-methods assessment involving dried blood spot sampling. Epilepsia 2013;54(6):1020–7.

[75] Wegner I, Edelbroek P, de Haan GJ, Lindhout D, Sander JW. Drug monitoring of lamotrigine and oxcarbazepine combination during pregnancy. Epilepsia 2010;51(12):2500–2.

[76] Lawson G, Cocks E, Tanna S. Quantitative determination of atenolol in dried blood spot samples by LC-HRMS: a potential method for assessing medication adherence. J Chromatogr B 2012;897:72–9.

[77] Lawson G, Cocks E, Tanna S. Bisoprolol, ramipril and simvastatin determination in dried blood spot samples using LC-HRMS for assessing medication adherence. J Pharm Biomed Anal 2013;81–82:99–107.

[78] Patteet L, Maudens KE, Stove CP, Lambert WE, Morrens M, Sabbe B, et al. The use of dried blood spots for quantification of 15 antipsychotics and 7 metabolites with ultra-high performance liquid chromatography-tandem mass spectrometry. Drug Test Anal 2015;7(6):502–11.

[79] Salimo AT, Ledwaba J, Coovadia A, Abrams EJ, Technau KG, Kuhn L, et al. The use of dried blood spot specimens for HIV-1 drug resistance genotyping in young children initiating antiretroviral therapy. J Virol Methods 2015;223:30–2.

[80] Jimmerson LC, Zheng JH, Bushman LR, MacBrayne CE, Anderson PL, Kiser JJ. Development and validation of a dried blood spot assay for the quantification of ribavirin using liquid chromatography coupled to mass spectrometry. J Chromatogr B 2014;944:18–24.

[81] Hatami M, Karimnia E, Farhadi K. Determination of salmeterol in dried blood spot using an ionic liquid based dispersive liquid-liquid microextraction coupled with HPLC. J Pharm Biomed Anal 2013;85:283–7.

[82] Leichtle AB, Ceglarek U, Witzigmann H, Gäbel G, Thiery J, Fiedler GM. Potential of dried blood self-sampling for cyclosporine c(2) monitoring in transplant outpatients. J Transplant 2010;2010:201918.

[83] Cheung CY, van der Heijden J, Hoogtanders K, Christiaans M, Liu YL, Chan YH, et al. Dried blood spot measurement: application in tacrolimus monitoring using limited sampling strategy and abbreviated AUC estimation. Transpl Int 2008;21(2):140–5.

[84] Ansari M, Uppugunduri CR, Déglon J, Théorêt Y, Versace F, Gumy-Pause F, et al. A simplified method for busulfan monitoring using dried blood spot in combination with liquid chromatography/tandem mass spectrometry. Rapid Commun Mass Spectrom 2012;26(12):1437–46.

[85] Antunes MV, Raymundo S, de Oliveira V, Staudt DE, Gössling G, Peteffi GP, et al. Ultra-high performance liquid chromatography tandem mass spectrometric method for the determination of tamoxifen, N-desmethyltamoxifen, 4-hydroxytamoxifen and endoxifen in dried blood spots—Development, validation and clinical application during breast cancer adjuvant therapy. Talanta 2015;132:775–84.

[86] Contreras Zavala L, Rivera Espinosa L, Ángeles Moreno AP, Ramírez San-Juan E, Zamudio Hernández S, García González A, et al. Development of a UPLC-MS/MS method for the quantification of sildenafil by DBS, and its use on pediatric pulmonary hypertension. Bioanalysis 2014;6(21):2815–24.

[87] Patel P, Mulla H, Kairamkonda V, Spooner N, Gade S, Della Pasqua O, et al. Dried blood spots and sparse sampling: a practical approach to estimating pharmacokinetic parameters of caffeine in preterm infants. Br J Clin Pharmacol 2013;75(3):805–13.

[88] Nirogi R, Ajjala DR, Kandikere V, Aleti R, Srikakolapu S, Vurimindi H. Dried blood spot analysis of an iron chelator—Deferasirox and its potential application to therapeutic drug monitoring. J Chromatogra B 2012;907:65–73.

[89] la Marca G, Malvagia S, Materazzi S, Della Bona ML, Boenzi S, Martinelli D, et al. LC-MS/MS method for simultaneous determination on a dried blood spot of multiple analytes relevant for treatment monitoring in patients with tyrosinemia Type I. Anal Chem 2012;84(2):1184–8.

[90] Rattenbury JM, Tsanakas J. Acceptance of domiciliary theophylline monitoring using dried blood spots. Arch Dis Child 1988;63(12):1449–52.

[91] Vu DH, Alffenaar JW, Edelbroek PM, Brouwers JR, Uges DR. Dried blood spots: a New tool for tuberculosis treatment optimization. Curr Pharm Des 2011;17(27):2931–9.

[92] Alffenaar JW. Dried blood spot analysis combined with limited sampling models can advance therapeutic drug monitoring of tuberculosis drugs. J Infectious Diseases 2012;205(11):1765–6.

[93] Bartelink IH, Savic RM, Mwesigwa J, Achan J, Clark T, Plenty A, et al. Pharmacokinetics of Lopinavir/Ritonavir and Efavirenz in food insecure HIV-infected pregnant and breastfeeding women in Tororo, Uganda. J Clin Pharmacol 2014;54(2):121–32.

[94] van Boekel GA, Aarnoutse RE, Hoogtanders KE, Havenith TR, Hilbrands LB. Delayed trough level measurement with the use of prolonged-release tacrolimus. Transpl Int 2015;28(3):314–18.

[95] Rauh M, Stachel D, Kuhlen M, Gröschl M, Holter W, Rascher W. Quantification of busulfan in saliva and plasma in haematopoietic stem cell transplantation in children. Clin Pharmacokinet 2006;45(3):305–16.

[96] Stifft F, Stolk LM, Undre N, van Hooff JP, Christiaans MH. Lower variability in 24-hour exposure during once-daily compared to twice-daily tacrolimus formulation in kidney transplantation. Transplantation 2014;97(7):775–80.

[97] Jones A, Beisty J, McKenna D, Clough D, Webb K, Morris J, et al. Monitoring of tobramycin levels in patients with cystic fibrosis by finger-prick sampling. Eur Respir J 2012;39(6):1537–9.

[98] Struthers SL, Nicholson T, Jones G, Connett GJ. Falsely elevated serum tobramycin levels in a patient receiving nebulised tobramycin. J Cyst Fibros 2002;1(3):146–7.

[99] De Kesel PM, Lambert WE, Stove CP. Does volumetric absorptive microsampling eliminate the hematocrit bias for caffeine and paraxanthine in dried blood samples? A comparative study. Anal Chim Acta 2015;881:65–73.

[100] Hill J., Taylor R., Clear C., Hutchinson K. An Improved Platform for the Recovery and Analysis of Cannabinoids from dried blood samples. Presented at the 7th annual MSACL US conference, San Diego, US. 2015.

[101] Christianson C.D., Johnson C.J., Sheaff C.N., Laine D.F., Zimmer J.S.D., Needham S.R. Dried matrix spot analysis—a novel dye to analyze drugs from clear fluids: tobramycin from tears. Poster consulted at <http://www.alturasanalytics.com/media/1209/alturas-analytics-dried-matrix-spot-tears-apa-poster-2010.pdf> (27.11.15).

[102] Abdel-Rehim A, Abdel-Rehim M. Dried saliva spot as a sampling technique for saliva samples. Biomed Chromatogr 2014;28(6):875–7.

[103] Zimmer JS, Christianson CD, Johnson CJ, Needham SR. Recent advances in the bioanalytical applications of dried matrix spotting for the analysis of drugs and their metabolites. Bioanalysis 2013;5(20):2581–8.

[104] Crawford E, Gordon J, Wu JT, Musselman B, Liu R, Yu S. Direct analysis in real time coupled with dried spot sampling for bioanalysis in a drug-discovery setting. Bioanalysis 2011;3(11):1217–26.

[105] Wiseman JM, Evans CA, Bowen CL, Kennedy JH. Direct analysis of dried blood spots utilizing desorption electrospray ionization (DESI) mass spectrometry. Analyst 2010;135 (4):720–5.

[106] Manicke NE, Abu-Rabie P, Spooner N, Ouyang Z, Cooks RG. Quantitative analysis of therapeutic drugs in dried blood spot samples by paper spray mass spectrometry: an avenue to therapeutic drug monitoring. J Am Soc Mass Spectrom 2011;22(9):1501–7.

[107] Lin CH, Liao WC, Chen HK, Kuo TY. Paper spray-MS for bioanalysis. Bioanalysis 2014;6 (2):199–208.

[108] Espy RD, Manicke NE, Ouyang Z, Cooks RG. Rapid analysis of whole blood by paper spray mass spectrometry for point-of-care therapeutic drug monitoring. Analyst 2012;137 (10):2344–9.

[109] Zhang Z, Xu W, Manicke NE, Cooks RG, Ouyang Z. Silica coated paper substrate for paper-spray analysis of therapeutic drugs in dried blood spots. Anal Chem 2012;84 (2):931–8.

[110] Shen L, Zhang J, Yang Q, Manicke NE, Ouyang Z. High throughput paper spray mass spectrometry analysis. Clin Chim Acta 2013;420:28–33.

[111] Li L, Chen TC, Ren Y, Hendricks PI, Cooks RG, Ouyang Z. Mini 12, miniature mass spectrometer for clinical and other applications—introduction and characterization. Anal Chem 2014;86(6):2909–16.

[112] Shi RZ, El Gierari el TM, Manicke NE, Faix JD. Rapid measurement of tacrolimus in whole blood by paper spray-tandem mass spectrometry (PS-MS/MS). Clin Chim Acta 2015;441:99–104.

[113] Siebenhaar M, Küllmer K, Fernandes NM, Hüllen V, Hopf C. Personalized monitoring of therapeutic salicylic acid in dried blood spots using a three-layer setup and desorption electrospray ionization mass spectrometry. Anal Bioanal Chem 2015;407 (23):7229–38.

[114] Aps JK, Martens LC. Review: the physiology of saliva and transfer of drugs into saliva. Forensic Sci Int 2005;150(2–3):119–31.

[115] Gallardo E, Barroso M, Queiroz JA. Current technologies and considerations for drug bioanalysis in oral fluid. Bioanalysis 2009;1(3):637–67.

[116] Langel K, Engblom C, Pehrsson A, Gunnar T, Ariniemi K, Lillsunde P. Drug testing in oral fluid—Evaluation of sample collection devices. J Anal Toxicol 2008;32(6):393–401.

[117] Crouch DJ. Oral fluid collection: the neglected variable in oral fluid testing. Forensic Sci Int 2005;150(2–3):165–73.

[118] Lomonaco T, Ghimenti S, Piga I, Biagini D, Onor M, Fuoco R, et al. Influence of sampling on the determination of warfarin and warfarin alcohols in oral fluid. PLoS One 2014;9 (12):e114430.

[119] Thomson WM. Dry mouth and older people. Aust Dent J 2015;60:54–63.

[120] Wintergerst U, Kurowski M, Rolinski B, Müller M, Wolf E, Jaeger H, et al. Use of saliva specimens for monitoring indinavir therapy in human immunodeficiency virus-infected patients. Antimicrob Agents Chemother 2000;44(9):2572–4.

[121] Berkovitch M, Goldman M, Silverman R, Chen-Levi Z, Greenberg R, Marcus O, et al. Therapeutic drug monitoring of once daily gentamicin in serum and saliva of children. Eur J Pediatr 2000;159(9):697–8.

[122] Wiesen MH, Farowski F, Feldkötter M, Hoppe B, Müller C. Liquid chromatography-tandem mass spectrometry method for the quantification of mycophenolic acid and its phenolic glucuronide in saliva and plasma using a standardized saliva collection device. J Chromatogr A 2012;1241:52–9.

[123] McAuliffe JJ, Sherwin AL, Leppik IE, Fayle SA, Pippenger CE. Salivary levels of anticonvulsants—practical approach to drug monitoring. Neurology 1977;27(5):409–13.

[124] Gröschl M, Köhler H, Topf HG, Rupprecht T, Rauh M. Evaluation of saliva collection devices for the analysis of steroids, peptides and therapeutic drugs. J Pharm Biomed Anal 2008;47(3):478–86.

[125] Drummer OH. Introduction and review of collection techniques and applications of drug testing of oral fluid. Ther Drug Monit 2008;30(2):203–6.

[126] Chang K, Chiou WL. Interactions between drugs and saliva-stimulating parafilm and their implications in measurement of saliva drug levels. Res Commun Chem Pathol Pharmacol 1976;13(2):357–60.

[127] Mendonza AE, Gohh RY, Akhlaghi F. Analysis of mycophenolic acid in saliva using liquid chromatography tandem mass spectrometry. Ther Drug Monit 2006;28(3):402–6.

[128] Liu H, Delgado MR. Therapeutic drug concentration monitoring using saliva samples—Focus on anticonvulsants. Clin Pharmacokinet 1999;36(6):453–70.

[129] Bolhuis MS, van Altena R, van Hateren K, de Lange WC, Greijdanus B, Uges DR, et al. Clinical validation of the analysis of linezolid and clarithromycin in oral fluid of patients with multidrug-resistant tuberculosis. Antimicrob Agents Chemother 2013;57(8):3676–80.

[130] Das R, Agrawal YK. Simultaneous monitoring of selective serotonin reuptake inhibitors in human urine, plasma and oral fluid by reverse-phase high performance liquid chromatography. J Chromatogr Sci 2013;51(2):146–54.

[131] Herkes GK, Eadie MJ. Possible roles for frequent salivary antiepileptic drug monitoring in the management of epilepsy. Epilepsy Res 1990;6(2):146–54.

[132] Reynolds F, Ziroyanis PN, Jones NF, Smith SE. Salivary phenytoin concentrations in epilepsy and in chronic renal-failure. Lancet 1976;2(7982):384–6.

[133] Gbotosho GO, Happi CT, Lawal O, Sijuade A, Sowunmi A, Oduola A. A high performance liquid chromatographic assay of Mefloquine in saliva after a single oral dose in healthy adult Africans. Malar J 2012;11:59.

[134] Rault JP, Hasselot N, Renard C, Cheminel V, Chaulet JF. Unreliability of saliva samples for monitoring chloroquine and proguanil levels during anti-malarial chemoprophylaxis. Gen Pharmacol 1996;27(1):65–7.

[135] van der Elst KC, van Alst M, Lub-de Hooge MN, van Hateren K, Kosterink JG, Alffenaar JW, et al. Clinical validation of the analysis of fluconazole in oral fluid in hospitalized children. Antimicrob Agents Chemother 2014;58(11):6742–6.

[136] Koks CH, Crommentuyn KM, Hoetelmans RM, Mathôt RA, Beijnen JH. Can fluconazole concentrations in saliva be used for therapeutic drug monitoring? Ther Drug Monit 2001;23(4):449–53.

[137] Patteet L, Maudens KE, Morrens M, Sabbe B, Dom G, Neels H. Determination of common antipsychotics in QuantisalTM-collected oral fluid by UHPLC-MS/MS: method validation and applicability for therapeutic drug monitoring. Ther Drug Monit 2015; Epub ahead of print.

[138] Shetty SJ, Desai PB, Patil NM, Nayak RB. Relationship between serum lithium, salivary lithium, and urinary lithium in patients on lithium therapy. Biol Trace Elem Res 2012;147(1–3):59–62.

[139] Rosman AW, Sczupak CA, Pakes GE. Correlation between saliva and serum lithium levels in manic-depressive patients. Am J Hosp Pharm 1980;37(4):514–18.

[140] El-Mallakh RS, Linder M, Valdes R, Looney S. Dialysis of saliva improves accuracy of saliva lithium determinations. Bipolar Disord 2004;6(1):87–9.

[141] Lamorde M, Fillekes Q, Sigaloff K, Kityo C, Buzibye A, Kayiwa J, et al. Therapeutic drug monitoring of nevirapine in saliva in Uganda using high performance liquid chromatography and a low cost thin-layer chromatography technique. BMC Infect Dis 2014;14:473.

[142] George L, Muro EP, Ndaro A, Dolmans W, Burger DM, Kisanga ER. Nevirapine concentrations in saliva measured by thin layer chromatography and self-reported adherence in patients on antiretroviral therapy at Kilimanjaro Christian Medical Centre, Tanzania. Ther Drug Monit 2014;36(3):366–70.

[143] Mortier KA, Renard V, Verstraete AG, Van Gussem A, Van Belle S, Lambert WE. Development and validation of a liquid chromatography-tandem mass spectrometry assay for the quantification of docetaxel and paclitaxel in human plasma and oral fluid. Anal Chem 2005;77(14):4677–83.

[144] Gorodischer R, Burtin P, Hwang P, Levine M, Koren G. Saliva versus blood sampling for therapeutic drug monitoring in children—patient and parental preferences and an economic analysis. Ther Drug Monit 1994;16(5):437–43.

[145] de Wildt SN, Kerkvliet KT, Wezenberg MG, Ottink S, Hop WC, Vulto AG, et al. Use of saliva in therapeutic drug monitoring of caffeine in preterm infants. Ther Drug Monit 2001;23(3):250–4.

[146] Umstead GS, Morales M, McKercher PL. Comparison of total, free, and salivary phenytoin concentrations in geriatric patients. Clin Pharm 1986;5(1):59–62.

[147] Perucca E, Crema A. Plasma protein binding of drugs in pregnancy. Clin Pharmacokinet 1982;7(4):336–52.

[148] Patteet L, Cappelle D, Maudens KE, Crunelle CL, Sabbe B, Neels H. Advances in detection of antipsychotics in biological matrices. Clin Chim Acta 2015;441:11–22.

[149] Tennison M, Ali I, Miles MV, D'Cruz O, Vaughn B, Greenwood R. Feasibility and acceptance of salivary monitoring of antiepileptic drugs via the US postal service. Ther Drug Monit 2004;26(3):295–9.

[150] Jones MD, Ryan M, Miles MV, Tang PH, Fakhoury TA, Degrauw TJ, et al. Stability of salivary concentrations of the newer antiepileptic drugs in the postal system. Ther Drug Monit 2005;27(5):576–9.

[151] Madsen V, Lind A, Rasmussen M, Coulthard K. Determination of tobramycin in saliva is not suitable for therapeutic drug monitoring of patients with cystic fibrosis. J Cyst Fibrosis 2004;3(4):249–51.

[152] Spencer H, et al. Measurement of tobramycin and gentamicin in saliva is not suitable for therapeutic drug monitoring of patients with cystic fibrosis. J Cystic Fibros 2005;4(3):209.

[153] Posa A, Bräuer L, Schicht M, Garreis F, Beileke S, Paulsen F. Schirmer strip vs. capillary tube method: non-invasive methods of obtaining proteins from tear fluid. Anal Anat Anat Anz 2013;195(2):137–42.

[154] Vanhaeringen NJ. Clinical biochemistry of tears. Surv Ophthalmol 1981;26(2):84–96.

[155] Pichini S, Altieri I, Zuccaro P, Pacifici R. Drug monitoring in nonconventional biological fluids and matrices. Clin Pharmacokinet 1996;30(3):211–28.

[156] Tondi M, Mutani R, Mastropaolo C, Monaco F. Greater reliability of tear versus saliva anticonvulsant levels. Ann Neurol 1978;4(2):154–5.

[157] Vanhaeringen NJ. Secretion of drugs in tears. Curr Eye Res 1985;4(4):485–8.

[158] Stoehr GP, Venkataramanan R, Dauber JH. Comparison of methods for determining unbound theophylline concentrations. Ther Drug Monit 1986;8(1):42–6.

[159] Dumortier G, Chaumeil JC. Lachrymal determinations: methods and updates on biopharmaceutical and clinical applications. Ophthalmic Res 2004;36(4):183–94.

[160] Stuchell RN, Feldman JJ, Farris RL, Mandel ID. The effect of collection technique on tear composition. Invest Ophthalmol Vis Sci 1984;25(3):374–7.

[161] Monaco F, Mutani R, Mastropaolo C, Tondi M. Tears as the best practical indicator of the unbound fraction of an anticonvulsant drug. Epilepsia 1979;20(6):705–10.

[162] Stuchell RN, Farris RL, Mandel ID. Basal and reflex human tear analysis II. Chemical analysis—lactoferrin and lysozyme. Ophthalmology 1981;88(8):858–62.

[163] Monaco F, Piredda S, Mastropaolo C, Tondi M, Mutani R. Diphenylhydantoin and primidone in tears. Epilepsia 1981;22(2):185–8.

[164] Monaco F, Piredda S, Mutani R, Mastropaolo C, Tondi M. The free fraction of valproic acid in tears, saliva, and cerebrospinal fluid. Epilepsia 1982;23(1):23–6.

[165] Nakajima M, Yamato S, Shimada K, Sato S, Kitagawa S, Honda A, et al. Assessment of drug concentrations in tears in therapeutic drug monitoring: I. Determination of valproic acid in tears by gas chromatography/mass spectrometry with EC/NCI mode. Ther Drug Monit 2000;22(6):716–22.

[166] Piredda S, Monaco F. Ethosuximide in tears, saliva, and cerebrospinal fluid. Ther Drug Monit 1981;3(4):321–3.

[167] Nakajima M, et al. Assessment of tear concentrations on therapeutic drug monitoring. III. Determination of theophylline in tears by gas chromatography/mass spectrometry with electron ionization mode. Drug Metab Pharmacokinet 2003;18(2):139–45.

[168] Steele WH, Stuart JF, Whiting B, Lawrence JR, Calman KC, McVie JG, et al. Serum, tear and salivary concentrations of methotrexate in man. Br J Clin Pharmacol 1979;7(2):207–11.

[169] Farandos NM, Yetisen AK, Monteiro MJ, Lowe CR, Yun SH. Contact lens sensors in ocular diagnostics. Adv Healthc Mater 2015;4(6):792–810.

[170] Kiang TK, Schmitt V, Ensom MH, Chua B, Häfeli UO. Therapeutic drug monitoring in interstitial fluid: a feasibility study using a comprehensive panel of drugs. J Pharm Sci 2012;101(12):4642–52.

[171] Donnelly RF, Mooney K, Caffarel-Salvador E, Torrisi BM, Eltayib E, McElnay JC. Microneedle-mediated minimally invasive patient monitoring. Ther Drug Monit 2014;36 (1):10–17.

[172] Müller M. Monitoring tissue drug levels by clinical microdialysis. Altern Lab Anim 2009;37:57–9.

[173] Combest A.J., Zamboni W.C. Microdialysis. In: Rudek MA, Chau CH, Figg WD, McLeod HL, editors. Handbook of Anticancer Pharmacokinetics and pharmacodynamics, 2nd ed. 2014. p. 477–98.

[174] Caricato A, Pennisi M, Mancino A, Vigna G, Sandroni C, Arcangeli A, et al. Levels of vancomycin in the cerebral interstitial fluid after severe head injury. Intensive Care Med 2006;32(2):325–8.

[175] Leboulanger B, Aubry JM, Bondolfi G, Guy RH, Delgado-Charro MB. Lithium monitoring by reverse iontophoresis in vivo. Clin Chem 2004;50(11):2091–100.

[176] Sekkat N, Naik A, Kalia YN, Glikfeld P, Guy RH. Reverse iontophoretic monitoring in premature neonates: feasibility and potential. J Control Release 2002;81(1–2):83–9.

[177] Donnelly RF, Moffatt K, Alkilani AZ, Vicente-Pérez EM, Barry J, McCrudden MT, et al. Hydrogel-forming microneedle arrays can be effectively inserted in skin by self-application: a pilot study centred on pharmacist intervention and a patient information leaflet. Pharm Res 2014;31(8):1989–99.

[178] Mooney K, McElnay JC, Donnelly RF. Children's views on microneedle use as an alternative to blood sampling for patient monitoring. Int J Pharm Pract 2014;22(5):335–44.

[179] Barbosa J, Faria J, Carvalho F, Pedro M, Queirós O, Moreira R, et al. Hair as an alternative matrix in bioanalysis. Bioanalysis 2013;5(8):895–914.

[180] Williams J, Patsalos PN, Mei Z, Schapel G, Wilson JF, Richens A. Relation between dosage of carbamazepine and concentration in hair and plasma samples from a compliant inpatient epileptic population. Ther Drug Monit 2001;23(1):15–20.

[181] Tracqui A, Kintz P, Mangin P. Hair analysis: a worthless tool for therapeutic compliance monitoring. Forensic Sci Int 1995;70(1–3):183–9.

[182] Villain M, Cirimele V, Kintz P. Hair analysis in toxicology. Clin Chem Lab Med 2004;42 (11):1265–72.

[183] Olds PK, Kiwanuka JP, Nansera D, Huang Y, Bacchetti P, Jin C, et al. Assessment of HIV antiretroviral therapy adherence by measuring drug concentrations in hair among children in rural Uganda. Aids Care 2015;27(3):327–32.

[184] Coetzee B, Kagee A, Tomlinson M, Warnich L, Ikediobi O. Reactions, beliefs and concerns associated with providing hair specimens for medical research among a South African sample: a qualitative approach. Future Virol 2012;7(11):1135–42.

[185] Gandhi M, Yang Q, Bacchetti P, Huang Y. A low-cost method for analyzing nevirapine levels in hair as a marker of adherence in resource-limited settings. AIDS Res Hum Retroviruses 2014;30(1):25–8.

[186] Huang Y, Yang Q, Yoon K, Lei Y, Shi R, Gee W, et al. Microanalysis of the antiretroviral nevirapine in human hair from HIV-infected patients by liquid chromatography-tandem mass spectrometry. Anal Bioanal Chem 2011;401(6):1923–33.

[187] Liu AY, Yang Q, Huang Y, Bacchetti P, Anderson PL, Jin C, et al. Strong relationship between oral dose and tenofovir hair levels in a randomized trial: hair as a potential adherence measure for Pre-Exposure Prophylaxis (PrEP). PLoS One 2014;9(1):e83736.

[188] Kintz P. Analytical and practical aspects of drug testing in hair. Boca Raton: CRC Press; 2007.

[189] Kronstrand R, Förstberg-Peterson S, Kågedal B, Ahlner J, Larson G. Codeine concentration in hair after oral administration is dependent on melanin content. Clin Chem 1999;45 (9):1485–94.

[190] Cooper GAA. Hair testing is taking root. Ann Clin Biochem 2011;48(6):516–30.

[191] Müller A, Jungen H, Iwersen-Bergmann S, Sterneck M, Andresen-Streichert H. Analysis of cyclosporin a in hair samples from liver transplanted patients. Ther Drug Monit 2013;35 (4):450–8.

[192] Poetzsch M, Baumgartner MR, Steuer AE, Kraemer T. Segmental hair analysis for differentiation of tilidine intake from external contamination using LC-ESI-MS/MS and MALDI-MS/MS imaging. Drug Test Anal 2015;7(2):143–9.

[193] Kintz P. Issues about axial diffusion during segmental hair analysis. Ther Drug Monit 2013;35(3):408–10.

[194] Williams J, Myson V, Steward S, Jones G, Wilson JF, Kerr MP, et al. Self-discontinuation of antiepileptic medication in pregnancy: detection by hair analysis. Epilepsia 2002;43 (8):824–31.

[195] Williams J, Lawthom C, Dunstan FD, Dawson TP, Kerr MP, Wilson JF, et al. Variability of antiepileptic medication taking behaviour in sudden unexplained death in epilepsy: hair analysis at autopsy. J Neurol Neurosurg Psychiatry 2006;77(4):481–4.

[196] Uematsu T, Sato R, Suzuki K, Yamaguchi S, Nakashima M. Human scalp hair as evidence of individual dosage history of haloperidol—method and retrospective study. Eur J Clin Pharmacol 1989;37(3):239–44.

[197] Sato R, Uematsu T, Sato R, Yamaguchi S, Nakashima M. Human scalp hair as evidence of individual dosage history of haloperidol—prospective study. Ther Drug Monit 1989;11 (6):686–91.

[198] Matsuno H, Uematsu T, Nakashima M. The measurement of haloperidol and reduced haloperidol in hair as an index of dosage history. Br J Clin Pharmacol 1990;29(2):187–94.

[199] Papaseit E, Marchei E, Mortali C, Aznar G, Garcia-Algar O, Farrè M, et al. Development and validation of a liquid chromatography-tandem mass spectrometry assay for hair analysis of atomoxetine and its metabolites: application in clinical practice. Forensic Sci Int 2012;218(1–3):62–7.

[200] Marchei E, Muñoz JA, García-Algar O, Pellegrini M, Vall O, Zuccaro P, et al. Development and validation of a liquid chromatography-mass spectrometry assay for hair analysis of methylphenidate. Forensic Sci Int 2008;176(1):42–6.

[201] Baxi SM, Liu A, Bacchetti P, Mutua G, Sanders EJ, Kibengo FM, et al. Comparing the novel method of assessing prep adherence/exposure using hair samples to other pharmacologic and traditional measures. J Acquir Immune Defic Syndr 2015;68(1):13–20.

[202] Bernard L, Vuagnat A, Peytavin G, Hallouin MC, Bouhour D, Nguyen TH, et al. Relationship between levels of indinavir in hair and virologic response to highly active antiretroviral therapy. Ann Intern Med 2002;137(8):656–9.

[203] Duval X, Peytavin G, Breton G, Ecobichon JL, Descamps D, Thabut G, et al. Hair versus plasma concentrations as indicator of indinavir exposure in HIV-1-infected patients treated with indinavir/ritonavir combination. Aids 2007;21(1):106–8.

[204] Huang Y, Gandhi M, Greenblatt RM, Gee W, Lin ET, Messenkoff N. Sensitive analysis of anti-HIV drugs, efavirenz, lopinavir and ritonavir, in human hair by liquid chromatography coupled with tandem mass spectrometry. Rapid Commun Mass Spectrom 2008;22(21):3401–9.

[205] van Zyl GU, van Mens TE, McIlleron H, Zeier M, Nachega JB, Decloedt E, et al. Low lopinavir plasma or hair concentrations explain second-line protease inhibitor failures in a resource-limited setting. J Acquir Immune Defic Syndr 2011;56(4):333–9.

[206] Gandhi M, Ameli N, Bacchetti P, Anastos K, Gange SJ, Minkoff H, et al. Atazanavir concentration in hair is the strongest predictor of outcomes on antiretroviral therapy. Clinical Infect Dis 2011;52(10):1267–75.

[207] Beck O, Stephanson N, Sandqvist S, Franck J. Determination of amphetamine and methylphenidate in exhaled breath of patients undergoing attention-deficit/hyperactivity disorder treatment. Ther Drug Monit 2014;36(4):528–34.

[208] Berchtold C, Bosilkovska M, Daali Y, Walder B, Zenobi R. Real-time monitoring of exhaled drugs by mass spectrometry. Mass Spectrom Rev 2014;33(5):394–413.

[209] Gamez G, Zhu L, Disko A, Chen H, Azov V, Chingin K, et al. Real-time, in vivo monitoring and pharmacokinetics of valproic acid via a novel biomarker in exhaled breath. Chem Commun 2011;47(17):4884–6.

[210] Erhart S, Amann A, Haberlandt E, Edlinger G, Schmid A, Filipiak W, et al. 3-Heptanone as a potential new marker for valproic acid therapy. J Breath Res 2009;3(1):016004..

[211] Morey TE, Wasdo S, Wishin J, Quinn B, van der Straten A, Booth M, et al. Feasibility of a breath test for monitoring adherence to vaginal administration of antiretroviral microbicide gels. J Clin Pharmacol 2013;53(1):103–11.

[212] Morey TE, Booth M, Wasdo S, Wishin J, Quinn B, Gonzalez D, et al. Oral adherence monitoring using a breath test to supplement highly active antiretroviral therapy. AIDS Behav 2013;17(1):298–306.

[213] Kadehjian LJ. Specimens for drugs-of-abuse testing. In: Wong RC, Tse HY, editors. Drugs of abuse: body fluid testing, 2005. p. 11–28.

[214] Phillips M. An improved adhesive patch for long-term collection of sweat. Biomater Med Devices Artif Organs 1980;8(1):13–21.

[215] De Giovanni N, Fucci N. The current status of sweat testing for drugs of abuse: a review. Curr Med Chem 2013;20(4):545–61.

[216] Kuwayama K, Tsujikawa K, Miyaguchi H, Kanamori T, Iwata YT, Inoue H. Time-course measurements of caffeine and its metabolites extracted from fingertips after coffee intake: a preliminary study for the detection of drugs from fingerprints. Anal Bioanal Chem 2013;405(12):3945–52.

[217] Parnas J, Flachs H, Gram L, Würtz-Jørgensen A. Excretion of antiepileptic drugs in sweat. Acta Neurol Scand 1978;58(3):197–204.

[218] Cirimele V, Kintz P, Gosselin O, Ludes B. Clozapine dose-concentration relationships in plasma, hair and sweat specimens of schizophrenic patients. Forensic Sci Int 2000;107 (1–3):289–300.

[219] Marchei E, Farré M, Pardo R, Garcia-Algar O, Pellegrini M, Pacifici R, et al. Usefulness of sweat testing for the detection of methylphenidate after fast- and extended-release drug administration: a pilot study. Ther Drug Monit 2010;32(4):508–11.

[220] Marchei E, Papaseit E, Garcia-Algar OQ, Farrè M, Pacifici R, Pichini S. Determination of atomoxetine and its metabolites in conventional and non-conventional biological matrices by liquid chromatography-tandem mass spectrometry. J Pharm Biomed Anal 2012;60:26–31.

[221] Marchei E, Papaseit E, Garcia-Algar O, Bilbao A, Farré M, Pacifici R, et al. Sweat testing for the detection of atomoxetine from paediatric patients with attention deficit/hyperactivity disorder: application to clinical practice. Drug Test Anal 2013;5(3):191–5.

[222] Henkin RI. Comparative monitoring of oral theophylline treatment in blood serum, saliva, and nasal mucus. Ther Drug Monit 2012;34(2):217–21.

[223] Johnson CJ, Christianson CD, Sheaff CN, Laine DF, Zimmer JS, Needham SR. Use of conventional bioanalytical devices to automate DBS extractions in liquid-handling dispensing tips. Bioanalysis 2011;3(20):2303–10.

[224] Yuan L, Zhang D, Aubry AF, Arnold ME. Automated dried blood spots standard and QC sample preparation using a robotic liquid handler. Bioanalysis 2012;4(23):2795–804.

[225] Oliveira RV, Henion J, Wickremsinhe ER. Fully-automated approach for online dried blood spot extraction and bioanalysis by two-dimensional-liquid chromatography coupled with high-resolution quadrupole time-of-flight mass spectrometry. Anal Chem 2014;86(2):1246–53.

[226] Oliveira RV, Henion J, Wickremsinhe ER. Automated high-capacity on-line extraction and bioanalysis of dried blood spot samples using liquid chromatography/high-resolution accurate mass spectrometry. Rapid Commun Mass Spectrom 2014;28(22):2415–26.

[227] Oliveira RV, Henion J, Wickremsinhe ER. Automated direct extraction and analysis of dried blood spots employing on-line SPE high-resolution accurate mass bioanalysis. Bioanalysis 2014;6(15):2027–41.

[228] Ingels A.S., Ramirez Fernandez M.D.M., Di Fazio V., Will S.M., Samyn N. Optimization of an automated solid phase extraction to determine drugs in preserved oral fluid using ultra performance liquid chromatography tandem mass spectrometry. Presented at the 53rd TIAFT meeting, Firenze, Italy, 2015.

[229] Annesley TM, McKeown DA, Holt DW, Mussell C, Champarnaud E, Harter L, et al. Standardization of LC-MS for therapeutic drug monitoring of tacrolimus. Clin Chem 2013;59(11):1630–7.

[230] Society of Hair Testing. Recommendations for hair testing in forensic cases. Forensic Sci Int 2004;145(2–3):83–84.

[231] Robijns K, Koster RA, Touw DJ. Therapeutic drug monitoring by dried blood spot: progress to date and future directions. Clin Pharmacokinet 2014;53(11) 1053–1053.

[232] De Jesús VR, Mei JV, Cordovado SK, Cuthbert CD. The newborn screening quality assurance program at the centers for disease control and prevention: thirty-five year experience assuring newborn screening laboratory quality. Int J Neonatal Screen 2015;1 (1):13–26.

[233] Dantonio PD, Stevens G, Hagar A, Ludvigson D, Green D, Hannon H, et al. Comparative evaluation of newborn bloodspot specimen cards by experienced laboratory personnel and by an optical scanning instrument. Mol Genet Metab 2014;113 (1–2):62–6.

[234] van der Elst KC, Span LF, van Hateren K, Vermeulen KM, van der Werf TS, Greijdanus B, et al. Dried blood spot analysis suitable for therapeutic drug monitoring of voriconazole, fluconazole, and posaconazole. Antimicrob Agents Chemother 2013;57(10):4999–5004.

[235] Baumann RJ, Ryan M, Yelowitz A. Physician preference for anti-epileptic drug concentration testing. Pediatr Neurol 2004;30(1):29–32.

[236] Marchei E, Farrè M, Pellegrini M, Rossi S, García-Algar O, Vall O, et al. Liquid chromotography-electrospray ionization mass spectrometry determination of methlyphenidate and ritalinic acid in conventional and non-conventional biological matrices. J Pharm Biomed Anal 2009;49(2):434–9.

[237] Marchei E, Farré M, Pardo R, Garcia-Algar O, Pellegrini M, Pacifici R, et al. Usefulness of sweat testing for the detection of methylphinidate after fast and extended-release drug administration: a pilot study. Ther Drug Monit 2010;32(4):508–11.

CHAPTER 14

Integrating Therapeutic Drug Monitoring in the Health Care Environment: Therapeutic Drug Monitoring and Pharmacists

William Clarke
Department of Pathology, Johns Hopkins University School of Medicine, Baltimore, MD, United States

CONTENTS

14.1 Introduction 337
14.2 Team-Based Care in the Modern Health Care Environment......337
14.3 Hypertension.....338
14.4 Diabetes and Hypertension 338
14.5 Cardiovascular Diseases 339
14.6 Pharmacists and Cost-Effective Care....339
14.7 Impact of Health Care Information Technology on TDM ...340
14.8 Practical Examples of Integrating Pharmacy for TDM and Patient Care 341
14.8.1 Anticoagulation...342
14.9 Epilepsy.............344
14.9.1 Antiinfective Management....................346
14.10 Conclusions 349
References 349

14.1 INTRODUCTION

Early in the development of health care systems, physicians were both prescribers and dispensers of therapeutic drugs. However, when the pharmacy discipline was established as an independent medical specialty in its own right, the need for credentialed professionals with specialized knowledge, skills, and responsibilities was recognized [1]. Community pharmacists, or retail pharmacists, primarily verify the appropriateness of prescriptions, dispense medications, counsel patients, and are available for consultation as needed. Pharmacists in hospitals and health systems are often integrated into a multiprofessional team, attending patient care rounds and advising the care team on the selection, dosages, potential drug–drug interactions, and side effects of prescribed medications. They also monitor the health and progress of patients to ensure that the drug is working properly, and when necessary, they function as a liaison between physician and patient to explain the details and precautions of the prescribed therapeutic regimen. In many states, pharmacists now have limited prescribing rights, working collaboratively with physicians through physician–pharmacist agreements. Although pharmacists are available to interpret therapeutic drug monitoring (TDM) testing and consult based on the results, they typically have not been the traditional drivers of TDM, although that is changing.

14.2 TEAM-BASED CARE IN THE MODERN HEALTH CARE ENVIRONMENT

In today's health care environment, it is widely recognized that a multidisciplinary team is needed for optimal patient care. With the increasing complexity of

W. Clarke & A. Dasgupta (Eds): Clinical Challenges in Therapeutic Drug Monitoring. DOI: http://dx.doi.org/10.1016/B978-0-12-802025-8.00014-3

drug therapy and an emphasis on individualized or precision medicine, pharmaceutical interventions cannot be implemented in their full breadth by only physicians [2]. Pharmacists have collaborated closely with physicians in the clinical setting for many years, and it is not uncommon, especially in specialized settings such as transplantation, infectious disease, and oncology, for pharmacists to be interwoven into the patient care team. In these settings, the pharmacist is not just providing basic dispensing instructions and drug information services but also involved in solving patient- and medication-related problems and making clinical decisions regarding drug prescribing, therapeutic monitoring, and making dosage adjustments. Pharmacist intervention impact can be measured in terms of economics, health-related quality of life, medication appropriateness, and adverse events related to drug administration (adverse drug events and adverse drug reactions).

14.3 HYPERTENSION

Hypertension is a significant worldwide health care problem, responsible for as many as 7 million deaths per year [3]. Controlling blood pressure can lead to major reductions in heart failure (50%), stroke (35–40%), and myocardial infarction (20–25%) [4]. In 2008, Carter et al. [5] investigated the impact of enhanced pharmacist input on the management of patients treated for hypertension in general medicine clinics. The study was a prospective, cluster randomized control trial of patients aged 21–85 years. In the study, pharmacists made recommendations to the physicians caring for patients at clinics designated as intervention clinics, but they did not directly participate in care decisions at the control clinics. The impact of increased physician–pharmacist collaboration in the intervention clinics resulted in significantly better mean blood pressure values and control rates, primarily through intensified medication therapy and improved patient adherence. A comparable randomized controlled trial (RCT) by Hunt et al. [6] demonstrated similar findings, in which 62% of intervention subjects achieved target blood pressure values compared to 44% of control subjects, with fewer physician visits necessary for the intervention group during the course of the study. These studies suggest that an active role for clinical pharmacists within the patient care team leads to improved clinical outcomes.

14.4 DIABETES AND HYPERTENSION

For patients with diabetes, the benefits of glycemic control, lipemic control, and blood pressure have been well established [7]. In efforts toward improving delivery of long-term care in diabetic patients, pharmacists are currently integrated with the care team [8–10]. Nkanash et al. reported findings of a retrospective, time-series, single-group study that investigated the impact of pharmacist medication management for patients treated by private practice physicians [11]. In this study, patients older than 18 years with either type 1 or type 2 diabetes treated in private practice clinics were referred to a pharmacy clinic within the practice and followed for 18 months. Compared to baseline (beginning of the study), at 18 months the investigators observed a significant reduction in hemoglobin A1c (primary marker for glycemic control in diabetes), with no significant changes in weight or number of patients at goal blood pressure. The study suggests that integration of clinical pharmacists into the patient care pathway significantly improves glycemic management and adherence to American Diabetes Association recommendations.

14.5 CARDIOVASCULAR DISEASES

Amiodarone is a drug commonly used for the treatment of life-threatening arrhythmias, as well as atrial fibrillation. Although it is an effective treatment option, it also has significant adverse effects, including liver, thyroid, and pulmonary toxicity. In addition, there are significant drug–drug interactions with other antiarrhythmics (eg, digoxin), warfarin, and statins. It is recognized that laboratory monitoring is necessary for effective management of amiodarone; some studies have reported that perhaps as many as half of patients taking amiodarone do not receive the necessary baseline monitoring test for thyroid and liver function [12]. In a 2011 study, Spence et al. [13] demonstrated the positive impact of a pharmacist-managed program for patients treated with amiodarone. In this retrospective cohort study, patients were identified using clinical and enrollment data with regard to whether they were in a pharmacist-managed program or the standard of care group. Laboratory test monitoring was recorded at baseline, 1–6 months after the index date, 7–12 months after the index date, and any time during the year (months 1–12). Monitoring rates between the two groups at any time during the year were compared, along with hospitalizations and emergency room (ER) visits. The study found that patients in the pharmacist-managed group were much more likely to get the needed laboratory tests, and there were significantly more hospitalizations and ER visits for the patients with abnormal lab tests in the standard of care group relative to the pharmacist-managed group. This study suggests that active involvement of pharmacists in the patient care team leads to greater compliance with safety measures and fewer adverse events.

14.6 PHARMACISTS AND COST-EFFECTIVE CARE

As the costs of providing health care strain an overburdened health care delivery system, there is an increased focus on the provision of cost-effective care. In theory, greater integration of pharmacy specialists into patient care teams results in more efficient utilization of resources and improved clinical outcomes. An early study (1988) by Clapham et al. [14] demonstrated that increased interaction with the care team by pharmacists compared to traditional models of hospital pharmacy reduced total average cost per patient by $1293. Additional studies [15,16] confirmed that inclusion of pharmacists on general medical teams resulted in reductions for length of stay and decreases in hospital and pharmacy costs. In 2001, Bond et al. [17] published a meta-analysis examining interrelationships and associations among mortality rates, drug costs, total cost of care, and length of stay in US hospitals; included in this analysis was an examination of these variables and outcomes relative to the presence of clinical pharmacy services. The study found that as clinical pharmacist staffing increased (measured in number of clinical pharmacists per occupied bed) from the 10th to the 90th percentile, the number of hospital deaths decreased from 113 to 64 per 1000 admissions (a 43% decrease). This translates to a reduction of approximately 400 deaths per hospital per year. In addition, the degree of clinical pharmacist staffing (clinical pharmacists/occupied bed), pharmacist participation in medical rounds, and pharmacist-driven drug protocol management were associated with decreased length of stay (thus decreasing inpatient costs per encounter). These studies suggest that an increased role for hospital pharmacists in patient care can reduce the overall cost of care and improve clinical outcomes.

14.7 IMPACT OF HEALTH CARE INFORMATION TECHNOLOGY ON TDM

One of the significant challenges for effective TDM is accurate capture of information regarding drug administration and specimen collection. A typical process flow for administration of a drug and TDM is given in Fig. 14.1. In most cases, information for drug dosing and administration is found in the clinical documentation section of an electronic patient record; it is manually entered and includes a wide window of acceptability (±1 h of the ordered and/or documented time). In many cases, the specimen collection time is also variable because it is dependent on the accuracy of information

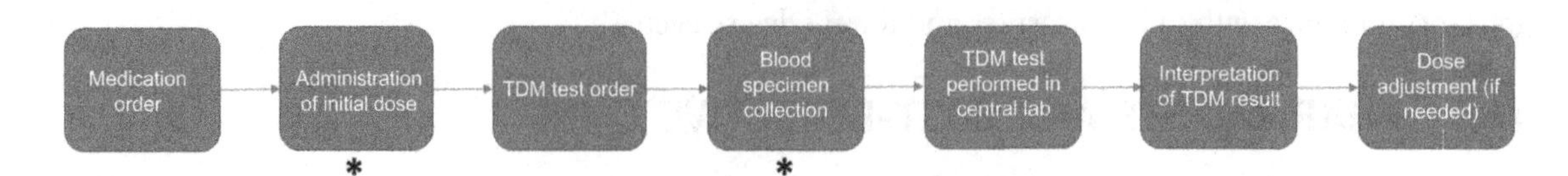

FIGURE 14.1
Typical process flow for TDM. *Steps with the most variability.

captured by the phlebotomy service or the nurse performing the specimen collection. With so much variability built into the information regarding dosing time and collection time, it can be very difficult to interpret TDM results because the physician or pharmacist cannot be sure whether the individual result is truly a peak or trough level.

However, within the past decade, point-of-collection bar code readers and label printers (eg, CareFusion) have been implemented for coordination of phlebotomy services. These devices and systems result in decreased specimen collection and labeling errors, and they have the additional benefit of obtaining accurate specimen collection times to within ±5 min. In addition, many newer-generation hospital information systems (HIS; eg, EPIC and Cerner Millennium) include bar code medication administration (BCMA) modules. This BCMA functionality assists with accurate administration of the drug by the provider, but it also allows for accurate collection of the time of dosing. Automation of collecting drug dosing and specimen collection data at the point of care removes uncertainty from the documented times in the patient record and improves interpretation of TDM levels collected for the patient.

In addition to improving interpretation of traditional TDM levels, the accurate information collection and automation available due to advances in hospital information system technology (IT) may allow approaches that are more advanced than TDM in the hospital. First, in simple cases, the automated collection of accurate data could lead to an automated interpretive report or decision support for TDM, allowing for dosing recommendations from a table embedded in the information system. Another potential application that could be built using the accurate information within the IT system is automated pharmacokinetic (PK) modeling from limited sampling data. As long as a robust PK model has been developed, the HIS will be able to combine the demographic data, clinical data, and blood level data (including the accurate dose administration and specimen collection data) to feed into the PK model. The calculated area under the curve or drug exposure can then be used to generate interpretive reports and perhaps even automated dosing recommendations. These types of applications are not widely implemented, but they are certainly possible with the newer HIS functionality.

14.8 PRACTICAL EXAMPLES OF INTEGRATING PHARMACY FOR TDM AND PATIENT CARE

Although automated interpretation of TDM is not yet standard of care, there are several examples in which pharmacy-driven TDM practice has been shown to improve patient care. These include the management of anticoagulants, in which pharmacist involvement has been in place for outpatients for many years. In addition, examples can be found in the treatment of epilepsy, in which TDM has been common for more than 40 years. In the treatment of infection, pharmacists have played an active role for many years as well.

14.8.1 Anticoagulation

Warfarin (Coumadin) is a commonly used anticoagulant drug used for prevention of thrombosis and thromboembolism. Common clinical indications for warfarin include patients with atrial fibrillation, patients with artificial heart valves, and patients diagnosed with deep vein thrombosis. It is a vitamin K epoxide reductase inhibitor, so it is subject to dietary interference from foods that are high in vitamin K, such as leafy green vegetables, in addition to potential drug–drug interactions. In addition, it demonstrates significant interindividual variability with respect to both PKs (CYP 2C9 variability) and pharmacodynamics (PDs; VKORC1 variability). Because of the significant bleeding risk associated with too much drug, and the risk of blood clots with too little drug, patients receiving warfarin must be carefully managed. The drug is not monitored via traditional TDM with PK measurements; it is instead managed using the PD marker of prothrombin time, which is a biomarker of coagulation. Prothrombin time is commonly expressed as international normalized ratio (INR) to account for assay variability between labs.

In a descriptive retrospective study published in 2012, Fowler et al. [18] examined the impact of implementing a pharmacist-led anticoagulation management service (AMS) for inpatients at a 380-bed tertiary care hospital in Minnesota. This investigation reviewed all hospitalizations for patients who received anticoagulation therapy with warfarin managed by the AMS between January 1, 2006, and August 31, 2007 (20 months). The primary endpoints for adverse events included thrombosis and bleeding complications during hospitalization. Of the 2794 hospitalizations for patients managed by the pharmacist-led AMS, only 59 complications were identified (2.1% complication rate). Of these, 14 were thrombosis events (0.5%) and 45 were bleeding events (1.6%).

In Singapore, Wong et al. [19] conducted a single-site, cohort study examining baseline data prior to implementation of an inpatient, pharmacist-managed anticoagulation service compared to postimplementation data. This study was conducted at a 1200-bed hospital in which pharmacist-led anticoagulation services had been available for outpatients since 1997 and were extended to inpatients in 2006. Baseline data were collected for 3 months prior to implementation, and then postimplementation data were collected for 12 months after the inpatient service was started. The endpoints for the study included percentage of INRs achieving therapeutic range within 5 days, INRs more than 4 during dose titration (typical target is 2 or 3), and subtherapeutic INRs on discharge. When comparing baseline data (26 patients) to postimplementation data (144 patients) for patients who had received pharmacist-managed anticoagulation, the investigators found that the provision of a pharmacist consult resulted in 88% of INR values achieving therapeutic range within 5 days compared to 38% for patients without the pharmacist consult in the baseline period. In addition, there was a reduction in INR values of more than 4 during dose titration from 27% in the baseline period to 2% with pharmacist-managed service; subtherapeutic INR values on discharge declined from 15% to 0%. The mean time to therapeutic INR was reduced from 6.5 to 3.9 days with pharmacist-managed anticoagulation, and the mean length of stay after initiation of warfarin was reduced from 11 to 7.7 days. These studies (among others) demonstrate the value of pharmacist involvement in therapeutic monitoring and management of warfarin for inpatient care.

Therapeutic management of warfarin in the outpatient setting can also benefit from incorporation of pharmacists into the care team. A 2011 study by Young et al. compared pharmacist-managed anticoagulation in a family medicine clinic with the standard of care [20]. This was a single-site retrospective cohort study in which an in-house pharmacist assumed anticoagulation management in 2006. During a 17-month period, the investigators compared a pharmacy-managed group of patients ($n = 112$) to a group of patients receiving the usual care that included physician-managed anticoagulation from a comparable period prior to 2006. The primary endpoint for the study was time in the therapeutic range for INR, and secondary outcomes included the percentage of time with ± 0.3 units of the target range (expanded therapeutic range) and percentage of time the INR was greater than 5.0 or less than 1.5. Upon comparison, the investigators found that the patients in the pharmacist-managed group were within the therapeutic range 73% of the time, whereas the usual care group was in the therapeutic range 65% of the time. When considering the expanded therapeutic range, the pharmacist-managed group was in the range 91% of the time compared to 85% for the usual care group. The occurrence of patients with INR <1.5 was twice as frequent in the usual care group compared to the pharmacist-managed group, but the occurrence of patients with INR >5.0 was slightly more frequent in the pharmacist-managed group. This is a good example of the benefit of incorporating pharmacists into a community practice clinical workflow for warfarin management.

Recently, a large meta-analysis was performed comparing pharmacist-led anticoagulation service versus standard medical care with physician-managed anticoagulation [21]. The study was conducted by performing a systematic search of PubMed, Scopus, Google Scholar, and the Cochrane Library over a period from database inception to January 1, 2014. After curating the search results for studies comparing pharmacist-driven anticoagulation to standard of care, the authors were able to incorporate 24 studies (4 RCTs and 20 non-RCTs) into their meta-analysis, covering 11,607 participants (662 in RCTs and 10,945 in non-RCTs). From the non-RCT studies, it was shown that the percentage of time in the therapeutic range was significantly increased for patients on pharmacist-managed anticoagulation services (72.1% vs 56.7%). In addition, major bleeding events (0.6% vs 1.7%), thromboembolic events (0.6% vs 2.9%), hospitalizations (3% vs 10%), and emergency department visits (7.9% vs 23.9%) all decreased for patients treated with pharmacy-managed anticoagulation compared to physician-managed anticoagulation. The evidence from RCTs was not as strong and did not replicate evidence from non-RCTs in some cases. As a result, the authors correctly note some limitations: (1) The number of RCTs relative to non-RCTs is much smaller, both in terms of the number of studies and in terms of the number of study participants; and (2) the RCTs covered shorter periods of time, which could cause unrecognized anticoagulation-related adverse effects. In general, the evidence seems to point to improved patient care for both inpatient and outpatient anticoagulation when pharmacists are performing TDM (INR monitoring) for warfarin and managing the therapeutic intervention.

14.9 EPILEPSY

Epilepsy is a significant health problem throughout the world, but pharmacotherapy for treatment of seizures is well described and used widely. TDM for antiepileptic drugs (AEDs) is the standard of care (see chapter: Overview of Therapeutic Drug Monitoring), but criteria of TDM are not firmly established for monitoring AEDs. In some cases, TDM is used to titrate the dose upon initiation of therapy, and in other cases, TDM is used to determine a baseline level after the patient is determined to be clinically stable at a particular dose. The challenges for TDM in the setting of epilepsy include ensuring that the testing is performed for the appropriate indication, that the specimens are collected at the appropriate times, and that the TDM results are interpreted and applied correctly so that dose adjustments are appropriate. One way to address these challenges is to drive TDM practice through a clinical pharmacist or pharmacy specialist.

Ratanajamit et al. [22] investigated the impact of pharmacist intervention on TDM utilization for AEDs. In this study, the investigators performed a baseline evaluation of the appropriateness of TDM utilization at a teaching hospital in southern Thailand, including indication, sampling time, and application of the measured AED levels—specifically phenytoin, carbamazepine, and valproic acid. The medical residents were primarily involved in TDM request, interpretation, and dosage adjustment recommendations during the baseline evaluation. During the intervention period, each of these steps was evaluated by a pharmacist on site for appropriateness, and if needed, a suggestion was provided by the pharmacist before the resident made the final recommendation. The criteria for TDM appropriateness were developed and validated by two independent neurologists. During the baseline evaluation, the TDM indication was appropriate 63.6% of the time, the sampling time was appropriate 47.7% of the time, and the treatment decisions based on TDM results were appropriate 63.6% of the time. Implementation of pharmacist intervention significantly improved the quality of TDM services: After implementation, the indication for TDM was appropriate 97.7% of the time, sampling time was appropriate 79.1% of the time, and application of the results was appropriate 83.7% of the time. During baseline evaluation, 16 TDM requests without indication were found compared to only one during the intervention period, reducing the unnecessary cost by 90%. The study found that pharmacist intervention significantly improved appropriateness of TDM use and substantially reduced unnecessary costs.

A recent study by Lertsinudom et al. [23] asserted that just over 39% of patients with epilepsy demonstrated drug–drug interactions between phenytoin and other medications, and almost 29% of these patients suffered from adverse drug reactions from AEDs. The authors described a system of multidisciplinary TDM composed of attending physicians, pharmacists, and nurses. In their system, the team discussed drug-related issues and seizure control, in which pharmacists assist with drug level interpretations and provide suggestions for the management of drugs; however, the physicians make the final treatment decision after group discussion. The study covered all consecutive patients with epilepsy in the authors' institution for the period of 1 year (January 1 to December 31), and data were collected regarding reasons for TDM, interpretation of AED drug levels, pharmacists' recommendations, and clinical decisions of the physician. In reviewing the data, the investigators found that drug levels were ordered for compliance slightly more than 1% of the time, suspected subtherapeutic levels 44.5% of the time, suspected toxicity almost 14% of the time, and for routine monitoring 40.5% of the time. Once drug levels were obtained, the pharmacist recommended adjustment of medication dose slightly less than 40% of the time and recommended a recheck of the drug levels on the next visit more than 50% of the time; it was rare that the pharmacist recommended changing the medication altogether. With regard to physicians incorporating pharmacist recommendations, they completely followed the recommendations more than 80% of the time and partially followed them 11.5% of the time. They disregarded pharmacy recommendations only 7.75% of the time. This study suggests that the pharmacist is an important facilitator for TDM in the clinical management of patients with epilepsy.

14.9.1 Antiinfective Management

Gentamicin is an aminoglycoside antibiotic used for treatment of bacterial infections caused by Gram-positive and Gram-negative organisms. Aminoglycoside antibiotics, including gentamicin, have a narrow therapeutic index and significant interindividual PK variability, which necessitates TDM for safe and effective therapy. The short half-life of gentamicin, coupled with the challenges of accurately collecting dosage and specimen collection information, can make interpretation of gentamicin difficult. In many institutions, the pharmacy is responsible for providing a PK consult service for managing aminoglycoside antibiotics.

An interesting study by Murphy et al. [24] examined the impact of removing a pharmacist-initiated TDM consult service for children treated with gentamicin. At the Hospital for Sick Children in Toronto, Canada, a pharmacist-initiated TDM consult service provided for patients admitted to a nursing unit that did not have a clinical pharmacist from 1988 to 2003, at which time it was withdrawn from the institution. While the service was active, pharmacists automatically assessed all serum drug concentrations outside of accepted therapeutic ranges when TDM testing was requested, and then they supplied individualized dosing regimens for the patients and coordinated future TDM services for those patients. When the service was withdrawn, the assessment of children with subtherapeutic or supertherapeutic serum drug concentrations became the responsibility of the attending physician. For this study, a chart review was conducted for patients treated with gentamicin for 6 months prior to removal of the pharmacist-based TDM consult service and for 6 months thereafter. All children admitted to general surgery units who received gentamicin and who had reported serum gentamicin concentrations outside of the therapeutic target interval were included in the study. For each TDM level included in the study, data regarding age, sex diagnosis, indication for therapy, gentamicin dose, duration of therapy, serum creatinine result, and serum gentamicin concentration were recorded. Based on these data, the clinician's actual response to the TDM result in terms of dose adjustment was compared to the ideal calculated response in order to determine appropriateness. A physician's response was considered appropriate if a dose adjustment was unnecessary and no dose adjustment was made or if a dose adjustment was necessary and the revised regimen was correct. A physician's response was considered inappropriate if a dose adjustment was needed but none was made, if a dose adjustment was necessary but the revised regimen was incorrect, or if a dose adjustment was unnecessary but a dose adjustment was made by the physician. Based on these criteria, the actions taken in response to serum gentamicin concentrations were appropriate in 99% of instances (93 of 94) with the TDM consultation service in place and in only 64% of instances (64 of 100) after removal of the TDM consultation service. This study demonstrates the value of a pharmacist-based TDM consult service for patient care.

Vancomycin is another antibiotic with a short half-life that requires TDM for effective management and avoidance of toxicity. It is a glycopeptide antibiotic used for treatment of many Gram-positive bacterial infections, and it is most commonly used for the treatment of proven or suspected methicillin-resistant *Staphylococcus aureus* infections. Most TDM for vancomycin is focused on whether the target level for effectiveness versus the microorganism is reached and whether the drug is cleared, but there is some interest in the time to reach a therapeutic trough level [25]. When vancomycin is not adequately cleared, it can have significant nephrotoxicity associated with high blood concentrations of the drug. Given the challenges in managing this drug, there is significant interest in finding the best way to safely achieve target vancomycin trough concentrations while avoiding permanent toxic effects.

Cardile et al. [26] investigated the impact of implementing a pharmacist-guided vancomycin TDM program (high trough therapy) compared to a standard physician-guided vancomycin therapy protocol. This was a single-center, pre- and postintervention observational study conducted at a 250-bed teaching hospital in Hawaii. The control group included patients treated with vancomycin from July 2007 to March 2008 (9 months), whereas the interventional group (pharmacist-guided therapy) included patients treated with vancomycin from July 2009 to March 2010 (9 months). In order to reduce the time to target trough, multiple changes were made, including placing a pharmacist in charge of the drug management, eliminating batching of vancomycin TDM with morning phlebotomy rounds (instead of providing TDM service "on demand"), and utilizing a standard formula for dose adjustment based on the vancomycin TDM results. Primary outcome measures for the study included time to target trough, time to clinical stability, all-cause mortality, and inpatient length of stay. Secondary outcomes included vancomycin treatment failure, time to normalization of white blood cell count, and inpatient length of time of vancomycin therapy. Clinical outcomes analysis was conducted only for patients with culture-confirmed Gram-positive infections sensitive to vancomycin and with normal renal function, which left only 145 patients eligible for the study (79 patients in the control group and 66 in the intervention group). More patients in the pharmacist-guided TDM group reached the initial target trough compared to the control group (80% vs 42%) at the first TDM draw; for the control group, there were 233 total TDM trough levels drawn with 47% being mistimed, whereas in the pharmacist-guided TDM group there were 185 total TDM trough levels drawn with only 32% being mistimed. The time to trough target was shorter for the pharmacist-guided TDM group compared to the control group (3 vs 5 days). Compared to the control group, patients in the pharmacist-guided TDM group were discharged from the hospital more rapidly, reached clinical stability quicker, and had shorter courses of inpatient vancomycin treatment. The study was able to demonstrate that a pharmacist-driven TDM program for vancomycin can improve patient outcomes with minimal vancomycin-derived nephrotoxicity.

Cystic fibrosis (CF) is an autosomal recessive genetic disorder that causes normally thin secretions in the body to become thick, particularly sweat, digestive fluids, and mucus. Initial presentation of the disorder is typically poor growth and weight gain, but patients with CF tend to have an accumulation of thick and sticky mucus, frequent chest infections, and coughing or shortness of breath. Patients with CF are often treated with aminoglycoside antibiotics for treatment of a variety of infections, both intravenously and via inhalation. Because the aminoglycoside antibiotics can cause hearing loss, damage to the balance system, or damage to the kidneys, they are managed carefully through TDM. However, due to the changes in mucus and secretions in CF patients, the distribution for the aminoglycosides as well as the target ranges for treatment in CF are quite different from those of the general population and require people with specialized knowledge of CF patients to manage the treatment.

A 2012 study by Cies and Varlotta [27] examined the impact of implementing a pharmacist-driven aminoglycoside TDM service in CF patients. The study was a single-center, retrospective cohort study conducted at St. Christopher's Hospital for Children in Philadelphia. In this setting, a practice change was implemented in September 2008 to include active, contemporaneous monitoring of serum aminoglycoside levels under the oversight of a dedicated clinical pharmacist. The medical records of all pediatric CF patients admitted to the study institution between January 2007 and May 2009 for acute CF pulmonary exacerbation meeting the criteria for aminoglycoside treatment (tobramycin ≥ 48 h) were included in the study. Patients meeting the inclusion criteria from January 2007 to August 2008 (20 months) were designated as the control group (standard of care), and patients meeting the inclusion criteria from September 2008 to May 2009 (9 months) were designated as the intervention group (pharmacist-driven TDM). The primary endpoint of this study was to compare the number of pediatric CF patient achieving the PK/PD target. The secondary endpoints included a comparison of the total number of orders and dose changes required, total number of days required to achieve the PK/PD target, the number of TDM tests required to achieve the PK/PD target, and length of stay. Based on the inclusion criteria, there were 29 patients (52 courses of therapy) in the intervention group and 22 patients (42 courses of therapy) in the control group. The investigators found that 98% of courses in the intervention group (51 of 52) reached the PK/PD target compared to 69% of courses in the control group (29 of 42). Patients in the intervention group reached the target in a mean of 1.9 days compared to a mean of 4.8 days for the control group. The mean length of stay for the group of patients with pharmacist-managed TDM was 9 days ($\pm$ 5 days), whereas that for patients receiving standard of care was 12 days ($\pm$ 7.5 days). For the pharmacist-managed group, the mean number of TDM tests needed was 2.7, whereas that required for the control group was 5.2. According to the study, the use of a pharmacist-managed TDM program resulted in a 14.5% reduction in resource utilization (based on patient charges). The study suggests that implementation of pharmacist-driven TDM services can improve clinical outcomes while more efficiently utilizing health care resources.

14.10 CONCLUSIONS

There are many opportunities for pharmacists to positively impact patient care, in both hospital and outpatient environments. As management of drug therapy increases in complexity, with more patients on polypharmacy and an increasing emphasis on individualized care, specialists such as pharmacists are needed to adequately address the accompanying issues. Information tools and the capability for decision support are rapidly advancing, allowing pharmacists an increasing capability for precise management of medication. Where the impact of integrated pharmacy services has been studied, the results almost uniformly demonstrate that patient care is improved, often with more efficient and cost-effective delivery. That paradigm extends to TDM as well, in which pharmacists can be an effective link between physicians and optimal TDM for patient care.

References

[1] Concepts and medicaments become modern. In: Sonnedecker G Kremers and Urdang's history of pharmacy. 3rd ed. Philadelphia, PA: J.B. Lippincott Co.; 1963.

[2] Schellens JHM, Grouls R, Guchelaar HJ, Touw DJ, Rongen GA, de Boer A, et al. The Dutch model for clinical pharmacology: collaboration between physician- and pharmacist-clinical pharmacologists. Br J Clin Pharmacol 2008;66:146–7.

[3] World Health Organization. World health report 2002—reducing risks, promoting healthy life. Geneva: World Health Organization; 2002<http://www.who.int/whr/2002/>.

[4] Chobanian AV, Bakris GL, Black HR, Cushman WC, Green LA, Izzo JL Jr, Jones DW, Materson BJ, Oparil S, Wright JT Jr, Roccella EJ; Joint National Committee on Prevention, Detection, Evaluation, and Treatment of High Blood Pressure. National Heart, Lung, and Blood Institute; National High Blood Pressure Education Program Coordinating Committee. Seventh report of the Joint National Committee on prevention, detection, evaluation, and treatment of high blood pressure. Hypertension 2003;42:1206–52.

[5] Carter BL, Bergus GR, Dawson JD, Farris KB, Doucette WR, Chrischilles EA, et al. A cluster randomized trial to evaluate physician/pharmacist collaboration to improve blood pressure control. J Clin Hypertens 2008;10:260–71.

[6] Hunt JS, Siemienczuk J, Pape G, Rozenfeld Y, MacKay J, LeBlanc BH, et al. A randomized controlled trial of team-based care: impact of physician-pharmacist collaboration on uncontrolled hypertension. J Gen Intern Med 2008;23:1966–72.

[7] American Diabetes Association. Standards of medical care in diabetes—2007. Diabetes Care 2007;20:S4–41.

[8] Cranor CW, Christensen DB. The Asheville Project: short-term outcomes of a community pharmacy diabetes care program. J Am Pharm Assoc 2003;43:149–59.

[9] Cioffi ST, Caron MF, Kalus JS, Hill P, Buckley TE. Glycosylated hemoglobin, cardiovascular, and renal outcomes in a pharmacist-managed clinic. Ann Pharmacother 2004;38:771–5.

[10] Coast-Senior EA, Kroner BA, Kelley CL, Trill LE. Management of patients with type 2 diabetes by pharmacists in primary care clinics. Ann Pharmacother 1998;32:636–41.

[11] Nkanash NT, Brewer JM, Connors R, Shermock KM. Clinical outcomes of patients with diabetes mellitus receiving medication management by pharmacists in an urban private physician practice. Am J Health-Syst Pharm 2008;65:145–9.

[12] Thompson PD, Clarkson P, Karas RH. Statin-associated myopathy. JAMA 2003;289:1681–90.

[13] Spence MM, Polzin JK, Weisberger CL, Martin JP, Rho JP, Willick GH. Evaluation of a pharmacist-managed amiodarone monitoring program. J Manag Care Pharm 2011;17:513–22.

[14] Clapham CE, Hepler CD, Reinders TP, Lehman ME, Pesko L. Economic consequences of two drug-use control systems in a teaching hospital. Am J Hosp Pharm 1988;45:2329–40.

[15] Haig GM, Kiser LA. Effect of pharmacist participation on a medical team on costs, charges, and length of stay. Am J Hosp Pharm 1991;48:1457–62.

[16] Boyko Jr WL, Yurkowski PJ, Ivey MF, Armistead JA, Roberts BL. Pharmacist influence on economic and morbidity outcomes in a tertiary care teaching hospital. Am J Health-Syst Pharm 1997;54:1591–5.

[17] Bond CA, Raehl CL, Franke T. Interrelationships among mortality rates, drug costs, total cost of care, and length of stay in United States hospitals: summary and recommendations for clinical pharmacy services and staffing. Pharmacotherapy 2001;21:129–41.

[18] Fowler S, Gulseth MP, Renier C, Tomsche J. Inpatient warfarin: experience with a pharmacist-led anticoagulation management service in a tertiary care medical center. Am J Health-Syst Pharm 2012;69:44–8.

[19] Wong YM, Quek Y-N, Tay JC, Chadachan V, Lee HK. Efficacy and safety of a pharmacist-managed inpatient anticoagulation service for warfarin initiation and titration. J Clin Pharm Ther 2011;36:585–91.

[20] Young S, Bishop L, Twells L, Dillon C, Hawboldt J, O'Shea P. Comparison of pharmacist managed anticoagulation with usual medical care in a family medicine clinic. BMC Family Practice 2011;12:88.

[21] Entezari-Maleki T, Dousti S, Hamishehkar H, Gholami K. A systematic review on comparing 2 common models for management of warfarin therapy; pharmacist-led service versus usual medical care. J Clin Pharm 2016;56(1):24–38.

[22] Ratanajamit C, Kaewpibal P, Setthawacharavanich S, Faroongsarng D. Effect of pharmacist participation in the health care team on therapeutic drug monitoring utilization for antiepileptic drugs. J Med Assoc Thai 2009;92:1500–7.

[23] Lertsinudom S, Chaiyakum A, Tuntapakul S, Sawanyawisuth K, Tiamkao S, Tiamkao S. Therapeutic drug monitoring in epilepsy clinic: a multi-disciplinary approach. Neurol Int 2014;6:5620.

[24] Murphy R, Chionglo M, Dupuis LL. Impact of a pharmacist-initiated therapeutic drug monitoring consult service for children treated with gentamicin. Can J Hosp Pharm 2007;60:162–8.

[25] Li J, Udy AA, Kirkpatrick CM, Lipman J, Roberts JA. Improving vancomycin prescription in critical illness through a drug use evaluation process: a weight-based dosing intervention study. Int J Antimicrob Agents 2012;39:69–72.

[26] Cardile AP, Tan C, Lustik MB, Stratton AN, Madar CS, Elegino J, et al. Optimization of time to initial vancomycin target trough improves clinical outcomes. SpringerPlus 2015;4:364.

[27] Cies JJ, Varlotta L. Clinical pharmacist impact on care, length of stay, and cost in pediatric cystic fibrosis (CF) patients. Pediatr Pulmonol 2013;48:1190–4.

Index

Note: Page numbers followed by "*f*" and "*t*" refer to figures and tables, respectively.

A

Abacavir (ABC), 137–138, 138*f*
Abbott AxSYM analyzer, 38
Acetazolamide, 33
Acetonitrile, 107–108
ACS:180, 32
Adenosine, 137–138
ADVIA Centaur cyclosporine immunoassay, 26–27, 33–34
Agranulocytosis, 180
AIDS, 76*t*. *See also* Antiretroviral therapy (ART)
  concentrations of phenytoin in, 82
Albumin, 73
Aldosteronism, 24
Alfentanil, 86–87
α_1 acid glycoprotein (AGP)
  drug–protein binding of, 73
  free drug monitoring of drugs bound to, 86–87
Alternative matrices for TDM, 280–282
  alternative matrix/systemic compartment correlation, 284
  correlation between systemic plasma or whole blood concentration and, 282–284, 282*f*
  criteria, 281*t*
  dried blood spots (DBS). *See* Dried blood spots (DBS)
  Pearson correlation coefficient, 283
Amikacin, 10, 198*t*, 239–240
Aminoglycosides, 173–176, 239–240, 255
  antibiotics, 10, 19
Amiodarone, 254
Amitriptyline, 9, 35–36, 202*t*
Amoxapine, 35, 178
Amprenavir, 91–92
Analgesics, 255
Anemia, 180
Antiarrhythmics, 197, 197*t*, 255
Antibiotics
  aminoglycoside, 10, 19
  macrolide, 33, 252–253
  pharmacokinetic profile, 173–176, 189–190
  in pregnancy, 198, 198*t*
  TDM measurements for, 61
Anticoagulants, 199*t*, 255
Anticoagulation management service (AMS), 342–344
Anticonvulsants, 57–58, 221*t*, 255
  pharmacokinetic properties of, 176–177
Antidepressants, 221*t*, 255
  pharmacokinetic profile, 178, 189–190
  in pregnancy, 202*t*
  tricyclic, 241
Antiepileptic drugs (AEDs), 57, 101, 198–200, 200*t*
  newer generation of, 112–123
  pharmacist medication management of, 344–349
  pharmacogenetics of, 111–112
  pharmacokinetic profile, 189–190
  serum concentration ratio, 102*t*
  TDM of, 104–112
    analytical methods used, 108–109
    factors to consider, 105*f*
    free drug fractions, 110
    in infants and newborns, 111
    in obese patients, 111
    in pregnancy, 110
    in renal failure and hepatic dysfunction patients, 110
    in special populations, 110–111
    therapeutic ranges and specimen types, 104–108
  therapeutic ranges and methodologies for measuring concentrations of, 106*t*
  for treatment of seizure disorders, 101–104
Antiinfective management, 346–349
Antimicrobial drugs, 11
Antineoplastic drugs, 180–181, 255
Antiretroviral therapy (ART), 135
  agents, 135–136
    analytical methodologies for, 148–149
    approved by U.S. Food and Drug Administration (FDA), 136*t*
    clinical trials, 149–152
    evaluation in The Netherlands (ATHENA) prospective, 149
    GENOPHAR trial, 150–151
    Pharmacologic Optimizations of PIs and NNRTIs (POPIN) clinical study, 151

Antiretroviral therapy (ART) (*Continued*)
Resistance and Dosage Adapted Regimens (RADAR) trial, 150–151
classes, 137–147
highly active antiretroviral therapy (HAART), 251–252
in pregnancy, 203*t*
regimens for treatment naive patients, 143*t*
TDM for, 136
challenges and limitations, 152–153
rationale for, 137
recommendations, 152–153
Anxiety, 4
ARCHITECT analyzer, 26–28
ARCHITECT clinical chemistry platforms, 25
Aripiprazole, 9
Association for Quality Assessment in Therapeutic Drug Monitoring and Clinical Toxicology, 321
Atazanavir, 145*f*, 203*t*
Atmospheric pressure chemical ionization (APCI), 53
Atmospheric pressure photoionization (APPI), 53
AxSYM TCA assay, 36–37
Azathioprine, 203, 204*t*
Azithromycin, 254

B

Bariatric surgery, 240–241
Beckman Synchron LX analyzer, 20
Benzhydryl piperazine drugs, 34
Bilirubin, interference in TDM assays, 20–21
Biosite Triage method, 36
Bipolar disorder, 4
Bronchodilators, 179–180, 202–203
Brugada syndrome, 7–8
Buflomedil, 38
Bulimia, 178
Bumetanide, 249–250
Bupivacaine, 86–87
Bupropion, 201, 202*t*
Burn patients, 76*t*
Busulfan, 12, 180–181, 290–292
2-Butanol, 317–318

C

Calcineurin inhibitors, 255
Candida albicans, 255–256
Can Resistance Enhance Selection of Therapy (CREST) study, 144–147
Carbamazepine, 3, 35–37, 73, 75*t*, 116–117, 120–121, 176, 200*t*, 221*t*, 247, 249, 306, 308–309, 314, 344–345
Carbamazepine-10,11-epoxide, 19, 33–34, 36–37
Carbamazepine immunoassays, 33–34
3-Carboxy-4-methyl-5-propyl-2-furanpropionate, 81
Cardiac arrhythmias, 7–9
Cardiology, TDM in, 7–8
Cardiovascular drugs, 178–179
Cefpiramide, 249–250
Cetirizine, 34
Chemical ionization (CI) technique, 52–53
Chemotherapy, 237–239
Chloramphenicol, 252–254
Chloroquine, 305
Chlorpropamide, 249*t*
Chromatography combined with mass spectrometry in TDM
high-performance liquid chromatography (HPLC), 48–49, 172–173, 203–204
combined with ultraviolet detection (HPLC-UV) method, 28–29, 35
hydrophilic interaction chromatography (HILIC), 50
liquid, 48–51
detector systems for, 50–51
normal-phase liquid chromatography (NPLC), 48–50, 49*f*
reversed-phase liquid chromatography (RPLC), 48–50, 49*f*
stationary phase (SP) and mobile phase (MP), 48–50
liquid chromatographic–high-resolution mass spectrometric (HRMS) assay, 148
liquid chromatographic methods combined with mass spectrometry (LC-MS), 17, 28, 46–47, 55–63, 91–92, 148, 152–153, 172–173, 203–204
antibiotics, TDM measurements for, 61
anticonvulsants, TDM measurements for, 57–58, 59*t*
antifungal drugs, TDM measurements for, 59–60, 61*t*
application of, 56–61
heterogeneity, 62
HPLC, 17, 94
human errors in, 63
limitations, 61–63
matrix effects, 63
schematic representation of, 52*f*
tricyclic antidepressants (TCAs), TDM measurements for, 58–59, 60*t*
validation guidelines, 62
mass, 51–55
detectors, 55
ion source, 51–53
with ion trap analyzers, 54
mass analyzers, 53–55
quadrupole analyzers, 54
TOF analyzers, 55
preanalytical stage, 55–56
liquid–liquid extraction (LLE), 55–56
solid phase extraction (SPE), 55–56
Chronic kidney disease epidemiology collaboration (CKD-EPI) equation, 219–220
Chronic liver disease, 76*t*
Chronic myeloid leukemia (CML), 12
Cimetidine, 33, 122–123, 254
Ciprofloxacin, 221*t*, 236–237
Cisplatin, 236–237
Citalopram, 178, 201, 202*t*, 223–224

Clarithromycin, 305
Clinical Pharmacogenetics Implementation Consortium, 111–112
Clobazam, 113–114, 113*f*
Clomipramine, 9, 201, 202*t*
Clostridium difficile, 173
Clozapine, 9, 201, 202*t*
Cobicistat, 153
Cockcroft–Gault formula, 219–220, 248
Colchicine, 249*t*, 254
Competition-based immunoassay, 18
Congestive heart failure, 24
Corticosteroids, 204*t*
Critically ill patients, 76*t*
  altered drug disposition in, 252–254
    causes, 254*t*
  application of TDM in, 255–257
    of antimicrobial drugs, 255–256
    of digoxin, 256
    of sedatives and analgesics, 257
  enterohepatic circulation (EHC), impact of, 254
  monitoring of free drug concentrations in, 246–247
Cyclobenzaprine, 35–36
Cyclophosphamide, 237–238
Cyclosporine, 17, 19, 26, 87, 203, 204*t*, 221*t*, 254
Cyclosporine A (CyA), 5–6, 56–57, 177–178
Cyclosporine immunoassays, 26–27
CYP3A4, 5–6
CYP2B6, genetic polymorphism in, 141
Cyproheptadine, 37
Cystic fibrosis (CF), 348
  dried blood spots (DBS) analysis, 296
  oral fluid (OF)-based TDM method for, 307
Cytidine, 137–138
Cytochrome P450 (CYP) enzymes, 3, 10, 141, 144, 195, 199, 235–236, 241, 249–250

D

Dabigatran, 274
Danazol, 33
Darunavir, 145*f*
Delavirdine, 140*f*
Desipramine, 9, 36
Diabetes, pharmacist medication management for, 338–339
Diacerein, 249–250
Diarrhea, 180
Dicloxacillin, 86
Didanosine, 138*f*
Dietary Supplement Health and Education Act of 1994, 25
Digibind, 8, 23–24, 90–91
DigiFab, 23–24, 90–91
Digoxin, 8, 21–26, 22*t*, 178–179, 197*t*, 221*t*, 249
  in critically ill patients, 256
  Digibind and DigiFab, effects of, 23–24
  digoxin-like immunoreactive substances (DLIS) interferences in, 22–23
  herbal supplements and, 25–26
  monitoring free, 90–91
  potassium-sparing diuretics and, 24–25
Diltiazem, 33
Dimension digoxin assays, 24–25
Dimension RXL analyzer, 26–27
Diphenhydramine, 38
Disopyramide, 86–87, 179, 197*t*
Diuretics, 255
Division of AIDS (DAIDS), 148–149
Dolutegravir, 142*f*
Donepezil, 219, 221*t*
Doxepin, 36, 252
Doxorubicin hydrochloride, 237–238
Dried blood spots (DBS), 106–107, 279–280, 284–296
  advantages, 291*t*
  analysis of capillary, 287–288
    correlation between capillary and venous drug levels, 287–288
    Hct effects, 287–288
  analysis of dried spots, 298–299
  classes of therapeutic drugs determined in, 291*t*
  contamination issues, 288
  correlation between plasma levels and, 287
  in developing countries, 289–290
  Hct effect of, 286–287
  homogeneity, 288–289
  preparation, 285
  quality criteria, 321–322
  sampling, 284–286
  self-sampling in, 284
  TDM applications, 290–296
    advantage of home-based sampling, 295
    anemic patients, 293
    cystic fibrosis patients, 296
    elderly, 292–293
    home-based DBS sampling, 294–296
    pediatric patients, 290–292
    pregnant women, 293
    psychiatric patients, 293–294
    remote and resource-limited settings, 294
    tuberculosis (TB) treatment, 294
  tedious punching step, 320–321
  volumetric, 296–297
Dried matrix spots (DMS), 279–280
Dried OF spots, analysis of, 297–307
Dried plasma spots (DPS), 279–280, 297
Drug–drug interactions, 75–76
  elevated free anticonvulsant due to, 83–86
    from protein binding sites, 83–86
Drug–protein binding, 72–73, 81, 246–247
  of AGP, 73
  in albumin, 73
  of commonly monitored therapeutic drugs, 74*t*
  equilibrium between free drug and, 72–73
  normal molar concentration of binding protein albumin, 73
Duloxetine, 201, 202*t*, 249*t*

E

Efavirenz, 91, 140, 140*f*, 153
Electron ionization (EI) technique, 52–53
Electrospray ionization (ESI) technique, 53
Elevated free anticonvulsant levels, mechanism of, 80–83
  in AIDS, 82
  in liver diseases, 81–82
  in pregnancy, 83
  in uremia, 80–81
Elvitegravir, 142*f*, 153
Emtricitabine, 138*f*, 153
Enfuvirtide, 147, 147*f*
Enterobacter species, 173–175
Enterococcus faecium, 255–256
Epilepsy treatment, TDM application and, 3–5
Eplerenone, 24
Escherichia coli, 173–175
Escitalopram, 178
Eslicarbazepine acetate, 113*f*, 114
Etavirine, 140*f*
Ethosuximide, 176
Etravirine, 153
Everolimus, 6, 17, 19, 27–28, 56–57, 87
Exhaled breath for TDM, 279–280, 316–318
  accessibility, 317
  application of, 317–318
    compliance monitoring in patients receiving oral therapy for ADHD, 317
    monitoring of VPA, 317
    using extractive electrospray ionization (EESI), 317
Ezogabine (retigabine), 113*f*, 114–115
  clearance of, 115
  serum half-life of, 114–115

F

Fab antibody fragments, 179
Felbamate, 113*f*, 115, 176–177, 199, 200*t*
Fenoprofen, 85–86
Fexofenadine, 250
Fisher Bioservices, 148–149
Flecainide, 179
Fluconazole, 306
Fluoroquinolones, 252–253
Fluorouracil, 252
5-Fluorouracil (5-FU) management, 12
Fluoxetine, 178, 201, 202*t*
Fluvoxamine, 178
Fosamprenavir, 145*f*
Fosinopril, 249–250
Fosphenytoin, 3, 31–32
Free anticonvulsant monitoring, 75–80, 76*t*
  drug–drug interactions and, 83–86
Free carbamazepine concentrations, monitoring of, 80
  in pregnancy, 83
Free drug concentrations, monitoring of, 73–75
  of anticonvulsants, 75–80
    for special patient populations, 76*t*
  of carbamazepine concentrations, 80
  methods, 92–94
  of phenytoin concentrations, 76–78
  using saliva and tears for determination of, 92
  of valproic acid concentrations, 78–80
Free fraction of a drug, 72–73
Free mycophenolic acid, 87–90
  clinical conditions, 88*t*
  in uremic patients, 89
Free phenytoin concentration
  in AIDS, 82
  in hepatic disease, 81–82
  monitoring of, 76–78
  nonsteroidal antiinflammatory drugs and, 85–86
Free valproic acid concentration, monitoring of, 78–80
Furosemide, 81

G

Gabapentin, 4, 107–108, 113*f*, 116, 176–177
  clearance of, 116
  concentration in saliva, 116
γ-aminobutyric acid (GABA), 101
Gastrointestinal (GI) inflammation, 2
Gentamicin, 10, 175–176, 198*t*, 300, 346
Gentamycin, 239–240
GI stromal tumors, 12
GI toxicity of mycophenolic acid (MPA), 6
Glucuronide (HPPG), 29–31
Glyburide, 249*t*
Greiner Bio-One Saliva Collection System (SCS), 301–302
Guanidine, 81
Guanidinosuccinic acid, 81
Guanosine, 137–138

H

Hair analysis for TDM, 279–280, 312–316. *See also* Oral fluid (OF)-based TDM method; Sweat for TDM; Tear fluid for TDM
  analytical techniques, 313
  application of, 314–316
    antiretroviral treatment, monitoring in, 316
    atomoxetine and methylphenidate metabolites, in children and adolescents, 315
    correlation between haloperidol hair and plasma levels, 315
    correlation with cyclosporine A hair levels, 315
    tenofovir and emtricitabine hair levels monitoring, 315–316
    tenofovir compliance monitoring, 315–316
  interpretation of quantitative and qualitative hair analysis results, 313
  sample collection, 312–313
Haloperidol, 201, 202*t*
Hard ionization technique, 51–52
  chemical ionization (CI), 52–53
  electron ionization (EI), 52–53
Hct effect, 286–287
Health care environment, team-based care in, 337–338

cardiovascular diseases, 339
cost-effectiveness, 339–340
for hypertension, 338
for patients with diabetes, 338–339
practical examples, 341–344
anticoagulation management service (AMS), 342–344
antiepileptic drugs (AEDs), TDM utilization for, 344–349
antiinfective management, 346–349
Health care information technology, impact on TDM, 340–341
Hemoglobin, interference in TDM assays, 20–21
Hepatic disease
altered drug disposition in, 250–252
free phenytoin concentration in, 81–82
hypoalbuminemia in, 81–82
monitoring free drug concentrations, 246–247
pharmacokinetic parameters of valproic acid and, 82
questions to manage medications in patients with, 252*t*
Hepatotoxicity, 11–12
Hippuric acid, 81
Hitachi 917, 20
HIV infection
concentrations of phenytoin in, 82
*HLA-B*15:02* allele, 111–112
Home-based DBS sampling, 294–296. *See also* Dried blood spots (DBS)
abbreviated area under the curve (AUC) estimation, 295
advantage of home-based sampling, 295
Hydrophilic interaction chromatography (HILIC), 50
10-Hydroxycarbazepine, 118–119
drug clearance of, 119
serum/plasma concentration of, 119
Hydroxychloroquine, 203, 204*t*
10-Hydroxy-10,11-dihydrocarbamazepine, 4–5
Hydroxyzine, 34
Hyperbilirubinemia, 180
Hypercholesterolemia, 6
Hypercholesterolemia patients, 76*t*
Hypertension, 338
diabetes and, 338–339
Hypoalbuminemia, 80, 247
elevated free mycophenolic acid and, 88–89
in hepatic disease, 81–82
phenytoin–oxacillin interaction in, 86
in pregnancy, 83
in uremia, 81
Hypocalcemia, 178–179
Hypoglycemics, 255
Hypokalemia, 178–179
Hypomagnesemia, 178–179
Hypotension, 179–180

I

Ibuprofen, 83, 85–86
Idarubicin, 252
Imipramine, 9, 36, 201, 202*t*, 254
Immune modulators, 203
Immunoassays, used in TDM, 18–19, 45–46
antibody-conjugated magnetic immunoassay (ACMIA), 18, 26–27
anticonvulsants and interferences in immunoassays, 29–35
chemiluminescent microparticle immunoassay (CMIA), 18, 26
cloned enzyme donor immunoassay (CEDIA), 18, 26–29
competition-based, 18
digoxin interferences, 21–26, 22*t*
enzyme-linked immunosorbent assay (ELISA), 18
enzyme-multiplied immunoassay technique (EMIT), 18, 20, 26, 29–31, 36, 94
fluorescence polarization immunoassay (FPIA), 18, 20, 23–25, 29–31
interferences
for anticonvulsants, 29–35, 31*t*
from bilirubin, hemoglobin, or lipids, 20–21
in carbamazepine immunoassays, 33–34, 36–37
of cyproheptadine and quetiapine in immunoassays for TCAs, 37
for immunosuppressants, 30*t*
issues of, 26–29
in lamotrigine immunoassays, 35
metabolite, 26–29
oxaprozin, 29–31
in phenobarbital and valproic acid immunoassays, 35
in phenothiazine and phenothiazine metabolites, 36
for TCAs, 32*t*, 35–38
microparticle enzyme immunoassay (MEIA), 18
monoclonal antibodies, use of, 19
particle-enhanced turbidimetric inhibition immunoassay (PETNIA), 18, 26, 33–34
Immunosuppressants, 56–57, 255
interferences for, 30*t*
monitoring free concentrations of, 87–90
pharmacokinetic properties of, 177–178
TDM measurements for, 56–57, 58*t*
Indinavir, 144, 145*f*, 203*t*
Indomethacin, 254
Indoxyl sulfate, 81
Infectious disease, TDM in, 9–11
Inflammatory bowel disease, 56–57
Insomnia, 4
Integrase strand transfer inhibitors (INSTIs), 135–136, 141–143
Interstitial fluid (ISF)-based TDM, 279–280, 310–312. *See also* Dried blood spots (DBS); Oral fluid (OF)-based TDM method
application of, 311–312
feasibility of using, 310
sampling method, 310–311
use of microneedles, 311–312
use of reverse iontophoresis, 311
Ionization technique, 51–52

Ionization technique (*Continued*)
APCI or APPI sources, 63
chemical ionization (CI), 52–53
electron ionization (EI), 52
electrospray ionization (ESI), 53
hard, 51–52
matrix-assisted laser desorption/ ionization (MALDI), 53
soft, 51–52
surface-enhanced laser desorption/ ionization (SELDI), 53
Ion trap analyzers, 54
Irinotecan, 2
Isoniazid, 33, 249–250

K

Kidney diseases, altered drug disposition in, 247–250
clearance of nonrenally cleared drugs, 249–250
clearance of renally cleared medications, 248–249, 249*t*
Klebsiella pneumoniae, 173–175

L

Lacosamide, 113*f*, 116
drug–drug interactions, 116
Lacrimal fluid. *See* Tear fluid for TDM
Lambert–Beer law of absorbance, 21
Lamivudine, 138*f*
Lamotrigine, 4, 29, 33, 35, 57, 107–108, 113*f*, 116–117, 176–177, 200*t*
drug–drug interactions, 117
half-life of, 117
metabolism of, 117
serum/plasma concentrations of, 117
L-asparaginase, 238
Lennox–Gastaut syndrome, 101–104, 113–114
Leucovorin, 11–12
Leukopenia, 6, 11–12
Levetiracetam, 4, 29, 107–108, 113*f*, 118, 176, 200*t*
drug clearance of, 118
serum half-life, 118
Lidocaine, 7, 86, 179
Linezolid, 305
Lipids, interference in TDM assays, 20–21
Lithium, 8, 201, 202*t*, 221*t*
Lopinavir, 91–92, 144, 145*f*, 203*t*
Lorazepam, 254
Low-molecular-weight heparin (LMWH), 198, 199*t*
Luminescent oxygen channeling technology (LOCI) digoxin assay, 25

M

Macrolide antibiotics, 33, 252–253
Maprotiline, 35, 178
Maraviroc, 147, 147*f*
Mass spectrum, 55
Matrix-assisted laser desorption/ ionization (MALDI), 53
Maximum tolerated dose (MTD), 11
Mefenamic acid, 85–86
MEIA digoxin assay, 23–25
Meperidine, 249*t*
Mercaptopurine, 252
Metabolite interferences
in cyclosporine immunoassays, 26–27
in mycophenolic acid immunoassays, 28–29
in sirolimus and everolimus immunoassays, 27–28
in tacrolimus immunoassays, 27
Methanol, 107–108
Methicillin-resistant *Staphylococcus aureus* infection, 10, 253
Methotrexate, 11–12, 180, 254
Methyl digoxin, 24
Methyl guanidine, 81
Methylxanthines, 179–180
Metronidazole, 33
Mexiletine, 179
Mianserin, 35
Minimal effective concentration (MIC), 9
Mirtazapine, 35, 178
Modification of diet in renal disease (MDRD) equation, 219–220
Monoamine reuptake inhibitors, 178
Monoclonal antibodies, 19
Monohydroxycarbamazepine (MHC), 4–5
Morphine, 249–250, 252, 254
Mycophenolate mofetil (CellCept), 6
Mycophenolate sodium (Myfortic), 6
Mycophenolic acid (MPA), 6, 17, 19, 26, 56–57, 87, 177–178, 247, 254, 306
free, 87–90
glucuronide, 89
immunoassays, 28–29

N

N-acetylprocainamide (NAPA), 7, 179
N-acetyltransferase 2 (NAT2), 196
Naproxen, 85–86
Nasal mucus for TDM, 279–280, 319
National Committee for Clinical Laboratory Standardization (NCCLS), 20
National Institute of Allergy and Infectious Diseases (NIAID), 148–149
National Institutes of Health (NIH) AIDS Reagent Program, 148–149
Nausea, 179–180
Nelfinavir, 145*f*, 203*t*
Neuromuscular blocking agents, 255
Neuropathic pain, 4
Neutropenia, 180
Nevirapine, 140, 140*f*, 203*t*
Nicardipine, 252
Nifedipine, 249–250
Nitrofurantoin, 249*t*
Nitroglycerine, 252
Non-nucleoside reverse transcriptase inhibitors (NNRTIs), 135–136, 140–141
CYP3A4–drug interactions of, 142
exposure–response relationships, 141
Pharmacologic Optimizations of, 151
toxicities and metabolic modulators, 140, 141*t*

Nonvitamin K antagonist oral anticoagulants (NOACs), 220, 261–262, 274
Nortriptyline, 9, 36, 201
Nucleoside/nucleotide reverse transcriptase inhibitors (NRTIs/NtRTIs), 135–140
  structures, chemical formulas, and molecular weights of approved, 138*f*
  toxicities, 138–139, 139*t*

O

Obesity, 231–232
  antimicrobial treatment in, 239–240
  body surface area (BSA), 232–233, 233*t*
  chemotherapy in, 237–239
  health complications, 231–232
  ideal body weight (IBW), 232–233
  lean body weight (LBW), 232
  optimal dosing in, 232–233
  pharmacokinetics after bariatric surgery, 240–241
  pharmacokinetics in, 234–237
    absorption, 234
    distribution, 234–235
    drug and metabolite elimination, 236–237
    drug clearance, 235
    drug metabolism, 235–236
    metabolic disruption and insulin resistance, 235
  total body weight (TBW), 232–233, 232*f*
  World Health Organization (WHO) defined, 231–232
Obsessive–compulsive disorder, 178
Olanzapine, 9, 201
Older people, TDM in
  of antiarrhythmic agents and anticonvulsants, 219
  clinical pharmacology of medicines, 214–217
    pharmacodynamics, age-related changes in, 217
    pharmacokinetics, age-related changes in, 215–217
  cost-effectiveness, 223–224
    using a smartphone, 223
  database and best evidence to support, 224
  dose adjustment, 219–220
  dried blood spots (DBS) method, 292–293
  factors influencing, 217*f*
  impact of pharmacokinetic changes, 215*t*
  living with Alzheimer's disease, 219
  main indications for, 218–220, 218*t*
  oral fluid (OF)-based TDM method, 303–304
  pharmacodynamic monitoring to guide dosing, 223
  rationale for using, 214
  studies investigating, 220–222, 221*t*
Oleate, 81
Oncology, TDM in, 11–12
Oral fluid (OF)-based TDM method, 279–280, 299–301, 320. *See also* Dried blood spots (DBS); Interstitial fluid (ISF)-based TDM; Sweat for TDM; Tear fluid for TDM
  applications, 302–307
    cystic fibrosis patients, 307
    elderly, 303–304
    home monitoring, 306
    patients with renal failure, 305
    pediatric patients, 303
    pregnant women, 304
    psychiatric patients, 304
    remote and resource-limited settings, 305
  availability, 299
  flow rate and/or composition, influencing factors, 300
  inter- and/or intraindividual variability, 300–301
  sample collection, 299
    challenges, 301–302
    collection method or collection device, 301
    contamination issues and, 302
  technical limitation, 302
Organ transplantation, TDM in, 5–6
  monitoring free mycophenolic acid in, 87–88
Oseltamivir, 287–288
Oxacillin, 86
Oxaprozin, 29–31
Oxatomide, 34
Oxcarbazepine, 4–5, 34–35, 113*f*, 118–119, 176, 200*t*
Oxprenolol, 249–250

P

Panic disorder, 178
Paper spray ionization (PSI), 298–299
Paroxetine, 178, 201
Pediatrics, TDM in
  basic concepts, 166–168
    absorption, 166
    distribution, 166–167
    drug metabolism, 167–168
    excretion, 168
  dried blood spots (DBS) analysis, 290–292
  drug dosing regimens, 171–172
  hair analysis for TDM, 315
  oral fluid (OF)-based TDM method, 303
  pharmacokinetic calculations, 168–171
    drug clearance, 170
    drug concentration–time relationship for a two-compartment model system, 170*f*
    elimination rate constant, 169
    first-order kinetics, 169
    half-life, 169–170
    minimum effective and toxic concentrations, 171*f*
    steady-state concentrations, equations for, 170
    volume of distribution, 169
  pharmacokinetic properties of drug classes, 174*t*
    antibiotics, 173–176
    anticonvulsants, 176–177
    antidepressants, 178
    antineoplastic drugs, 180–181

Pediatrics, TDM in (*Continued*)
bronchodilators, 179–180
cardiovascular drugs, 178–179
immunosuppressants, 177–178
sample collection, analysis, and interpretation, 171–173
using chromatographic techniques, 172–173
Perampanel, 113*f*, 119
Perphenazine, 252
P-glycoprotein, 140
Pharmacist medication management
cardiovascular diseases, 339
cost effectiveness, 339–340
diabetes, 338–339
hypertension, 338
practical examples, 341–344
anticoagulation management service (AMS), 342–344
antiepileptic drugs (AEDs), TDM utilization for, 344–349
antiinfective management, 346–349
Pharmacokinetic–pharmacodynamic relationship, 1
Pharmacometrics, 205–206
PharmAdapt study, 150–151
Phenobarbital, 3–4, 35, 83, 116–117, 120–121, 176, 199, 200*t*, 306, 308–309
Phenobarbitone, 319
Phenothiazine, 35, 38
Phenytoin, 3–4, 29–31, 57, 73, 75*t*, 116–117, 120–121, 176, 199, 200*t*, 247, 249–250, 306, 319, 344–345
cytochrome P450 (CYP) isozyme 2C9 and, 112
immunoassays, 19
issues of interferences with, 29–32
in patients with renal failure, monitoring in, 305
5-(*p*-Hydroxyphenyl)-5-phenylhydantoin (HPPH), 29–31, 176–177
Physician–pharmacist collaboration in intervention clinics
cardiovascular diseases, 339
cost effectiveness, 339–340
for hypertension, 338
for patients with diabetes, 338–339
practical examples, 341–344
anticoagulation management service (AMS), 342–344
antiepileptic drugs (AEDs), TDM utilization for, 344–349
antiinfective management, 346–349
Polypharmacy, 219
Potassium canrenoate, 24
Potassium-sparing diuretics, 24–25
Pregabalin, 29, 113*f*, 119–120
plasma/serum concentrations of, 120
Pregnancy, TDM during, 185–186
of antibiotics, 198, 198*t*
of antidepressants, 202*t*
of antiepileptic drugs (AEDs), 110
of antiretrovirals, 203*t*
criteria of drugs and conditions that warrant, 186–187, 187*f*
dried blood spots (DBS) method, 293
drug classes used, 196–203
antiarrhythmics, 197, 197*t*
antibiotics, 198, 198*t*
anticoagulants, 198, 199*t*
antidepressants/antimanics/antipsychotics, 201, 202*t*
antiepileptics, 198–200, 200*t*
antiretrovirals, 201–202, 203*t*
bronchodilators, 202–203
immune modulators, 203, 204*t*
elevated free anticonvulsant levels, mechanism of, 83
free carbamazepine concentration, 83
limitations of, 203–206
oral fluid (OF)-based TDM method, 304
pharmacoeconomics impact, 186
pregnancy-induced changes, 192–196
in drug metabolism, 195–196
impact on drug transporters, 196
metabolic changes, 195*t*
pharmacokinetic/pharmacodynamic changes, 195*t*
physiological, 193–194, 193*t*
schedule for performing, 186
steps involved in, 187–192, 188*f*
clinical interpretation of drug levels, 191–192
communication of lab result, 191
decision to request a drug concentration, 188
drug concentration *vs* time profile, 189*f*
implementation and therapeutic management, 192
laboratory measurement of drug concentrations in specimens, 190–191
sample collection, 188–190
Primidone, 3, 83, 176
Probenecid, 249*t*
Procainamide, 7, 179, 197*t*, 236–237, 249–250
Prodrug, 2
Proficiency testing (PT) program for TDM, 321
Progabide, 33
Proguanil, 305
Propoxyphene, 33, 249*t*
Propranolol, 252
Protease inhibitors (PIs), 135–136, 143–147
CYP3A4, 144
drug–drug interactions associated with, 144
free concentrations of, 91–92
Pharmacologic Optimizations of, 151
structures, chemical formulas, and molecular weights, 145*f*
toxicities, mechanism of metabolism, 144, 146*t*
Protein-free ultrafiltrate, 23
Proteus mirabilis, 173–175
Proton pump inhibitors, 255
Pseudomonas aeruginosa, 173–175
Psoriasis, 56–57
Psychiatry, TDM in, 8–9

Q

QMS everolimus assay, 28
Quadrupole analyzers, 54
Quetiapine, 33, 37, 201, 252
Quinidine, 7–8, 86, 178–179

R

Raltegravir, 141–142, 142*f*
Refractory-schizoaffective disorder, 37
Remacemide, 33
Restless leg syndrome, 4
Rheumatoid arthritis, 56–57
Rifampicin, 254
Rifampin, 120–121, 252–253
Rilpivirine, 140*f*
Risperidone, 9, 201
Ritonavir, 91–92, 144, 145*f*, 203*t*
Ritonavir-boosted darunavir therapy, 142
Roux-en-Y gastric bypass, 241
Roxithromycin, 221*t*
Rufinamide, 107–108, 113*f*, 120
  serum/plasma concentrations of, 120

S

Salicylate, 85–86
Saquinavir, 91, 145*f*, 203*t*
Selective serotonin reuptake inhibitors (SSRIs), 178, 201, 221*t*, 241
Seradyn FPIA, 28
Seradyn QMS lamotrigine assay, 35
Serotonin reuptake inhibitors, 178
Sertraline, 178, 201
Siemens Diagnostics, 25
Sinusoidal obstructive syndrome, 11–12
Sirolimus, 6, 17, 19, 26–28, 56–57, 87, 203, 204*t*
SN-38, 2
Soft ionization technique, 51–52
  electrospray ionization (ESI), 53
Spironolactone, 24, 90, 249*t*
St. John's wort, 120
Staphylococcus aureus, 173
Stavudine, 138*f*
Stevens–Johnson syndrome, 111–112
Stiripentol, 33, 113*f*, 120–121
  serum/plasma concentrations of, 121
Structural analogs, 56
Succinylacetone, 290–292
Sudlow's site I and site II, 73
Sulfadimidine, 249–250
Sulfamethoxazole/trimethoprim, 249*t*
Sumatriptan, 252
Surface-enhanced laser desorption/ ionization (SELDI), 53
Sweat for TDM, 279–280, 318–319
  application of, 319
    antiepileptic drugs, monitoring of, 319
    noninvasive compliance monitoring in ADHD patients, 319
Synchron analyzers, 20
Syva RapidTest, 36

T

Tachycardia, 179–180
Tacrolimus (Tac), 5, 17, 19, 26–27, 56–57, 87, 177–178, 203, 204*t*, 252
Tamoxifen, 294–295
Target concentration intervention (TCI), 213–214
TDx Phenytoin II, 32
Tear fluid for TDM, 279–280, 307–310. *See also* Hair analysis for TDM; Oral fluid (OF)-based TDM method; Sweat for TDM
  advantages, 307
  applications of, 308–310
  basal tears, 307–308
  correlation between cerebrospinal fluid (CSF) concentrations and, 308–309
  limitations, 308
  sample collection, 307–308
  tear:free plasma concentration ratio, 309
Tenofovir disoproxil fumarate, 139–140, 153
Tenofovir (TFV), 137–138, 138*f*
Tertiary amine TCA, 35–36
Tetracyclines, 252–253
Tetrahydrofolate, 180
Theophylline, 179–180, 221*t*, 224, 309
Therapeutic drug monitoring (TDM), 71
  in antiretroviral management, 2
  clinical conditions applying, 2–12
    in cardiology, 7–8
    epilepsy, 3–5
    infectious disease, 9–11
    in oncology, 11–12
    organ transplantation, 5–6
    in psychiatry, 8–9
  criteria for, 1–2
  immunoassay platforms used in, 18–19
  primary function of, 2
  principles of, 1–2
  process of, 46*f*
  results, uses of, 2
  routine implementation of, 319–322
  serum or plasma use in, 17
Therapeutic index, 1–2
Therapeutic interval, 3–4
  for sirolimus, 6
Thrombocytopenia, 6, 180
Thymidine, 137–138
Tiagabine, 113*f*, 121
  bioavailability of, 121
  protein binding capability of, 121
  serum half-life of, 121
Tipranavir, 144, 145*f*
Titration to clinical effect, 1
Tobramycin, 10, 198*t*, 239–240, 296, 309–310
TOF analyzers, 55
Tolmetin, 85–86
Topiramate, 29, 107–108, 113*f*, 122
  half-life of, 122
  serum/plasma concentrations of, 122
Total drug concentration, 72
Trazodone, 178
Triamterene, 249*t*

Tricyclic antidepressants (TCAs), 9, 19
immunoassays for, 32*t*, 35–38
Triple quadrupole (QQQ) instruments, 54
Tswett, Michael, 48
Tyrosine, 290–292

U

Ulcerative stomatitis, 11–12
Unfractionated heparin, 198, 199*t*
Uremia, 76*t*
free drug concentrations in, 81
free fraction of valproic acid in, 80–81
hypoalbuminemia in, 81
monitoring free drug concentrations, 246–247
uremic toxins, 81
Uridine diphosphate glucuronosyltransferase (UGT) enzymes, 236

V

Valnoctamide, 33
Valproate, 57
Valproic acid, 4, 35, 73, 75*t*, 83, 116–117, 122–123, 176, 199, 200*t*, 221*t*, 247, 249, 308–309, 344–345
Valpromide, 33
Vancomycin, 10, 19, 173, 176, 198*t*, 240, 253, 347–348
Vasopressors, 255
Venlafaxine, 178, 201, 252
Venoocclusive disease, 11–12
Verapamil, 33, 179, 252
Vigabatrin, 113*f*, 122, 199, 200*t*
bioavailability, 122
in epilepsy treatment, 122
excretion of, 122
R-enantiomer of, 109
S-enantiomer of, 109
Viral fusion inhibitors, 135–136
Vitros assays, 32
Volumetric absorptive microsamples (VAMS), 279–280
Volumetric DBS, 296–297
Vomiting, 179–180

W

Warfarin (Coumadin), 199*t*, 254, 261–262, 342–343
chemical structure of *R*- and *S*-, 263, 263*f*
clinical benefit, 270–272
cost-effectiveness of pharmacogenomics testing in, 272–273
factors associated with higher requirements of, 267*t*
factors associated with lower requirements of, 267*t*
genetic mutations and, 265–266
medications to replace, 273–274
nongenetic factors affecting dosing of, 266–267, 266*t*
pharmacology, 263–270
age effect on clearance, 268–269
anthropometric variables, 269–270
CYP2C9 status, 263–264, 264*t*, 271–272
dietary effects, 268
gender effect on clearance, 269
percentage of time in the therapeutic range (PTTR), 270–271
regeneration of vitamin K, 268
VKORC status, 264–265, 265*t*, 271–272
potential for pharmacogenomics, 262–263

Z

Zalcitabine, 138–139, 138*f*
Zidovudine (ZDV), 137–138, 138*f*, 249–250
Zonisamide, 29, 113*f*, 122–123, 176–177
in epilepsy treatment, 123
pharmacokinetic variability with, 122–123